TWELFTH INTERNATIONAL CONFERENCE ON ADAPTIVE STRUCTURES AND TECHNOLOGIES

University of Maryland
15-17 October 2001

Edited by

NORMAN M. WERELEY
INDERJIT CHOPRA
DARRYLL J. PINES

Sponsored by
Defense Advanced Research Projects Agency
A. James Clarke School of Engineering
Minta Martin Fund for Aeronautical Research
Alfred Gessow Rotorcraft Center
at the
University of Maryland at College Park

CRC Press
Taylor & Francis Group
Boca Raton London New York

CRC Press is an imprint of the
Taylor & Francis Group, an **informa** business

First published 2002 by CRC Press
Taylor & Francis Group
6000 Broken Sound Parkway NW, Suite 300
Boca Raton, FL 33487-2742

Reissued 2018 by CRC Press

© 2002 by Taylor & Francis
CRC Press is an imprint of Taylor & Francis Group, an Informa business

Publisher's Note
The publisher has gone to great lengths to ensure the quality of this reprint but points out that some imperfections in the original copies may be apparent.

Disclaimer
The publisher has made every effort to trace copyright holders and welcomes correspondence from those they have been unable to contact.

ISBN 13: 978-1-138-10500-3 (hbk)
ISBN 13: 978-1-138-56287-5 (pbk)
ISBN 13: 978-1-315-12133-8 (ebk)

Visit the Taylor & Francis Web site at http://www.taylorandfrancis.com and the CRC Press Web site at http://www.crcpress.com

Contents

HELICOPTER APPLICATIONS
Session Chair: Gary Anderson (Army Research Office)

ADVANCED ACTUATORS
Session Chair: Ephrahim Garcia (Defense Advanced Research Projects Agency)

SMART MATERIALS
Session Chair: Manfred Wuttig (University of Maryland)

MODELING AND IDENTIFICATION
Session Chair: Alison Flatau (National Science Foundation)

HIGH LOAD / HIGH SPEED SYSTEMS
Session Chair: Norman M. Wereley (University of Maryland)

Preface

The Twelfth International Conference on Adaptive Structures and Technologies (12th ICAST) was held on October 15-17, 2001 on the campus of the University of Maryland, College Park.

The objective of the 12th International Conference was to provide a forum for discussions of research in adaptive structures at an international level. Researchers from the USA, France, Japan, Germany, Hong Kong, and India, gathered in College Park. A key goal of this conference is to provide opportunities for international collaboration and interaction that are critically important to advancing the state of the art in the field of smart materials and structures. Several speakers provided notable invited lectures. Dr. Steve Wax of the Defense Advanced Research Projects Agency (DARPA), described smart structures research currently ongoing under DARPA support. Dr. Gary Anderson of the U.S. Army Research Office (ARO) described several initiatives of the U.S. Army's current program in adaptive structures. Dr. Jay Kudva (Northrop Grumman Corporation) spoke on smart wings and described new frontiers for adaptive wing structures. Dr. Peter Janker (Daimler-Chrysler) described new research in adaptive structures at the EADS Research Center in Germany. Dr. V.K. Aatre, the Scientific Advisor to the Defense Minister of the Indian Ministry of Defense, described ongoing research in adaptive structures in India. Finally, Prof. Richard James (University of Minnesota) spoke about recent developments on ferromagnetic shape memory alloy, a promising new smart material with the potential for numerous novel applications.

Two distinguished pioneers in the field of smart materials provided a historical perspective on the development of two classes of smart materials. Prof. Eric Cross (Pennsylvania State University) reflected on his 50-year career studying ferroelectricity, and the scientific advances that have lead to the popularity of piezoelectric material-based actuators widely used today. Dr. Kristl Hathaway, U.S. Office of Naval Research, provided an engaging perspective on long and short magnetostriction, and provided insights into the science and application of magnetostrictive actuators.

Six technical sessions were held:

1. Helicopter Applications, chaired by Dr. Gary Anderson (ARO), USA
2. Advanced Actuators, chaired by Dr. Ephrahim Garcia of DARPA, USA
3. Smart Damping, chaired by Prof. Yuji Matsuzaki of Nagoya University, JAPAN
4. Identification and Control, chaired by Prof. Roger Ohayon of the Conservatoire National des Arts et Metier, FRANCE
5. Smart Materials, chaired by Prof. Manfred Wuttig of the University of Maryland
6. Modeling and Identification, chaired by Dr. Alison Flatau of the National Science Foundation

We thank all of our invited speakers and sessions chairs for their efforts in making this international conference a success. In addition, we also thank all of the researchers who presented their work in the spirit of international cooperation and collaboration. This spirit of international cooperation was especially evident when the conference attendees enjoyed the conference banquet during an evening cruise down the Potomac River aboard the Spirit of Washington.

Key financial support for this conference was provided by the Defense Advanced Research Projects Agency. Thanks go also to the A. James Clarke School of Engineering, the Minta Martin Fund for Aeronautical Research, and the Alfred Gessow Rotorcraft Center, all of the University of Maryland College Park for providing additional financial support.

Norman M. Wereley
Inderjit Chopra
Darryll J. Pines

Participants in the International Conference on Adaptive Structures and Technologies
at the University of Maryland, October 15-17, 2001.

INTERNATIONAL ORGANIZING COMMITTEE

Inderjit Chopra, *University of Maryland, USA* (General Chairman)

Norman M. Wereley, *University of Maryland, USA* (Technical Chairman)

V. Baburaj, *Nagoya University, Japan*

R. Barboni, *University of Rome, Italy*

E. J. Breitbach, *German Aerospace Center, Germany*

M. Bernadou, *Pole Universitaire Leonard de Vinci and INRIA, France*

G. P. Carman, *University of California at Los Angeles, USA*

S. Chonan, *Tohoku University, Japan*

L. E. Cross, *The Pennsylvania State University, USA*

A. Flatau, *National Science Foundation, USA*

T. Fujita, *University of Tokyo, Japan*

E. Garcia, *DARPA Defense Sciences Office, USA*

P. Gaudenzi, *University of Rome, Italy*

N. W. Hagood, *Massachusetts Institute of Technology, USA*

H. Hanselka, *Otto-von-Guericke-Universitat Magdeburg, Germany*

A. Hariz, *University of South Australia, Australia*

D. J. Inman, *Virginia Polytechnic Institute and State University, USA*

S. J. Kim, *Seoul National University, Korea*

C. K. Lee, *National Taiwan University, Taiwan*

I. Lee, *Korea Advanced Institute of Science and Technology, Korea*

W. H. Liao, *Chinese University of Hong Kong, China*

D. Martinez, *Sandia National Laboratory, USA*

Y. Matsuzaki (Chair), *Nagoya University, Japan*

K. Miura, *Structural Morphology Research, Japan*

Y. Murotsu, *Osaka Prefectural College of Technology, Japan*

M. C. Natori, *Institute of Space and Astronautical Science, Japan*

R. Ohayon, *Conservatoire National des Arts et Metiers, France*

M. Regelbrugge, *Rhombus Consultants Group, Inc., USA*

S. W. Ricky Lee, *Hong Kong University of Science and Technology, China*

J. Sater, *Institute for Defense Analysis, USA*

C. Stavrinidis, *ESTEC European Space Agency, The Netherlands*

N. Takeda, *University of Tokyo, Japan*

J. Tani, *Tohoku University, Japan*

B. K. Wada, *Jet Propulsion Laboratory, Retired, USA*

T. Weller, *Israel Institute of Technology, Israel*

LOCAL COMMITTEE

Inderjit Chopra, *University of Maryland* (General Chairman)

Norman M. Wereley, *University of Maryland* (Technical Chairman)

Amr Baz, *University of Maryland, USA*

Christopher Cadou, *University of Maryland, USA*

Alison Flatau, *National Science Foundation, USA*

Ephrahim Garcia, *Defense Advanced Research Projects Agency, USA*

Darryll J. Pines, *University of Maryland, USA*

Manfred Wuttig, *University of Maryland, USA*

Jin Hyeong Yoo, *University of Maryland, USA*

SPONSORS

U.S. Defense Advanced Research Projects Agency

Minta Martin Fund for Aeronautical Research

Alfred Gessow Rotorcraft Center and the Alfred Gessow Fund for Rotorcraft Research

PARTICIPATING ORGANIZATIONS

American Institute of Aeronautics and Astronautics

American Society of Mechanical Engineers

American Astronautical Society

Japanese Society for Aeronautical and Space Sciences

Japanese Society of Mechanical Engineers

The American Institute of Aeronautics and Astronautics

Associazione Italiana di Aeronautica e Astronautica

Deutsche Gesellschaft fur Luft- und Raurnfahrt

The Royal Aeronautical Society

Helicopter Applications

FEASIBILITY OF ADAPTIVE MICRO AIR VEHICLES

Felipe Bohorquez, Chris Cadou, and Darryll Pines

Abstract

This paper reviews some of the important technical barriers that must be overcome to achieve truly efficient flying adaptive micro air vehicles (MAVs). As defined by the Defense Advanced Research Agency (DARPA), MAVs are vehicles with no length dimension greater than 6 inches. These vehicles typically weigh less than 100 grams and some can fly for approximately 30 minutes. Over the past decade significant progress has been made in developing these small-scale mechanical flying machines. However, there is still much work to be done if these vehicles are to approach the efficiency and performance of biological fliers. This paper reviews the status of current miniature mechanical flying machines and compares their performance with common biological flyers such as birds, and small insects. This comparison reveals that advances in aerodynamic efficiency, lightweight and adaptive wing structures, energy conversion/propulsion systems and flight control are required to match or exceed the performance of nature's great flyers.

1.0 Introduction

In the US, research on MAVs has been spearheaded by the Defense Advanced Research Projects Administration's (DARPA's) need to develop autonomous, lightweight (100g or less), small-scale flying machines (no length dimension greater than 15 cm) that are appropriate for reconnaissance over land, in buildings, and other confined spaces [3-7]. Toward this end, DARPA has supported numerous research activities including the development of vehicle configurations, energy sources, miniature electronics, and sensors. Figure 1 depicts some of the vehicles that have been developed. While these represent substantial progress in the field, the fact that none has been able to achieve true long-loiter times (>100 minutes) or efficient hovering flight is a testament to the difficulty of flying extended missions with small vehicles. Careful inspection of these vehicles reveals a variety of challenges. For example, a detailed breakdown of the mass fractions of three of these vehicles reveals a number of shortcomings when compared to

Felipe Bohorquez, Graduate Research Assistant, Department of Aerospace Engineering, University of Maryland.
Chris Cadou and Darryll Pines, Associate Professors, Department of Aerospace Engineering, University of Maryland.

full-scale systems. Figure 2 displays the mass fractions of three micro flyers and a full scale Boeing 767 commercial jetliner. Notice that for the small-scale flyers the mass fraction of the propulsion system (batteries and motor/transmission) is in excess of 60% of the total vehicle mass. In contrast the jetliner has a propulsion/fuel mass fraction of approximately 40%. It appears that this 20% savings at full scale is used entirely for payload since the payload mass fraction is 29% for the 767 and just 9% for the University of Maryland's MICOR (Micro Coaxial Rotorcraft) and CalTech's Microbat respectively. Additionally, there is a wide variation in the mass fraction of the structure required to support flight of the three small-scale vehicles compared in figure 2.

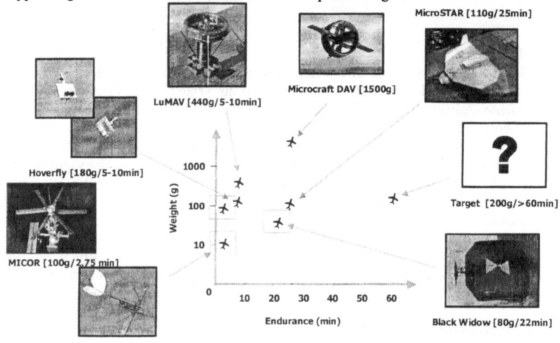

Figure 1: Existing MAV Designs and Endurance

Table 1: Design and Performance Parameters of some current MAVs [8-10]

	Black Widow (Aerovironment)	LUMAV (Auburn U.)	MicroStar (Lockheed-M)	Microbat {CalTech}	MICOR (UMD)
GTOW grams	80	440	110	10.5	103
Cruise Speed m/sec	13.4	5	13.4 to 15.6	5	5
Wing/Disk Loading-N/m2	40.3	185	70.9	40	25
Endurance minutes	30	20	25	2 min 16sec	3
Power Source	Lithium Ion Batteries	2-stroke IC Engine	Lithium Ion Batteries	NiCad Batteries	Lithium Ion Batteries
EnergyDensity W-hr/kg	140	5500 methanol	150	45	150
Power (hover) Watts	N/A	70	N/A	N/A	11
Hover FM	N/A	45%	N/A	N/A	42%

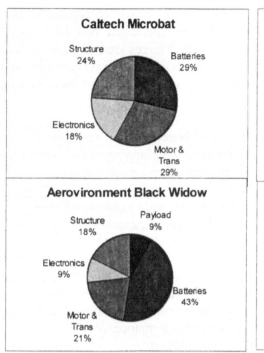
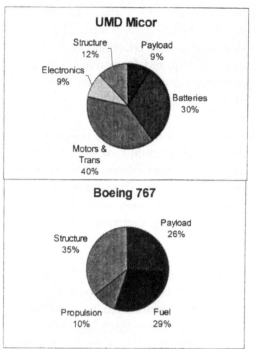

Figure 2: MAV Subsystem Mass Fractions

Table 1 summarizes some of the performance parameters of recent MAV designs flying under their own power. The values have been either estimated by the authors or obtained from data available in the literature [8-10]. A glance at this table reveals that the majority of current fixed wing and rotary wing designs rely on battery power for energy, and conventional airfoil shapes for achieving aerodynamic lift. With the exception of Microbat, all rely on conventional aerodynamic mechanisms for generating thrust and lift. Similar to small insects and birds, CalTech's Microbat uses flapping of its wings via an electric motor to generate thrust and lift, suggesting possibly a new *paradigm shift in the design* and development of future Micro Aerial Vehicles. Is it possible that MAV designers can improve the efficiency and performance of MAVs by studying biological flyers? Unlike subsonic fixed wing aircraft with their steady, almost inviscid flow dynamics, biological flyers such as insects and small birds fly in a sea of vortices when they flap their wings. These vortices can be used to keep MAVs aloft, especially in the case of hovering flight. Recently, it has been demonstrated by CFD analysis that bumblebees flap their wings in a complex kinematic pattern to generate lift and thrust. Similarly, hummingbirds are well-known masters of hovering flight by flapping their wings in excess of 20 Hz. This paper reviews the design and performance of current MAVs and compares this performance with biological flyers. This paper attempts to summarize technology advances that are required to accomplish efficient adaptive MAV flight.

2.0 Fundamental Technical Barriers Limiting Performance
Closer examination of biological fliers reveals that existing MAVs cannot match the aerodynamic performance (maneuverability and efficiency) of insects and small birds. However, the underlying physics that are responsible for these achievements is not well understood. For example, how an insect can take off backwards, fly sideways, and land

upside down [11] cannot be explained using conventional aerodynamic theory. Moreover, when insect wings are placed in a wind tunnel and tested over the range of air velocities that they encounter when flapped, the measured forces are substantially smaller than those required for active flapping flight [12-15]. Said another way, it appears that insect wings produce lift more efficiently than one would expect based on conventional aerodynamic theory.

The reasons for this remain research topics but some conclusions are emerging. Recent work by experimental biologists indicates that the pitching/plunging motion of the insect wing may improve efficiency by enabling the recovery of wake vorticity [16-17]. Other studies indicate that birds may increase their aerodynamic efficiency via large-scale morphing of their wing geometries [17]. A remarkably wide range of changes can be affected including variations in anhedral, dihedral, planform, camber, wing sweep and wing warping.

Given that the performance of the current generation of MAVs is vastly inferior to that of birds and insects, it seems logical that subsequent generations of adaptive MAVs could improve their performance by mimicking at least some aspects of biological flight. Thus, if MAVs are to approach and possibly exceed the performance of biological flyers, advances are required in several fundamental areas including:

-low Reynolds number aerodynamics
-lightweight and adaptive structures and materials
-energy storage/conversion to useful power/propulsion
-flight control

The status of each of these areas is reviewed below:

2.1 Low Reynolds Number Aerodynamics
Probably the greatest challenge for researchers is determining how insects and small birds can generate forces that at times are twice their body weight. Conventional steady-state aerodynamic theory is unable to explain this phenomenon. When insects are placed in a wind tunnel and tested over the range of air velocities that they encounter, the measured forces are substantially smaller than those required for active flight [12]. Thus, something about the complexity of the wing pitching/plunging/lagging motion increases the lift produced by a wing above and beyond that which it could generate under steady flow conditions or that can be predicted by conventional steady-state aerodynamic theory.

2.1.1 Unsteady Aerodynamic Mechanisms
This limit in conventional steady-state aerodynamic theory has prompted many researchers to search for the unsteady aerodynamic mechanisms that might explain the high forces produced by insects and small birds [16-31]. Pioneering research by Lighthill [20], Pennycuick [23-24], and Rayner [21-22] has provided some insight on avian flight. However, until recently little was known regarding the complex kinematics of insect flight. According to Ellington [16] (See Figure 3), the wingstroke of an insect is typically divided into four kinematic portions: two translational phases (upstroke and downstroke), when the wings sweep through the air with a high angle of attack, and two rotational

phases (pronation and supination), when the wings rapidly rotate and reverse direction. The unsteady mechanisms that have been proposed to explain the elevated performance of insect wings typically emphasize either the translational or rotational phases of wing motion [16-19]. The first unsteady effect to be identified was a flapping mechanism termed the "clap and fling" [18], a close apposition of the two wings preceding pronation that accelerates the development of circulation during the downstroke [19]. Although the clap and fling may be important, especially in small species, it is not used by all insects [16] and thus cannot represent a general solution to the phenomenon of force production. Recent studies by Liu [29] (See Figure 4) and Dickinson [26,27], suggests that "delayed stall and wake capture," might explain how insect wings generate such large aerodynamic forces.

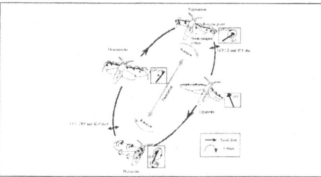 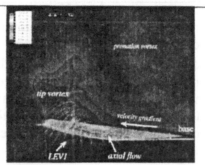

Figure 3: Diagram of the vortex system during the complete wingbeat cycle. The shaded area at pronation denotes the morphological lower wing surface on the insect diagram (insets). A large leading-edge vortex *(LEV)* with strong axial flow is observed during the downstroke. This *LEV* is still present during supination, but turns into a hook-shaped vortex. A small *LEV* is also detected during the early upstroke, and gradually grows into a large vortex in the latter half of the upstroke. This *LEV* is still observed closely attached to the wing during the subsequent pronation, where a trailing-edge vortex *(TEV)* and a shear-layer vortex *(SLV)* are also formed, together forming a complicated vortex system. (Ellington, 99) [26]	Figure 4: Top view of iso-velocity contours in a plane cutting the leading edge of the wing together with instantaneous streamlines at f=0 °. There is a spanwise velocity gradient from the base to the tip of the wing, but the flow becomes more complicated near the point of breakdown of the leading-edge vortex. The shed vortex of the early downstroke is still visible above the wing. A strong tip vortex with high velocity is detected where the leading-edge vortex *(LEVI)* breaks down and is shed. (Liu et al, 1998) [29]

A more detailed look at these aerodynamic mechanisms can be explained by examining the flow physics associated with any airfoil with increasing angle of attack. The high adverse pressure gradients that build up near the leading edge under dynamic translating (flapping) conditions cause flow separation to occur there. Experimental evidence suggests the formation of a shear layer that forms just downstream of the leading edge, which quickly rolls up and forms a vortex. Not long after it is formed, this vortex leaves the leading-edge region and begins to convect over the upper surface of the airfoil. This induces a pressure wave that sustains lift and produces airloads well in excess of those obtained under steady conditions. While delayed stall might account for enough lift to keep an insect aloft, it cannot easily explain how many insects can generate aerodynamic forces that exceed twice their body weight while carrying loads. Several additional unsteady mechanisms have been proposed, mostly based on wing rotation. Depending on the Reynolds number, these mechanisms include delayed stall, wake capture, rotational circulation and bound circulation. Recently, Dickinson [28] has shown that the flight of a drosophilia fly relies on complex kinematic motion of the insect's wings. This kinematic motion gives rise to unsteady lift and drag forces that exceed lift forces under steady aerodynamic loads at the same Reynolds number. While these recent advances in understanding aerodynamic physics have given researchers a clearer picture of low Reynolds number flight, few if any of the current micro air vehicle designs exploit these aerodynamic mechanisms to achieve efficient lift and thrust.

Thus, the major obstacle in realizing truly efficient micro air vehicles in the <100-200g class, are the complex unsteady aerodynamic mechanisms that contribute to the efficient lift and flight maneuvering capability of insects and small birds in the Reynolds number range 50< Re <10,000. A recent study by Baxter [30] in this Reynolds number regime using steady-state aerodynamic analysis reveals that the minimum drag-minimum power configuration of MAVs requires vehicles with lift coefficients in excess of 3. Baxter results indicate that as the profile drag coefficient increases relative to the induced drag coefficient, the operating C_L at which minimum drag and minimum power are obtained are significantly higher than those required at more conventional flight Reynolds numbers. This conclusion is largely due to the fact that the flow in this regime is highly viscous. Thus, it will be necessary to develop a new set of analytical and computational tools to compliment experimental investigations to study the complex aerodynamic flow behavior at Low Reynolds numbers for biological flyers.

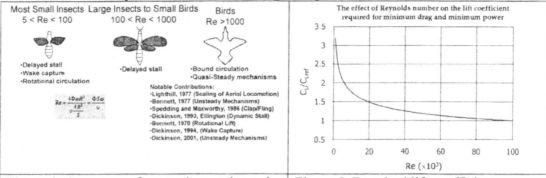

Figure 6: Summary of unsteady aerodynamic mechanisms governing low Reynolds flight regime of insects and small birds.	Figure 5: Required lift coefficient versus Reynolds Number for minimum drag minimum power.

2.2 Lightweight and adaptive structures and materials

While it is important to understand the aerodynamic physics associated with MAV flight, one must not neglect the contribution of lightweight, flexible and adaptive wing structures to the overall system performance. To determine if current MAV designs are structurally efficient, it is useful to compare MAV structural design parameters such as wingspan, aspect ratio, and wing loading to biological creatures. To enable this comparison, Rayner [21-22] has developed a number of geometric scaling relationships for birds. These approximate relationships are given below as a function of total body mass, m:

$$Wingspan = 1.17m^{0.39} \tag{1}$$

$$Aspect\ Ratio = 8.56m^{0.06} \tag{2}$$

$$WingLoading = 62.2m^{0.28} \tag{3}$$

2.2.1 Geometric Wing Scaling: Wingspan, Aspect Ratio and Wing Loading

Figures 6 thru 9 illustrate these geometric scaling relationships for small birds and how they compare to three current MAV designs. Although approximate, these scaling relationships illustrate that most MAV designs have shorter wingspans and lower aspect ratios than their biological counterparts, suggesting a higher maneuverability than the equivalent size bird. However, higher maneuverability for the same mass also implies higher bandwidth control for these systems. Another important property of winspan and aspect ratio is its connection to the aerodynamic properties of an aircraft. As the wingspan and aspect ratio increases, the lift to drag ratio also increases affecting the glide ratio of the aircraft. Thus, birds with long wingspans and high aspect ratios are more akin to dynamic soaring. Since most MAVs have short wingspans and low aspect ratios, one would not expect for these vehicles to have great glide or soaring properties. Finally, Figure 8 displays the average wing loading (N/m2) values for birds as a function of body mass. Specific birds are displayed by the 'x' symbol. The wing loading for three MAVs is also displayed in this figure with the 'o' symbol. Notice that the wing loading for MAVs is significantly higher than the equivalent size bird. This suggests that MAVs must fly faster in comparison to birds of comparable geometric size, aerodynamic properties and weight to stay aloft. To accomplish this goal MAVs must expend more power to overcome the induced aerodynamic drag. Another interesting aspect of nature is that the wing and aspect ratio for hummingbirds tends to be independent of body mass while their wingspan increases monotonically with mass. Figure 9 summarizes these geometric scaling laws for birds on a single chart and indicates that nature has figured out more efficient ways of achieving flight at low Reynolds number than humans.

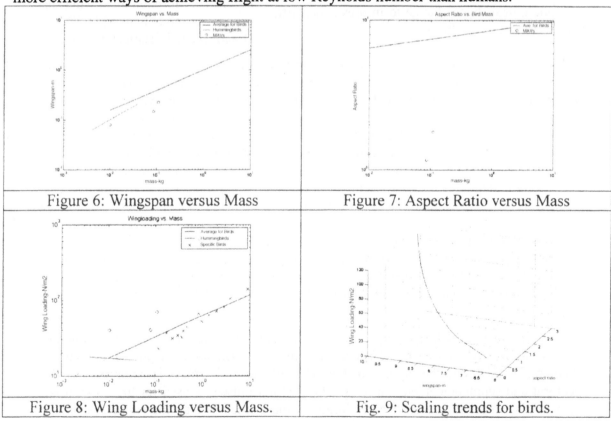

Figure 6: Wingspan versus Mass

Figure 7: Aspect Ratio versus Mass

Figure 8: Wing Loading versus Mass.

Fig. 9: Scaling trends for birds.

Finally, and probably most intriguing to the adaptive structures community is the fact that it has been observed that both insects and birds undergo significant morphological shape change in the geometry/stiffness (compliance) of their wings during flapping flight. This morphological shape change for some species is believed to occur passively as a built in mechanism that enables the animal to reduce its drag profile during the upstroke motion of its wing. To illustrate this phenomenon, Figure 10 displays the cross sections of a dragon fly wing undergoing flapping flight. Notice that the chord dimension reduces during the upstroke. In addition, one can

also see some changes in camber of the wing at various cross-sections. Similarly, wing shape changes occur in-flight for birds such as the trumpeter swan illustrated in the bottom part of Figure 10. Notice that during translational (flapping) flight that wings undergo planform, sweep and other deformations. Currently, none of the MAVs displayed in figure 1 undergo morphological shape change to improve their lift, thrust, or maneuverability.

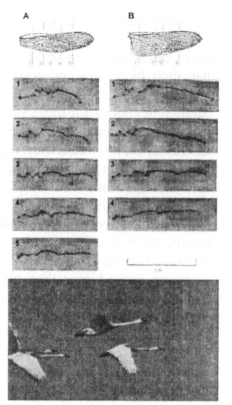

Figure 10: Morphological wing shape changes found in nature [31,32]. Dragonfly and Trumpeter Swan.

2.3 Energy storage/conversion to useful power/propulsion

While maximizing the power-to-weight ratio is important for all aero power systems, two factors make it absolutely critical to micro-air vehicles. First, the overall aerodynamic efficiencies of conventional fixed-wing vehicles using steady-state analysis tools decreases with size. This difference is reflected in the fact that the Reynolds numbers associated with micro-air vehicle flight are several orders of magnitude smaller than those associated with conventional flight making the aerodynamic efficiency of lift-generating systems for micro-air vehicles substantially less than their conventional-scale counterparts. Finally, the efficiency of the power/propulsion system appears to degrade with decreasing size. Together, these factors conspire to make the power/weight ratio and efficiency of the power system critical and enabling for hovering as well as fixed-wing mechanical flying vehicles. Moreover, while improving efficiency at small scale is important, one must also achieve mass fractions of the propulsion that are less than 50% of the total MAV mass. Current MAVs exhibit propulsion system mass fractions of approximately 60%.

In comparison, a large fraction of the weight of small biological flyers is concentrated in muscle matter used to generate large-scale complex translational (flapping) kinematics. Rayner [21-22] found that approximately 16% of the bird's mass is comprised of the pectoral and supracoracoideus muscles. The pectoral muscle is used for the downstroke of a bird's wing and is significantly larger in mass fraction than the supracoracoideus muscle that is used for the upstroke motion of the wing. In comparison, birds have approximately 3 times the mass fraction of muscle found in humans. Thus, this large muscle mass fraction coupled with the elasticity/flexibility and low inertia of a bird's wing provides the necessary power and lift to weight ratios required for efficient low Reynolds number flight.

Figure 11 shows the advantage that hydrocarbon fuels enjoy over other energy sources for hovering micro-air vehicles. It plots the mass fraction of fuel required for a 'typical' hovering vehicle weighing 100g and flying for 30 minutes against the energy density of the fuel. Assuming that the fuel mass fraction of a hovering micro-air vehicle will need to be the same or smaller than that of a conventional hovering vehicle (10%) in order to accommodate a payload, the figure shows that only hydrocarbon fuels and hydrogen are capable of meeting the requirement. Therefore, the challenge becomes how to build a high power/weight ratio engine that releases the energy stored in the fuels *efficiently*.

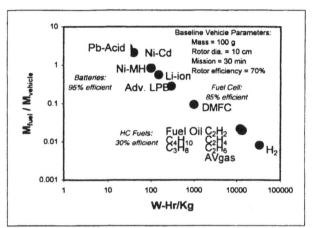

Figure 11. Required fuel mass fraction as a function of the energy content of the fuel for a generic micro-air vehicle in hover.

Unfortunately, building small *and* efficient engines is difficult because thermal and frictional losses scale directly with surface/volume ratio and therefore increase as the size of the engine is reduced [33,34]. Several research teams building micro-engines in the 10watt/5gram class have recognized this problem but exactly how these losses scale with device size is not well understood. While this information is important to future designers of micro-power systems, there has been very little work done in this area.

2.4 Summary of Status of Existing MAV Development

In summary, it appears that the major reason for the across-the board failure to achieve true long-loiter, highly maneuverable and hovering flight in a single configuration is that all of the configurations/designs pictured in figure 1 rely on conventional steady-state aerodynamic mechanisms that become inefficient at the low Reynolds numbers associated with the operation of MAVs. The inefficient utilization of aerodynamic lift coupled with relatively inefficient systems for storing and releasing energy (power systems) are the two factors inhibiting the development of efficient miniature mechanical flying

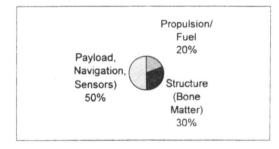

Figure 12. Estimate of Mass Fractions computed for birds.

machines. From a systems perspective, it appears that biological flyers achieve remarkable mass fractions (See Figure 12) when compared to miniature mechanical flyers. Thus, future generations of MAV will undoubtedly take lessons from nature to achieve efficient operating mechanical flying machines.

3.0 Opportunities for Adaptive Structures

It is likely that adaptive structures technology will play an important role in the next generation of bio-mimetic MAV designs inspired by nature. However, there are numerous technical questions that must be answered if truly adaptive bio-mimetic MAVs are to become practical. While there are numerous opportunities to affect shape change, modify aerodynamic properties and achieve large-scale kinematic motion of micro air vehicles, one question that arises is what smart material actuators will be used to achieve shape changes and kinematic motion at low and high bandwidth? Will new advances in actuator material/smart compliance mechanisms be required to improve efficiency? In addition, will it be feasible to integrate this technology in flying machines weighing less than 200 grams? To help answer these questions, Table 2 reviews some common properties of representative smart material actuators. Closer examination of the properties of these actuators may offer some clues to answer the questions raised above. Forces required for MAV flight are typically less than 2 N. Thus, most of the actuators in this table have the capacity to support such loads. However, if morphological shape changes are required to improve aerodynamic efficiency, large strains would be necessary. Shape memory alloys actuators activated by thermal temperature gradients have the capability of inducing very large strains (>4%). However, these actuators suffer bandwidth limitations, implying that shape changes cannot occur instantaneously. Recently, magnetic shape memory alloys have become available to the adaptive structures community. While these new actuators possess many desirable attributes such as large strains (>2%) and higher bandwidth response, they suffer from excessive power requirements due to losses in magnetic field intensity used to induce a shape and compliance transformation in the material. In addition, magnetic shape memory alloy devices tend to be heavy. In summary, there is no simple answer to the choice of a smart material actuator for a particular MAV application. The choice will depend on the system level requirements, vehicle design, mechanical transduction device, wing compliance, electronics and choice of the energy storage/power conversion system. The underlying theme of bio-mimetic adaptive MAVs will be achieving high system level efficiency.

Table 2: Properties of Smart Material Actuators

Property	Piezoelectric-PZT	Magnetostrictive (Terfenol-D)	Shape Memory Alloy (TiNi)	Magnetic SMA
Block Force	>100 N	>100N	10 N	10 to 100 N
Strain	0.3%	0.5%	6%	2 to 6%
Bandwidth	100 kHz	10 kHz	1 Hz	1 kHz
Elastic Storage Energy J/cm3	0.1	0.025	>100	>100
Energy Density (J/g)	0.013	0.0027	>15	

3.1 Application of Adaptive Structures Technology to Conventional MAVs

In order to improve the overall performance of MAVs the use of smart material actuators is very appealing. Adaptive structures can be used over a broad range of components and applications. For example, in rotorcraft and fixed wing MAVs modifying the twist distribution of the blades or wings requires different actuator characteristics. To actively modify the twist distribution in the rotor blades the basic requirement is to have the capability to induce enough twist to affect the aerodynamics. Mainly two smart materials can be used depending on the application. For large DC strains, and low bandwidth SMA (shape memory alloys) are well suited. When optimizing the twist distribution a relatively low bandwidth is necessary, since twist changes will be required mainly between hover and forward motion (where optimum hover twist is detrimental). The possible bandwidth of the actuation is around 1 Hz, which is an acceptable value for this application. A tube of SMA strained in torsion can accomplish the relatively large amplitude of the actuation. This same mechanism can be used for modifying the twist distribution of propellers. In this case also low bandwidth actuation with relatively large forces are necessary. The requirements for twisting the wings and modifying the airfoil shape of a fixed wing aircraft are similar, however the bandwidth required is higher if the actuation is used for control purposes and not only for aerodynamic optimization. Higher bandwidth actuators like piezo electric materials have considerably less free strain values and present drift of the actuator strain in response to a DC excitation. The drift phenomenon is a slow increase in the free strain with time after the application of a DC field. This makes them inappropriate for applications that need to maintain long-term static deflections.

However this is not the case of the twist changes required for control purposes where embedded Piezo actuators oriented at 45 degrees along the span of the blades generate pure twist. In order to create a control moment, cyclic actuation at one per revolution has to be possible. Micor's rotors turn at 4000 RPMs in hover, which is equivalent to an actuation frequency of 66 Hz. Impossible to reach for thermally induced SMA, but well within the capability of piezoelectric actuators.

In MICOR's case, smart materials could find use in the rotor blades (See Figure 13). The twist distribution could be adaptively modified in order to obtain a more uniform inflow in hover or at any forward speed. This is a similar strategy that has been investigated on Mach scale rotor blades to control transmitted vibration. The twist of the rotor blades is actively modified along the azimuthal angle to vary the lift distribution of the blades, in order to produce bending moments on the hub that can eliminate part of the undesirable vibrations. In the case of MICOR [10] the optimized twist distribution would increase the overall aerodynamic performance of the vehicle reducing the induced power requirements of the rotors, and thus extending the endurance of the vehicle.

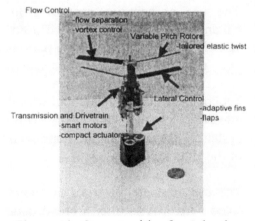

Figure 13: Opportunities for Adaptive Structures for a Hovering MAV

As shown earlier in Table 1 the performance, i.e. Figure of Merit (FM) of MICOR is considerably lower than that of a full-scale helicopter, where Figures of Merit close to 80% are common. This low performance is a direct result of the low Reynolds number regime. In this environment, the viscous effects take over the inertial effect in the flow, and as a consequence early flow separation and a reduction in the Clα slope occur, making it more difficult to the aerodynamic hover efficiency. The actual performance of the prototype can be considered as a benchmark that gives a hint about the achievable performance of a conventional small-scale rotor with and without twist. Figures 14 and 15 display hover efficiency results from experiments performed on two sets of uniform blades (10 degrees linear twist and no twist) and just one airfoil (8% arc) design were used. These blades were chosen based on their ease of manufacture and relatively good aerodynamic performance at low Re. Notice that only a 2.5% improvement is achieved with the blades incorporating 10 degree linear twist. Thus, the application of adaptive structures technology to conventional rotary wing configurations will depend on the improved aerodynamic performance gain versus the additional power and mission requirements. Better results can be obtained with improved aerodynamic tools and adaptive structures technology to deliver optimal airfoil properties at various flight conditions. Nevertheless, these results illustrate the potential benefits and challenges facing low Reynolds number flight.

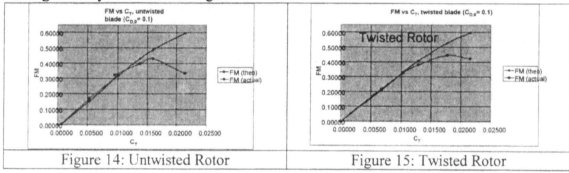

| Figure 14: Untwisted Rotor | Figure 15: Twisted Rotor |

The use of adaptive structures in the rotor is not restricted to an aerodynamic enhancement; they can also be used to provide alternative lateral control means to the vehicle. If the rotor blades are periodically twisted at one per revolution the lift produced over the disk could be varied azimuthally, creating the moments required to laterally control the rotorcraft. Such a system would be equivalent to the cyclic control of a helicopter with swashplate, but with the advantage of having no moving parts, hence with a lesser mechanical complexity. This system could replace the gimbaled drive train/rotor system or the aerodynamic flaps design in MICOR, reducing the overall weight and drag of the vehicle.

The use of smart materials in other MAV configurations can also be considered. The fixed wing MAVs offer a broad range of possible new concepts to be tested. Let's first think about the aerodynamic enhancement that could be obtained by controlling the boundary layer using smart synthetic microjets, which are small air pumps on the surface of the wing that inject and suck air, having a cero net mass flow in time. At low Reynolds numbers the adverse pressure gradients on the upper surface of the airfoils commonly result in the phenomenon known as laminar separation bubble, where the laminar boundary layer separates from the surface of the airfoil, only to be reattached further behind as a turbulent Boundary layer. The region between the separation and reattachment points is known as the separation bubble, here the flow may be circulating, the direction near the airfoil surface may even be the opposite of the direction of the outer flow. There is almost no energy exchange with the outer flow, which makes the laminar separation bubble quite stable. This phenomenon sharply reduces the lift and increases the drag of the wings.

One of the possible strategies that can be used to avoid the formation of the separation bubble is to create enough disturbances in the flow to cause transition into the turbulent state, before the laminar separation can occur. This can be achieved by the use of synthetic microjets. The use of smart materials to actuate the micro jets could result in an enhancement of the aerodynamic characteristics of the aircraft. Delaying the onset of stall in airfoils, especially in high angle of attack maneuvers. The consequences of this are a better performance and maneuverability, increased lift and decreased drag for airfoils operating at or near stall conditions.

Just like in MICOR's case not only an aerodynamic performance improvement can be sought, alternative control schemes can also be explored. Let's conceive a flying wing MAV that has no control surfaces, but that instead of using conventional elevons or ailerons, it actively modifies its fuselage shape to control its trajectory. Independent twist control of the wings to increase or reduce the lift along the span could accomplish this task.

3.2 Emerging Research Trends: Flapping Flight and Ornithopter Configurations

The next generation of MAV designs that shows the greatest promise for the use of adaptive structures technology are the ornithopter configurations. Up to now, there has been little research toward the development of these types of configurations. Recently, Caltech, with its prototype Microbat has placed its focus on the wing design, where flexible thin membranes, similar to bat wings are used. Since reproducing the complex kinematics of biological flyers is challenging, one of the primary problems faced when using simplified movements is the increased drag in the upstroke. To reduce this drag, MEMs electrostatic actuator valves have been incorporated to parylene wings in order to control the flow through the membrane [35-36]. Without electrostatic actuation air can move freely from one side of the skin to the other side trough the vent holes. With actuation, these vent holes are sealed and the airflow is controlled. The membrane behaves as a complete diaphragm. Actuation is done on the downstroke and stopped on the upstroke reducing the drag of the wings over the flapping cycles. To aid the design of ornithopters, one can again examine the characteristic of biological creatures. Studies of birds by Rayner [21] and Norberg [32] have revealed the following relationships characterizing the flapping wingbeat frequency used to acquire thrust and lift:

$$f_{min} = 2m^{\frac{-1}{6}} \tag{4}$$

$$f_{max} = 8m^{\frac{-1}{3}} \tag{5}$$

$$f_{hum\min gbird} = 1.32m^{-0.6} \qquad (6)$$

3.2.1 Kinematic Motion

Actuation of the flapping mechanism is another potential application for smart material, the use of conventional electric motors, requires the transformation of a rotating motion into a linear reciprocating movement. This involves a certain mechanical complexity, added weight, and more important, a fixed movement pattern, that makes difficult the change of flapping amplitude and hence restricts the control schemes and experimental degrees of freedom. Smart linear actuators can alleviate these problems by providing controllable stroke amplitude at high frequencies. In order to apply these concepts in prototypes or experimental setups certain physical characteristics are required from the smart actuators depending on its use.

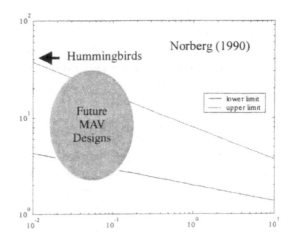

Figure 16: Wingbeat Frequencies Associated with Avian Flight

3.2.2 Morphological Shape Changes

But not only can the spanwise shape of the wings be modified, the airfoil itself can undergo radical transformations. Camber modification along the chord could be used to adapt the airfoil to different flight conditions and for controlling the pitch of the vehicle. A relatively small reshaping of the tail of the airfoil (see reflex airfoils for example) results in a large variation on the effective pitching moment on the fuselage. So the simultaneous use of these two shape transformations could provide enough control authority to command the MAV. Another key component of a fixed wing MAV that could be improved through the use of smart structures is the propeller. The twist distribution can actively be varied using a torsional actuator so that every section works at an optimal angle of attack at different advance ratios.

4.0 Summary and Conclusions

In summary, nature has evolved thousands of miniature flying machines (insects and small birds) that out perform man-made miniature flying machines routinely. While some of the details underlying the operational success of biological fliers remain a research topic, a general picture is emerging which indicates that the overwhelming superiority of biological fliers over existing MAVs stems from three fundamental factors: an ability to generate lift more efficiently than existing technologies, an ability to harness morphological changes in wing kinematics/structure and an ability to store and release energy more efficiently. The major obstacle in this endeavor is our limited understanding of how to build pitching/plunging aerodynamic structures that exploit unsteady aerodynamic effects to achieve high maneuverability and efficiency, and morphing structures to achieve high efficiency. Since morphing aerodynamic structures are simpler than pitching/plunging ones, developing MAVs that incorporate adaptive morphing aerodynamic surfaces is the logical first step along the path to developing fully bio-mimetic airframes. Adaptive structures technology will undoubtedly play a major role in future airframe designs.

REFERENCES

[1]   Wootton, R. J. (1981). Palaeozoic insects. *Annu. Rev. Ent.* 26, 319–344.

[2]   Ellington, C. P. (1991a). Aerodynamics. and the origin of insect flight. *Adv. Insect Physiol.* 23, 171–210.

[3]   Gallington et al., (1996),"Chapter 6: Unmanned Aerial Vehicles", Future Advances in Aeronautical Systems, AIAA, Reston, VA.

[4]   "Micro Spy Planes, Inside the World's Smallest Aircraft," Popular Science, Jan. 1998, pp. 53.

[5]   "Micro Air Vehicles Hold Great Promise, Challenges," Aviation Week & Space Technology, April 14, 1997, pp. 67.

[6]   Dr. J.M. McMichael and Col. M.S. Francis, "Micro Air Vehicles - Toward a New Dimension in Flight", http://www.aero.ufl.edu/~issmo/mav/info.htm.

[7]   Davis, Jr., W.R., Kosicki, B.B., Boroson, D.M., and Kostishack, D.F., (1996), "Micro Air Vehicles for Optical Surveillance", The Lincoln Laboratory Journal, Vol. 9, No. 2.

[8]   Joel M. Grasmeyer, Matthew T. Keenon (AeroVironment, Inc., Simi Valley, CA)," Development of the Black Widow Micro Air vehicle", AIAA Paper 2001-0127, AIAA, Aerospace Sciences Meeting and Exhibit, 39th, Reno, NV, Jan. 8-11, 2001.

[9]   T.N. Pornsin-Sirirak, S.W. lee, H.Nassef, J. Grasmeye, Y.C. Tai, C. M. Ho, M. Keennon, MEMS Wing Technology for a battery-powered Ornithopter", Proceedings of the 13 th IEEE Annual International Conference on MEMS 2000, Miyazaki, Japan, 1/23-27/00,pp. 779-804

[10]   Bohorquez, F., et al., (2000), "Design and Development of a Micro Coaxial Rotorcraft", AHS Design Conference, San Francisco, CA, Jan. 2000.

[11]   Dalton, S., Borne on the Wind, Reader's Digest Press, New York, 1975.

[12]   Ellington, C.P., (1984), "Morphological Parameters, The Aerodynamics of hovering insect flight", Philos. Trans. Roy. Soc. London Ser. B305:17-40..

[13]   Biological Fluid Dynamics, C. P. Ellington and T. J. Pedley, Eds. (Company of Biologists, London, 1995), pp. 109Ð129.

[14]   Nachtigall, W., Insects in Flight (McGraw-Hill, New York, 1974)

[15]   Ellington, C. P. (1991a), "Aerodynamics. and the origin of insect flight". *Adv. Insect Physiol.* 23, 171–210.

[16]   Dickinson, M.H. (1996). "Unsteady mechanisms of force generation in aquatic and aerial locomotion". Amer. Zool., 36

[17]   Rayner, (1988), "Form and Function in Avian Flight", Current Ornithology, editor, Johnston, R.F., vol. 5, New York, Plenum Press, pp.1-66.

[18]   Weis-Fogh, T., (1973), "Quick estimates of flight fitness in hovering animals, including novel mechanisms for lift p;roduction.. 59,169-230.

[19]   Spedding , G.R.,and Maxworthy, T., J.Fluid Mech. 165, 247 (1986).

[20]   Ligthhill, MJ (1977), "Introduction to the scaling of aerial locomotion", Scale Effects in Animal Locomotion , editor: Pedley, TJ, New York: Academic Press, pp.365-77.

[21]   Rayner, JMV, (1979), "A new Approach to Animal Flight Mechanics", J. of Exp. Bio, 80:pp. 17-54.

[22]   Rayner, JMV (1988), "Form and Function in Avian Flight," Current Ornithology, editor, Johnston, R.F., vol. 5, New York, Plenum Press, pp.1-66.

[23]   Pennycuick, CJ, (1989), Bird Flight Performance: a practical calculation manual, New York: Oxford University Press.

[24]   Pennycuick, CJ (1996), "Wingbeat frequency of birds in steady cruising flight: new data and improved predictions", J. of Exp. Bio, 199:pp. 1613-8.

[25]   Ellington, C.P., (1999), "The Novel Aerodynamics of Insect Flight: Applications to Micro-Air Vehicles", J. Exp. Bio., 202:3439-3448.

[26]   Dickinson, M.H., Lehmann, F.-O., and Sane, S., (1999), "Wing rotation and the aerodynamic basis of insect flight," Science, in press.

[27]   Dickinson, M.H., Lehmann, F.-O., and Chan, W.P., (1998), "The control of mechanical power in insect flight", Amer. Zool., 38, 718-728.

[28]   Sanjay, S. P., Dickinson, M. H., et al. (2001), "The control of flight force by a flapping wing: lift and drag production, J. Exp. Bio.204:2607-2626.

[29]   Liu, H, Ellington, C.P., and Kawachi, K., (1998), "A Computational Study of Hawkmoth Hovering", J. Exp. Bio, 201, 461-477.

[30]   Baxter, D., and East, R., "A Survey of Some Fundamental Issues in Micro-Air-Vehicle Design," *Proceedings of the 14th Bristol International Conference on UAV Systems*, April 1999, (page 34.2)

[31]   McMahon and Bonner, (1983), One Size One Life, Scientific American Books, Inc., New York.

[32]   Norberg, (1990), Vertebrate flight: mechanics, physiology, morphology, ecology and evolution. Springer, New York.

[33]   A. H. Epstein and S. D. Senturia, 'Macro Power from Micro Machinery', *Science*, Vol 276, May 1997, p. 1211.

[34]   K. Fu et al., 'Design and Experimental Results of Small-Scale Rotary Engines', *Proceedings 2001 (IMECE)*, New York, November 11-16, 2001.

[35]   T.N. Pornsin-Sirirak, Y.C. Tai H. Nassef, C.M.Ho; Flexible Parylene Actuator for Micro Adaptive Flow Control; 2001

[36]   T.N. Pornsin-Sirirak, Y.C. Tai H. Nassef, C.M.Ho, "Unsteady-State Aerodynamic Performance of Mems Wings", International Symposium on Smart Structures and Micro systems 2000, The Jockey Club, Honk Kong,10/19-21/00,pp. G1-2

DEVELOPMENT OF A STRAIN-RATE DEPENDENT MODEL FOR UNIAXIAL LOADING OF SMA WIRES

Harsha Prahlad and Inderjit Chopra

ABSTRACT

This paper describes a methodology to incorporate the effects of non-quasistatic loading on the extensional behavior of an SMA wire. An SMA shows considerable differences between quasistatic loading conditions and loading at higher strain rates. The model incorporates a coupled thermo-mechanical analysis in conjunction with the rate form of an SMA quasistatic model (Brinson model is used as a representative model). This model is quite generic and is equally applicable with any other quasistatic model describing SMA behavior. In this model, the temperatures and temperature rates are not prescribed, but are derived from energy conservation of the material. The stress rate is described in terms of the instantaneous stress, temperature and strain and also the strain and temperature rates. A coupled heat transfer analysis yields the temperature rate as a function of the state variables that include stress, temperature, strain and rates of strain and temperature. By solving these coupled pseudo-first order differential equations with Runge-Kutta method, the evolution of both the stress and temperature is simultaneously obtained for a given strain rate and environmental temperature. Important parameters governing heat transfer such as specific heat, heat transfer and latent heat are modeled and validated with experimental data. The predictions for the stress-strain curves show good agreement with the experimental data. The model is also shown to accurately predict the behavior under more complex loading conditions involving composite strain rates.

1. INTRODUCTION

Shape Memory Alloys (SMAs) are being viewed as one of the most promising among the emerging 'smart' materials. Their applications encompass a wide range of fields such as aerospace, civil, mechanical and biomedical engineering [1]. These wide range of applications are all derived from crystallographic changes that occur as the material transforms

[1]Graduate Research Assistant, Alfred Gessow Rotorcraft Center, Department of Aerospace Engineering, University of Maryland, College Park, MD, 20742, U.S.A.
[2]Alfred Gessow Professor and Director, Alfred Gessow Rotorcraft Center, Department of Aerospace Engineering, University of Maryland, College Park, MD, 20742, U.S.A.

from one equilibrium phase to another. The unique capacity of the SMA to develop very large plastic strains (6-8%) that are completely recoverable upon heat activation and the corresponding ability to transform this energy into useful work is used in a variety of actuators [2]. When the SMA is restrained such that it cannot recover this strain, a very large recovery force develops in the wire, which can then be used to tune the properties of a structure [3,4]. This phase transformation is also accompanied by a large change in resistivity, specific heat, young's modulus and other properties of the material.

The behavior of the material is primarily a function of three variables: stress, strain and temperature, and their associated rates. These variables are interdependent, and the material behavior is a non-linear function of these variables. A constitutive model attempts to describe the material behavior as a function of these variables. These models are based on thermo-mechanics, or a combination of thermo-mechanics and SMA phenomenology, and/or statistical mechanics. The properties of a particular alloy depend on the precise composition of the constituent materials, the processing technique and other factors involving the manufacturing and heat treatment of the alloy. To accommodate these variations, most of these constitutive models employ phenomenological parameters that are determined by experimentation for a particular material [5].

Most of these constitutive models [6–9] are developed for quasistatic loading only, and it is assumed that the material at each instant is in thermodynamic equilibrium. However experiments at higher strain rates have been observed to yield results that are quite different from those obtained with low strain rates [10,5]. The cause of the variation of the material characteristics with strain rate is a subject of active current research.

Most research in this area have fallen in two broad categories. One approach is to derive constitutive models with a fundamental dependance on strain rate. This approach has been extensively used in modelling high rate plastic behavior of metals [11]. This models the kinetics of the phase transformation using strain energy functions [12] and Phase Interaction Energy Functions (PIEF) [13], among others. The other approach is to use a heat transfer model on a macroscopic level to couple with the rate form of quasistatic models [14,15]. These models, however, have been applied only to the pseudoelastic regime in most cases [16].

In addition to these two approaches, phenomenological models [17] that estimate the effects of frequency and strain rate on the mechanical characteristics of the SMA in a limited temperature range have also been proposed. The more generalized models based on thermodynamics and kinetics, may, however, be effective in modeling some of these effects over the entire range of thermomechanical conditions.

Effects of Non-Quasistatic Loading in SMA Wires

Experimental investigations previously reported by the authors [5] demonstrated a number of deviations of the SMA response between quasistatic and non-quasistatic loading. These included :

1. The transformation stresses increased significantly as function of the strain rates above rates of $5 \times 10^{-4}/s$. Below this strain rate, there was no appreciable change in the mechanical properties of the material at different strain rates. In this paper, strain rates below

this value are referred to as "quasistatic", and strain rates above this as "non-quasistatic". The increase in transformational stresses during non-quasistatic loading was accompanied by a change in the path followed by the transformation, and also significant changes in the instantaneous temperature of the material.

2. When the material was loaded under non-quasistatic loading, the stresses reached an instantaneous value which was higher than the those reached during quasistatic loading for the same final strains. However, when the strain was subsequently held constant, the material returned to lower values of stress consistent with quasistatic loading. This phenomenon, known as "stress-relaxation", has also been reported before in different SMA alloy systems [18]. The magnitude of the decrease in stress due to the stress-relaxation was found to be significant (of the order of 30% of original stress).

3. The phenomenon of stress relaxation was also observed during loading patterns that combine quasistatic and non-quasistatic loading. When the material was loaded under non-quasistatic loading, stresses instantaneously reached high values, and returned to values consistent with quasistatic loading when the strain rate was stepped down to quasistatic values.

In each of these tests, non-quasistatic loading was consistently accompanied by increases in temperature. Shaw and Kyriakides [10] experimentally observed that the high strain rate behavior is accompanied by a significant change in material temperature, which in turn affects the mechanical behavior of the material. This is due to the origination of local nucleation sites with temperature differences along the wire [19]. It was shown that the material may momentarily reach higher temperatures locally, and then settle down to an equilibrium that is at the same temperature as that of its surroundings due to convective heat losses(i.e. the temperature controlled in the chamber). It has also been demonstrated that the dependance on strain rate disappears when the wire is placed in an effective heat sink [20], further indicating that the cause of the mechanical variations is the variations of the temperature associated with non-quasistatic loading.

However, it is still to be determined if this can explain the magnitude of the variations of stress reported experimentally [5]. The objective of the current research is to investigate whether a coupled material-heat transfer approach can adequately predict the effects of non-quasistatic loading in the entire thermomechanical range.

The coupled model has two parts; one describing the rate form of SMA constitutive models to prescribe the stress rate, and another an energy analysis to prescribe the temperature rates induced in the wire. Since the material properties that are used in this material have a primary influence on the model predictions, a detailed experimental validation of each of the parameters and a brief parametric study is also discussed in the following sections.

2. RATE FORM OF QUASISTATIC SMA CONSTITUTIVE MODELS

This section derives the rate form of a quasistatic SMA model. The Brinson model [8] is used as a representative model in this study, although the principles used are equally applicable to other quasistatic models that predict SMA behavior, such as Tanaka [6] model and Liang and Rogers [7] model. In the present formulation, the stress rate is determined as a function of not only the state variables - strain and temperature, but also their associated rates. The constitutive equation is represented in the following first-order form as

$$\dot{\sigma} = \sigma(\epsilon, T, \dot{\epsilon}, \dot{T}) \tag{1}$$

where $\dot{\sigma}$, $\dot{\epsilon}$, and $\dot{T}$ are the time rates of stress, strain and temperature respectively, and σ, ϵ and T are their instantaneous values. In the following analysis, the temperature rate and corresponding instantaneous temperature can be either prescribed externally, or determined from a coupled heat transfer analysis described in the following section.

The quasistatic formulation for Brinson model [8] has been shown to be

$$\sigma - \sigma_0 = E(\xi)\epsilon - E(\xi_0)\epsilon_0 + \Omega(\xi)\xi_s - \Omega(\xi_0)\xi_{s0} + \Theta(T - T_0) \tag{2}$$

A rate form of these equations can be derived by taking derivatives with respect to time

$$\dot{\sigma} = E\dot{\epsilon} + \Omega_s\dot{\xi}_s + \Theta\dot{T} \tag{3}$$

Here, it is assumed that the Young's modulus $E(\xi)$ of the linear elastic region is constant for a given starting conditions ($\dot{E} = 0$) and that the transformation constant (Ω) also does not vary with strain rate. Both these assumptions are made from experimental observations of the stress profiles at different rates of loading [5]. Neglecting the contribution of pure-phase thermal expansion ($\Theta\dot{T}$ term) and applying initial conditions to these equations, we see that the can derive the rate-form of the simplified Brinson equation to be

$$\dot{\sigma} = E(\xi)(\dot{\epsilon} - \epsilon_l\dot{\xi}_s) \tag{4}$$

The equations for the martensite volume fraction rates are obtained by taking time derivatives of the corresponding quasi-static equations. For the conversion to detwinned martensite, the martensitic volume fractions are given by For $T > M_S$ and $(\sigma_s^{cr} + C_M(T - M_s)) < \sigma < (\sigma_f^{cr} + C_M(T - M_s))$,

$$\dot{\xi}_S = -(\tfrac{1-\xi_{s0}}{2})(\tfrac{\pi}{\sigma_s^{cr}-\sigma_f^{cr}})(\dot{\sigma} - C_M\dot{T}) \tag{5}$$

$$\times sin\left[\tfrac{\pi}{\sigma_s^{cr}-\sigma_f^{cr}}(\sigma - \sigma_f^{cr} - C_M(T - M_s))\right](\dot{\sigma} - C_M\dot{T}) \tag{6}$$

$$\dot{\xi}_T = -(\tfrac{\xi_{T0}}{1-\xi_{S0}})\dot{\xi}_s \tag{7}$$

(8) For $T < M_S$ and $\sigma_s^{cr} < \sigma < \sigma_f^{cr}$,

$$\dot{\xi}_S = -(\tfrac{1-\xi_{s0}}{2})(\tfrac{\pi}{\sigma_s^{cr}-\sigma^{cr}r_f}) \ sin\left[\tfrac{\pi}{\sigma_s^{cr}-\sigma^{cr}r_f}(\sigma - \sigma_f^{cr})\right]\dot{\sigma} \tag{9}$$

$$\dot{\xi}_T = -(\tfrac{\xi_{T0}}{1-\xi_{s0}})\dot{\xi}_S + \dot{\Delta}_{T\epsilon} \tag{10}$$

(11) where, if $M_f < T < M_S$ and $T < T_0$

$$\dot{\Delta}_{T\epsilon} = -(\frac{1-\xi_{T0}}{2})a_M sin(a_M(T - M_f))\dot{T} \tag{12}$$

else

$$\dot{\Delta}_{T\epsilon} = 0 \tag{13}$$

For conversion to austenite, these variables then become: For $T > A_S$ and $C_A(T - A_f) < \sigma < C_A(T - A_S)$:

$$\dot{\xi} = -(\tfrac{\xi_0}{2})a_A \ \ sin\left[a_A(T - A_S - \tfrac{\sigma}{C_A})\right](\dot{T} - \tfrac{\dot{\sigma}}{C_A}) \tag{14}$$

$$\dot{\xi}_S = \tfrac{\xi_{so}}{\xi_0}\dot{\xi} \tag{15}$$

$$\dot{\xi}_T = \tfrac{\xi_{To}}{\xi_0}\dot{\xi} \tag{16}$$

(17)where the material constants are the same as defined in the quasi-static model, are obtained from a comprehensive experimental characterization of the SMA wire [5]. It is important to note that the development of the rate equations yield the same predictions as the quasistatic form if the temperature is held constant (temperature rate is zero).

Using the rate formulation and given the instantaneous temperature and rates of temperature and strain, we can calculate and solve this differential equation to solve for the instantaneous stresses. However, in reality, the temperature rise of the material is not an independent prescribed function, but is coupled to the loading pattern, material characteristics and heat transfer aspects of the test sample. Describing the instantaneous temperature rates in terms of these states requires an energy analysis of the SMA material [21]. The following section describes the development of this energy analysis and describes methods to validate each component of this analysis with experimental data.

3. THERMO-MECHANICAL ENERGY ANALYSIS

This section deals with a heat transfer analysis to determine the instantaneous temperature as a function of time for a prescribed mechanical loading. This is done by determining thermodynamic equilibrium between the input mechanical power, rate of heat loss to the surroundings through convective heat transfer, and rate of heat absorbed by the SMA wire. The heat absorbed by the SMA wire has two components - a specific heat component(related to a rise in temperature of the homogenous phase), and a latent heat component (related to the rate of transformation in the material). The equilibrium equation then becomes

$$\dot{E}_{mech} = \dot{E}_{loss} + \dot{E}_{specific} + \dot{E}_{latent} \tag{18}$$

Each of these quantities can be expressed as functions of the material states, and validated against experimental data.

3.1. Mechanical Energy Input

If we assume that the material is homogenous, i.e. stress, strain and temperature are constant throughout the volume of the material, then the instantaneous mechanical input energy and the corresponding power input to the material are given by

$$E_{mech} = \tfrac{1}{2} \int_0^x (PV\,dx) \tag{19}$$

$$\dot{E}_{mech} = \tfrac{1}{2} V(\sigma\dot{\epsilon} + \dot{\sigma}\epsilon) \tag{20}$$

(21) where P, x and V represent the force, deflection, and volume of the material. During the loading cycle (i.e. increasing stress and strain, positive time-derivatives), this quantity is positive, indicating work done on the system. During unloading (i.e. decreasing stress and strain, negative time-derivatives), this quantity becomes negative, indicating work done by the system. Note that the stress rates are related by the material behavior described in equation 4.

3.2. Convective Heat Transfer

A portion of the mechanical work input into the material is dissipated to the surrounding environment through convective heat transfer. The heat transfer comes due to the rise (or fall) in temperature of the sample relative to the environmental temperature due to work input to(or output from) the material. The heat output from the material is described by the heat transfer equation

$$\dot{E}_{loss} = hA\Delta T \tag{22}$$

where A is the surface area of the material, ΔT is the difference between sample and environmental temperatures, and h is the convective heat transfer coefficient. The convective heat transfer coefficient, h, is a function of the configuration, dimensions and environmental temperature. This can be estimated using empirical formulae for a given configuration. For a horizontal cylinder with free convection [22], the empirical relationship for heat transfer coefficient $\bar{h}$ is given by

$$\bar{h} = \frac{\overline{Nu_D}k}{D} \tag{23}$$

where $\overline{Nu_D}$ is the Nusselt number based on the cylinder diameter and ambient temperature, k is the thermal conductivity of air and D is the diameter of the cylinder. The Nusselt number $\overline{Nu_D}$ and thermal conductivity k of air are determined from empirical relationships by Churchill and Chu [23]

$$\overline{Nu_D} = \left[0.57 + \frac{0.377 Ra_D^{\frac{1}{6}}}{\left[1+(\frac{0.539}{Pr})^{\frac{9}{13}}\right]^{\frac{8}{57}}}\right]^2 \tag{24}$$

$$Ra_D = \frac{g\beta(T_s - T_{inf})D^3}{\nu\alpha} \tag{25}$$

(26) where Ra_D is Rayleigh number, which represents the degree of turbulence in the thermal boundary layer of the element. T_S and T_{inf} are the temperatures of the SMA surface (assumed homogenous throughout the SMA) and the environment respectively. The values of Prandtl number Pr, volumetric thermal expansion coefficient β, dynamic viscosity

ν and thermal diffusivity α are determined from look-up of tables of thermophysical properties of air [22]. It is useful to note that the Nusselt number and thus the heat transfer coefficient themselves are a function of the temperature difference between the sample and the ambient air, and therefore need to be updated constantly as the material temperature varies, reaching a converged solution for each time-step.

From this calculation, we can get an estimate of the heat transfer coefficient for a given material. Note that the heat transfer coefficient varies directly in proportion to the surface area and inversely with the volume, and is therefore more likely to affect the calculations for a thin wire. The heat transfer coefficient also varies greatly with surface finish and other properties of the material interface. The heat transfer coefficient obtained using the empirical formulae therefore should be validated with experimental cooling profiles of the material, as described in section 3.4.

3.3. Heat absorbed by material

The difference between the mechanical power input to the material and the heat loss to the surroundings represent the net heat absorbed (or released) in the material. This heat has two components.

The specific heat component is the portion of energy required (or released) to raise (or cool) the portion of the material that is in pure phase. Since this is associated with a single phase material, the net specific heat is the sum of the specific heats of the martensite and austenite components of the material. The net specific heat C_p and the heat rates to change the temperature of the material is given by

$$C_p = \xi_M * C_{pM} + (1 - \xi_M)C_{pA} \tag{27}$$
$$\dot{E}_{spec} = mC_p\Delta T \tag{28}$$

(29)where m is the total mass and C_{pM}, C_{pA} and C_p are the specific heats of pure martensite, pure austenite and the mixed phase SMA respectively.

The latent heat represents the heat required (or released) to transform the material between austenite and martensite phases. It has been shown [10] that the martensite to austenite transformation is exothermic (heat emitting), while the austenite to martensite transformation is endothermic (heat absorbing). Since this quantity is related with the transformation process, it is a function of the rate of change of martensite volume fraction and can be represented as

$$\dot{E}_{latent} = -mL\dot{\xi} \tag{30}$$

where L is the latent heat constant, i.e. the heat released (or absorbed) when a unit mass of SMA transforms completely from austenite to martensite (or vice versa). The latent heat raten

takes the opposite sign of $\dot{\xi}$, and is negative during transformation from austenite to martensite (exothermic), and positive during the reverse transformation (endothermic), accurately representing the physical nature of the latent heat.

Having obtained the individual components of heat flow rate into the material, the

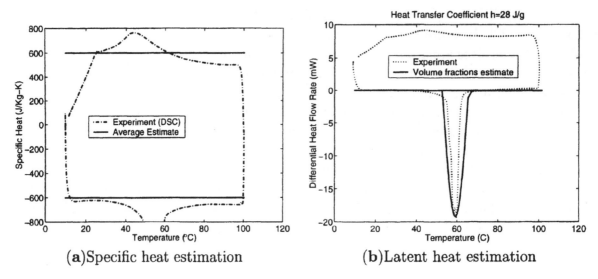

(a)Specific heat estimation   (b)Latent heat estimation

Figure 1. Obtaining thermal properties from experimental DSC characteristics

total heat flow rate is then

$$\dot{E}_{in} = \dot{E}_{specific} + \dot{E}_{latent} = mL\dot{\zeta} + mC_p\Delta T \tag{31}$$

This energy equilibrium equation is then used in conjunction with Equation 4 and solved simultaneously to determine the heat and temperature profiles as a function of time for a given mechanical loading.

The formulation for the energy analysis involves three important parameters that affect the magnitude of the temperature profiles (and thus the mechanical behavior of the SMA). These parameters (specific heat, latent heat and convective heat transfer coefficients) are obtained from experimental data that is independent of the mechanical loading of the material.

3.4. Determining Heat Transfer Parameters

The material parameters that are required for the heat transfer model are the specific heat C_p, heat transfer coefficient h and latent heat of transformation L. While the specific and latent heats are well-defined for a particular composition of the material, the heat transfer coefficient is a function of the surfaces and the dimensions of the material involved. These parameters were obtained experimentally in the current work. The specific heat and latent heats of the material are obtained from curvefits to the experimentally measured Differential Scanning Calorimeter (DSC) characteristics (figure 1(a)). The DSC provides the heat flow into the material as a function of temperature for a sample of a given weight of the material. Since the constant heat transfer rate to change the temperature of the material in pure phase is determined by the specific heat, this property of martensite and austenite can be obtained from the constant y-intercepts of the DSC(figure 1(a)). The low temperature phase (martensite) heat transfer rates vary only slightly from the high temperature phase (austenite), and are therefore assumed to be equal for simplicity.

The specific heat can be related to the heat flow rate obtained experimentally using the prescribed temperature rate in the experiment as

$$C_p = \frac{\dot{E}_{DSC}}{(m_{sample}\dot{T}_{DSC})} \tag{32}$$

where E_{DSC} is the experimentally measured heat flow rate for a sample of mass m_{sample} being heated (or cooled) at a temperature rate of $\dot{T}_{DSC}$. From this expression, an average specific heat for martensite and austenite can be found for the flat portion of the DSC.

The Latent heat of transformation, $L_{M\to A}$, is determined by curvefitting the DSC curve between the transformation temperatures to the estimate obtained in equation 16. The martensite volume fraction rate is obtained by numerical differentiation of the volume fraction estimated by Brinson's quasistatic model during thermal cycling at no stress -

$$\frac{d\xi}{dt} = \frac{d\xi}{dT} \times \frac{dT}{dt} \tag{33}$$

Using the appropriate values for the latent heat, this simple model shows good agreement to experimental data within the transformation temperatures (figure 1(b)). Although the experimental latent heats of transformation vary slightly between the $M \to A$ and $A \to M$ transformations, it is assumed constant in the analysis for simplicity, with a change in sign representing the direction of heat flow.

The convective heat transfer coefficient determined from equation 13 was also validated by comparing the estimate for the time history obtained from the calculated heat transfer coefficient to experimental time histories. For a material with a homogenous starting temperature of T_0 and environment temperature of T_∞, the variation of instantaneous temperature with time is given by

$$T = T_\infty - e^{-\frac{hAt}{mC_p}}(T_\infty - T_0) \tag{34}$$

In order to experimentally determine the time histories, an SMA rod of 0.25inch(0.006 m) diameter (used for torsional testing) and an SMA wire of 0.015inch (0.0003 m) diameter used for axial testing were mounted on a thermally insulating base 2(a). Two k-type thermocouples were mounted on each of the specimens, and the temperature histories were recorded. The initial and environmental temperatures of the sample were varied by moving the assembly into or out from a thermal chamber that was set at a fixed temperature. The cooling profile for both samples is compared with theoretical predictions in figure 2(b). Good agreement was observed between the two, with the empirical heat transfer coefficient slightly overpredicting the cooling rate of the material.

4. RESULTS AND DISCUSSION

Having obtained the coupled rate-dependant material equation(3) and the heat transfer equation(10) to determine the stress and temperature rates in first order form, the equations are solved simultaneously using a 4th order Runge-Kutta method with adaptive time-stepping. It may be noted that equation 10 describes the stress rate in terms of

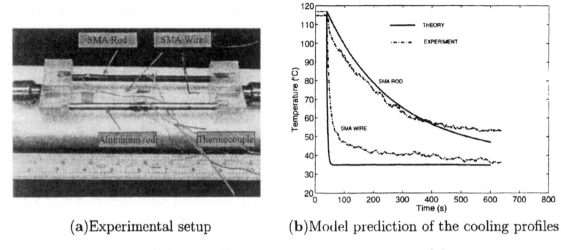

(a) Experimental setup   **(b)**Model prediction of the cooling profiles

Figure 2. Validation of heat transfer model with experimental data

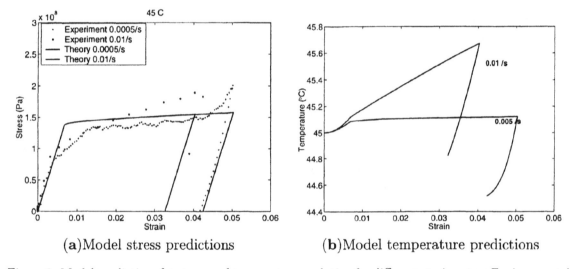

(a) Model stress predictions   **(b)**Model temperature predictions

Figure 3. Model predictions for stress and temperature evolution for different strain rates. Environmental temperature = 45 °C

not only the instantaneous stresses, strains and temperatures, but also the temperature rate, which violates first order representation. It therefore becomes necessary to solve the equations iteratively with the temperature rates of the previous time step being used to determine the stress rates in the current step.

Figure 3(a) shows the predicted trends as given by the strain rate model with the experimental data described previously at two different strain rates for a temperature of 45°C (below M_f). Figure 3(b) shows the corresponding theoretical temperature profiles. The model predicts a small rise and drop in temperature during straining of the material. The temperature rise (and fall) is proportional to the imposed strain rate, with about 2°C variation at the higher strain rate tested. The change in predicted temperature is small in magnitude due to the absence of any latent heat as predicted by the

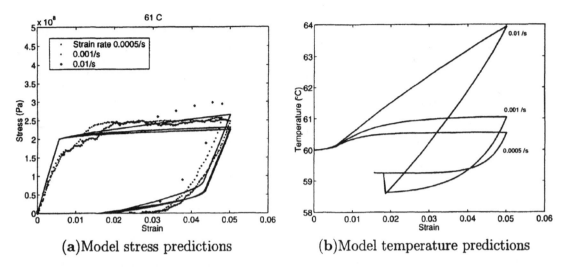

(a) Model stress predictions   **(b)**Model temperature predictions

Figure 4. Model predictions for stress and temperature evolution for different strain rates. Environmental temperature $= 60\ °C$

model (no austenite-martensite transformation) during detwinning. Further, since the original quasistatic model assumes that the transformational stresses are independent of the temperature below M_s, a temperature variation causes no change in the mechanical properties at low temperature. Due to these two factors, no significant strain rate variation is predicted at these temperatures (the thermal expansion term, $\Theta \dot{T}$, is negligible in comparison with the magnitudes of the variations in the stresses reported here). In reality, the transformational stresses vary linearly with temperature below M_S between , and thus we observe a dependance of the mechanical properties on the strain rate.

This discrepancy does not occur in some other models such as the Tanaka model [6], since the critical stresses are represented as a linear function of temperature until zero stress, and thus a significant temperature variation will manifest itself as variations in stress. However, the Tanaka model does not include variables to model the detwinning process in pure martensite. This suggests that significant fundamental drawbacks exist with the applicability of this analysis to both Tanaka and Brinson models, and this needs to be further addressed in future modeling approaches. Figure 4 shows the corresponding stress and temperature profiles for an environmental temperature of 60°C (slightly below A_f). From figure 4(b), we see that the change in temperature in this case is much more significant, and upto a 8°C variation is predicted at a high strain rate of 0.01 /s. This is due to the associated latent heat that is generated during the $A \to M$ transformation. Since this temperature is above M_s, these changes in temperature cause a large deviation in the mechanical behavior at higher strain rates(figure 4(a)), which appears to be in good agreement with experiments. The magnitude of the variation of the stresses with strain rate is, however, was found to be consistently underpredicted by the model for all the temperatures.

The model can also be applied to any arbitrary loading condition where the strain rate is prescribed as a function of time. Figure 5 shows the predicted stress and temperature profiles for a test involving composite strain rates (where the strain rate is stepped down

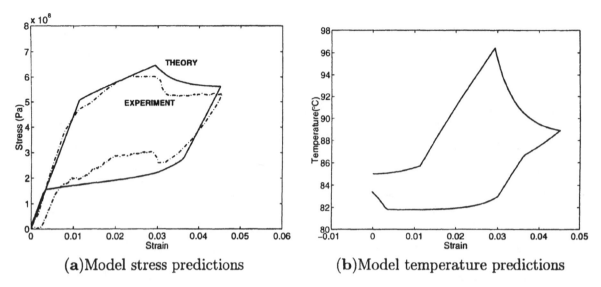

(a)Model stress predictions   (b)Model temperature predictions

Figure 5. Model predictions for stress and temperature evolution for composite strain rates, 0.01/s and 0.0005/s. Environmental temperature = 84 °C

from a value of 0.01/s to 0.005/s during loading). From figure 5(a), we observe good qualitative agreement with the experimental data. However, the rate of stress relaxation is underpredicted, possibly owing to the heat transfer coefficient being underpredicted. There is an instantaneous drop in the temperature rates when the loading condition is changed 5(b), further justifying this modeling approach to predict strain rate variations. The experimental heating characteristics of the wire consistently show a higher sensitivity to strain rate than those predicted by the model. This indicates that the heat transfer coefficient may be underpredicted by the empirical heat transfer coefficient. By modifying the heat transfer coefficients from those obtained by the thermal analysis by about 50%, it is possible to obtain better agreement with observed experimental data. From figure 6(a), we observe good agreement with the loading profile, and general agreement with the experimental unloading profiles. Further thermal analysis is required to fully validate the heat transfer coefficients used in the models so that better correlation can be achieved.

The results shown here serve to prove the validity of this modeling approach to predict various phenomena associated with non-quasistatic loading under different conditions. The model shows good qualitative agreement and fair quantitative agreement with experimental data at temperatures above M_f. Fundamental drawbacks in the Brinson model in temperature regimes below M_s were, however, found to manifest themselves in strain rate model. A more definitive conclusion regarding the quantitative agreement can be made after a more thorough approach to determine the heat transfer parameters of the material. However, the fair agreement with this method shows good promise for further refinement. It may be noted that although the present approach involves the rate form of the Brinson equation, the same approach can be used in conjunction with the rate forms of any other quasistatic model to effectively predict strain rate behavior in SMAs.

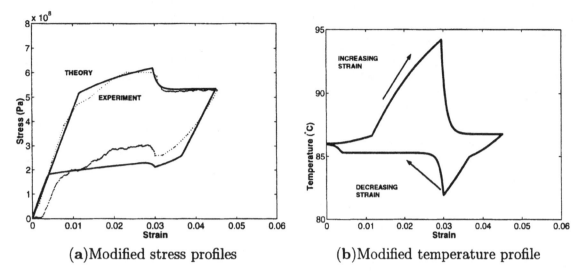

(a)Modified stress profiles   (b)Modified temperature profile

Figure 6. Model predictions for stress evolution for composite strain rates, modified heat transfer coefficient, 84 °C environmental temperature.

5. SUMMARY AND CONCLUSIONS

The paper described the development and validation of a strain-rate dependant model to capture the effects of non-quasistatic loading on an SMA wire. The rate form of the material thermomechanical model was developed from Brinson model and coupled with a macroscopic energy analysis to simultaneously describe the temperature and stress profiles in the material. The model shows good overall agreement with experimental data, and captures experimental phenomena such as strain-rate dependencies and stress-relaxation behavior in SMA wires. The parameters required for the model are obtained systematically from experiments not related to the mechanical properties of the wire. However, an optimization of these parameters with the observed mechanical properties shows the potential of yielding good agreement with observed test data involving complex loading profiles. The agreement of the model indicates that coupling a quasistatic model with an energy analysis is sufficient to predict most effects associated with non-quasistatic loading in an SMA wire. This also indicates that the temperature rise in the material when strained at high strain rates plays a seminal role in determining the mechanical stress developed in the material, and must be taken into account in applications involving continuous loading at higher strain rates.

REFERENCES

1. C.M. Wayman and T.W. Duerig. *Engineering Aspects of Shape Memory Alloys.* Butterworth-Heinemann Publishers, 1990.
2. L. Schetky. Shape memory alloy effect for robotic devices. *Robotics Age*, 6:13–17, 1984.
3. Imam K. Baz, A. and J. McCoy. Active vibration control of beams using shape memory actuators. *Journal of Sound and Vibration*, 140(3):437–456, 1990.

4. J. Epps and I. Chopra. Shape memory alloy actuators for in-flight tracking of helicopter rotor blades. *Smart Materials and Structures*, v10(1):104–111, 2001.

5. H. Prahlad and I. Chopra. Experimental characteristics of ni-ti shape memory alloys under uniaxial loading conditions. *Journal of Intelligent Materials and Structures*, 11(4):272–282, 1991.

6. K. Tanaka. A thermomechanical sketch of shape memory effect: One - dimensional tensile behavior. *Res. Mechanica*, 18:251–263, 1986.

7. C. Liang and C.A. Rogers. One- dimensional thermo mechanical constitutive relations for shape memory material. *Journal of Intelligent Materials and Structures*, 1:207–234, 1990.

8. L.C. Brinson. One-dimensional constitutive behavior of shape memory alloys: Thermo mechanical derivation with non-constant material functions. *Journal of Intelligent Materials and Structures*, 1:207–234, 1990.

9. Boyd J.G. and D.C. Lagoudas. A thermodynamic constitutive model for the shape memory materials part i. the monolithic shape memory alloys. *International Journal of Plasticity*, 12(6):805–842, 1998.

10. J.A. Shaw and S. Kyriakides. Thermomechanical aspects of niti. *Journal of the Mechanics and Physics of Solids*, 43(8):1243–1281, 1995.

11. Ravichandran G. Rosakis A.J. Hodowany, J. and P. Rosakis. Partition of plastic work into heat and stored energy in metals. *Experimental Mechanics*, 40(2):113–123, 2000.

12. R. Abeyaratne and S. Kim. Cyclic effects in shape-memory alloys: A one-dimensional continuum model. *International Journal of Solids and Structures*, 34(25):3273–3289, 1997.

13. Matsuzaki Y. Naito H. and Ikeda T. A unified model of thermomechanical behavior of shape memory alloys. In *Proceedings of SPIEs symposium on Smart Structures and Materials, Smart Structures and Integrated Systems, Newport Beach, CA*, pages 291–300, 2001.

14. Brinson L.C. Bekker, A. and K. Issen. Localized and diffuse thermoinduced phase transformation in 1-d shape memory alloys. *Journal of Intelligent Material Systems and Structures*, 9(5):355–365, 1998.

15. T.J. Lim and D.L. McDowell. Mechanical behavior of an ni-ti shape memory alloy under axial-torsional proportional and nonproportional loading. *Journal of Engineering Materials and Technology, Transactions of the ASME*, 121(1):9–18, 1998.

16. Tobushi H. Tanaka K. Hattori T. Lin, P.H. and M. Makita. Pseudoelastic behavior of niti shape memory alloy subjected to strain variations. *Journal of Intelligent Material Systems and Structures*, 5:694–710, 1994.

17. B. Malovrh and F. Gandhi. Mechanism-based phenomenological models for the pseudoelastic hysteresis behavior of shape memory alloys. *Journal of Intelligent Material Systems and Structures*, 12(1):21–30, 2001.

18. L.G. Chang. On diffusionless transformation in au-cd single crystals containing 47.5 atomic percent cadmium: Characteristics of single-interface transformation. *Journal of Applied Physics*, 23(7):725–728, 1952.

19. J.A. Shaw and S. Kyriakides. Initiation and propagation of localized deformation in elasto-plastic strips under uniaxial tension. *International Journal of Plasticity*, 13(10):837–871, 1997.

20. Yang F. Pu Z. Wu, K. and J. Shi. The effect of strain rate on the detwinning and superelastic behavior of niti shape memory alloys. *Journal of Intelligent Material Systems and Structures*, 7:138–144, 1996.

21. A. Bhattacharyya and Weng G.J. An energy criterion for the stress-induced martensitic transformation in a ductile system. *Journal of Mechanics and Physics of Solids*, 42(11):1699–1724, 1994.

22. F.P. Incorperra and D.P. DeWitt. *Fundamentals of Heat and Mass Transfer, 3rd ed.* John Wiley and Sons, 1990.

23. S.W. Churchill and H.H.S. Chu. Correlating equations for laminar and turbulent free convection from a horizontal cylinder. *International Journal of Heat and Mass Transfer*, 18:1049, 1975.

ADAPTIVE CONTROL OF SEMI-ACTIVE VARIABLE STIFFNESS DEVICES FOR NARROW-BAND DISTURBANCE REJECTION

Farhan Gandhi and Phuriwat Anusonti-Inthra[1]

ABSTRACT

The present study examines the potential of using a semi-active controllable stiffness device, whose spring coefficient can be modulated in real-time, for narrow-band disturbance rejection applications. A frequency-domain optimal control algorithm is developed for determining the controllable stiffness device input (of twice the disturbance frequency) that minimizes the force transmitted to the support at the disturbance frequency. The effectiveness of open-loop, closed-loop, and adaptive controllers in rejecting the transmitted disturbances are evaluated. The results of the study indicate that for physically achievable stiffness coefficient variations, the support force could be reduced by about an additional 30%, beyond the levels due to the passive isolation characteristics (no stiffness variation). When the disturbance phase changed during the simulation, the effectiveness of the open-loop controller rapidly degraded. While the closed-loop controller (with inputs based on current support vibration levels) performed better, there was still some degradation in performance, and support vibrations were not reduced to levels prior to change in disturbance phase. The results show that for the semi-active system to retain its effectiveness in rejecting disturbances, a closed-loop, adaptive controller (with on-line system identification) is required; even when there is only a change in disturbance, and no change in basic system properties.

1. INTRODUCTION

Undesirable vibrations in many engineering applications are often concentrated in a narrow frequency range. Such narrow-band vibrations could, for example, be produced by rotating machinery – turbines, engines, compressors, propellers, and helicopter rotors. Without any treatment, the vibrations could potentially result in both component as well as human fatigue, require increased maintenance, and reduce the effectiveness of sensitive equipment. Thus, considerable effort has been devoted in the past to suppress narrow-band vibrations, using both passive design concepts as well as active strategies (see, for example, Refs. 1-11).

Common passive techniques for reduction of narrow-band vibrations or rejection of tonal disturbances include the use of vibration absorbers and isolation schemes. While these concepts can be reasonably successful, there is usually a weight penalty involved, and their effectiveness can be significantly reduced with changes in operating conditions (for example, frequency of the disturbance) as well as changes in properties of the system. Active vibration suppression strategies such as active mounts, active truss modules, and Active Control of Structural Response or ACSR (used in helicopters) employ actuators to provide active forces for directly canceling the vibratory forces. Although active methods can be very effective and can be configured to cope with change in incoming disturbances (through closed-loop controllers) and variation in system properties (through adaptive controllers), they usually require significant power and actuator authority, and result in increased system complexity.

1Farhan Gandhi, Associate Professor, Department of Aerospace Engineering, The Pennsylvania State University, 229 Hammond Building, University Park, PA 16802, USA, e-mail: *fgandhi@psu.edu*
Phuriwat Anusonti-Inthra, Graduate Research Assistant.

In recent years, a new approach, known as *semi-active control*, has been receiving increasing attention for vibration reduction applications. In this approach, system properties such as damping or stiffness are controlled to modify the system response and reduce vibration. Since large active forces are not being directly applied to the system (as in the case of active vibration reduction schemes), semi-active control schemes are characterized by very low power requirements, while still retaining the ability to adapt to changes in conditions. Additionally, with semi-active approaches the risk for instability is eliminated, because unlike active approaches, energy is not being pumped into the controlled system. The majority of the studies on semi-active concepts are focused on *broad-band* vibration suppression applications such as controlling structural response due to earthquakes (see, for example, Refs. 12-15) and in the design of suspensions in the automobile industry (see, for example, Refs. 16-19). For practical implementation of the semi-active control concept, a wide range of discrete devices are available that can change the effective stiffness or damping characteristics of the system into which they are introduced. These include stiffness control devices, Electrorheological and Magnetorheological dampers, friction control devices, controllable orifice devices, and tuned mass/liquid dampers (see Ref. 20 for a detailed description of such devices).

Although vibration reduction using semi-active concepts are attractive for the reasons cited above, special control algorithms have to be developed, since the semi-active system is nonlinear or bilinear in nature. One of the most commonly used algorithms is based on the LQ clipped-optimal control scheme [14, 21-24]. Other algorithms include "bang-bang" or "on-off" controllers (see, for example, Refs. 15, 25-27), controllers with a more sophisticated scheduled input [28], and fuzzy-logic controllers [29]. Some researchers have also used nonlinear sliding-mode controllers [12, 23]. All of the above studies demonstrate that semi-active control systems are quite effective in reducing broad-band vibrations, but so far, semi-active control has not been exploited for narrow-band disturbance rejection.

2. FOCUS OF THE PRESENT STUDY

The primary focus of the present study is to examine the potential of using semi-active control for narrow-band disturbance rejection applications. The semi-active system considered is a variable stiffness device (modeled as a controllable spring) whose effective stiffness can be modulated in real-time. For a harmonic disturbance, a frequency-domain optimal control algorithm is developed that determines the stiffness variation for minimizing the forces transmitted to the support. The effectiveness of open-loop, closed-loop, and adaptive controllers in rejecting the transmitted disturbances are evaluated.

3. ANALYSIS

3.1 System Description

To evaluate the effectiveness of the semi-active variable stiffness device for rejecting narrow-band disturbances, a simple single-degree-of-freedom (SDOF) system is considered, as shown in Fig. 1. The mass, m, supported on the semi-active controllable spring and a parallel damper, is subjected to a harmonic disturbance force of amplitude, F_o, at frequency, Ω. The total stiffness of the semi-active spring is expressed as:

$$K(t) = K_o + K_1 u(t) \tag{1}$$

where K_o represents the baseline (passive) stiffness, and $K_1 u(t)$ represents the variation in stiffness due to the command input u. Rejection of the incoming disturbance (at frequency Ω) is achieved by optimally modulating the stiffness of the semi-active controllable spring. Without any modulation of the stiffness ($u = 0$), the baseline spring, K_o, in parallel with the damper, C, provides a passive isolation treatment between the disturbance and the support.

The equilibrium equation for the system in Fig. 1, and the corresponding force at the support, can be expressed as,

$$m\ddot{x} + C\dot{x} + (K_o + K_1 u)x = F = F_o \sin(\Omega t) \tag{2a}$$

$$F_s = C\dot{x} + (K_o + K_1 u)x \tag{2b}$$

The bilinear term, ux, which appears in the above equations (such a bilinear term appears in most semi-active systems), makes it difficult to apply conventional linear control theories.

3.2 Fundamentals of Controller Design

3.2.1 OVERVIEW OF THE SEMI-ACTIVE CONTROLLER FOR NARROW-BAND DISTURBANCE REJECTION

For narrow-band disturbance rejection using a semi-active variable stiffness device, an optimal controller is developed in the frequency-domain, as an adaptation of an approach previously used in Ref. 30 for disturbance rejection through pure active control. The vibratory force at the support, F_s, and the command input, u, are expressed in the frequency-domain (as cosine and sine components of specified harmonics), and are denoted as z and u_c, respectively. The harmonics of u_c are carefully selected to reject the incoming disturbance at frequency Ω (see Section 3.2.3 for detail). The control algorithm is based on the minimization of a quadratic objective function, J, defined as:

$$J = z^T W_1 z + u_c^T W_2 u_c \tag{3}$$

In the above equation, W_1 and W_2 represent penalty weighting corresponding to the vibratory force at the support, z, and the input, u_c, respectively.

Due to the bilinear term in Eqs. 2, the relationship between the input, u_c, and the support vibration, z, is not linear. However, the sensitivity of the vibrations, z, to perturbations in the frequency-domain inputs, u_c, are still expressed as:

$$z = z_o + Tu_c \tag{4}$$

where T is the system transfer matrix, and z_o represents the baseline support vibration levels without the input, u_c. The transfer matrix, T, can be calculated using both off-line and on-line approaches (detail discussion is presented in Section 3.2.4). A gradient-based method is used to minimize J and determine the optimal inputs, u_c. By substituting Eq. 4 into Eq. 3 and setting $\partial J/\partial u_c = 0$, the resulting optimal input may be obtained as:

$$u_c = \bar{T} z_o \tag{5a}$$

$$\text{with} \quad \bar{T} = -\left(T^T W_1 T + W_2\right)^{-1} T^T W_1 \tag{5b}$$

A Frequency-to-Time domain conversion (F/T) unit is used to obtain the optimal time-domain input, u, corresponding to the frequency-domain input, u_c (see Fig. 2). The amplitude of the time-domain input is also calculated during the conversion, and the input is modified if necessary to ensure that maximum or minimum values of physical achievable device stiffness are not exceeded. This is discussed in greater detail in the next section.

3.2.2 SEMI-ACTIVE DEVICE SATURATION CONSIDERATION

In the present study the semi-active controllable stiffness device has a baseline stiffness of, K_o, and it is assumed that the maximum and minimum physically achievable values of stiffness are $K_o + K_1$ and $K_o - K_1$, respectively. Typically K_1 would be some fraction of K_o, and is assumed to be $0.7 K_o$ in the present study. It can be deduced from Eq. 1 that the range of variation in u is:

$$-1 \le u \le 1 \tag{6}$$

If the actual input voltage required to produce a maximum possible increase of K_1 in stiffness is u_o volts (and the corresponding voltage to produce a maximum possible decrease is $-u_o$ volts), then u effectively represents a non-dimensional voltage input (non-dimensioanlized by u_o). In the F/T conversion unit (Fig. 2), the amplitude of the optimal frequency-domain input, u_{max}, is determined, and if it exceeds the maximum permissible value (of unity), then the converted time-domain input signal, $\hat{u}(t)$, can be "scaled down" as follows:

$$u(t) = \frac{\hat{u}(t)}{u_{max}} \tag{7}$$

However, with such a scaling-down there is a question regarding the optimality of the input signal. Alternately, the input can be reduced by increasing the input penalty weight, W_2.

3.2.3 FREQUENCY CONTENT OF THE SEMI-ACTIVE INPUT, u_c

In order to reject a tonal disturbance of frequency, Ω, higher harmonic semi-active inputs are required. This is different from a fully active system where the control input, u_c, would simply consist of cosine and sine components at the disturbance frequency, Ω, which would essentially minimize the support vibrations, z, at Ω. However, for a semi-active controller input at Ω (producing stiffness variation at Ω), it is seen from Eqs. 2 that the resulting semi-active force would be at 2Ω, due to the bilinear ux term. Thus, there would be no rejection of the disturbance at Ω, and additionally support vibrations would now be introduced at 2Ω. Instead, a semi-active controller input (stiffness modulation) at 2Ω would directly result in semi-active forces (proportional to ux) at Ω and 3Ω, with the component at Ω then canceling the incoming disturbance. Thus, for the present problem the semi-active input, u_c, and output, z (used in minimization of objective function, J, Eq. 3), are selected as:

$$u_c = \begin{bmatrix} u^{2\Omega c} & u^{2\Omega s} \end{bmatrix}^T \tag{8a}$$

$$\text{and} \quad z = \begin{bmatrix} F_s^{\Omega c} & F_s^{\Omega s} \end{bmatrix}^T \tag{8b}$$

In the above equations, the superscripts "c" and "s" represents cosine and sine components, respectively, at frequencies Ω or 2Ω. It should be noted that while the selected inputs will reduce the incoming disturbance at Ω, the support will now experience additional forces at 3Ω. These could in principle be reduced by expanding z to include these components and introducing additional harmonics in the input u_c.

3.2.4 IDENTIFICATION OF THE TRANSFER MATRIX, T

The system transfer matrix, T, can be identified using both off-line and on-line approaches. Off-line identification of the T matrix is achieved by perturbation of individual component of the input, u_c.

The first column of T matrix, which corresponds to the first input of u_c, is obtained by setting the first input $u^{2\Omega}$ to a non-zero value (while the other inputs are set to zero), and the column is calculated as:

$$\begin{bmatrix} t_{11} \\ \vdots \\ t_{1n} \end{bmatrix} = \frac{z - z_o}{u_c^1} \tag{9}$$

This process is repeated for all inputs of u_c to obtain all columns of the T matrix.

For the on-line identification of the T matrix, an initial estimate is obtained using the batch least square method [31], and it is updated using the recursive least square method (with variable forgetting factors) [32]. The on-line batch least square method yields an initial estimate of the T matrix from an array of inputs, u_c, and corresponding support vibration measurements, z, at $m+1$ time steps, as follows:

$$T = Z\Phi^T \left(\Phi\Phi^T \right)^{-1} \tag{10a}$$

$$\Phi = \begin{bmatrix} u_c(k) & u_c(k-1) & \cdots & u_c(k-m) \end{bmatrix} \tag{10b}$$

$$Z = \begin{bmatrix} z(k) & z(k-1) & \cdots & z(k-m) \end{bmatrix} \tag{10c}$$

where $u_c(k)$ and $z(k)$ represents the input and the corresponding support vibration level at the kth time step. It should be noted that the number of time steps used has to be greater than or equal to the number of inputs. For the simulations in the present study, a value of $m=4$ was used. For the batch least squares, an interval of six disturbance cycles was used between the successive inputs.

An on-line recursive least square method is implemented for introducing updates to the T matrix. A variable forgetting factor, λ, is used to prevent parameter estimation 'blow-up', which can occur when the estimation is running continuously for a long time without any change in parameters being estimated. The recursive least square identification is summarized as follows:

$$T(k) = T(k-1) + \varepsilon(k)K(k) \tag{11a}$$

$$\varepsilon(k) = z(k) - T(k-1)u_c(k) \tag{11b}$$

$$K(k) = \left[I + u_c^T(k)P(k-1)u_c(k) \right]^{-1} u_c^T(k)P(k-1) \tag{11c}$$

$$P(k) = \frac{P(k-1)}{\lambda(k)} \left[I - u_c(k)K(k) \right] \tag{11d}$$

$$\lambda(k) = 1 - \left[1 - K(k)u_c(k) \right] \frac{\varepsilon^T(k)\varepsilon(k)}{\Sigma_o} \tag{11e}$$

where Σ_o was chosen to be 0.0025, and lower limit of λ was set at 0.15, in the present study. Updates to $T(k)$ were carried out at intervals of four disturbance cycles.

3.3 Open-Loop Controller

An open-loop control scheme can in principle be effective for narrow-band disturbance rejection if the disturbance force and the system are not changing with time. In such a situation, the optimal control input in the frequency-domain, u_c, is based only on the *baseline (uncontrolled) support vibration levels*, z_o, as seen in Eq. 5, (and not on any measurements of "current" vibration levels). Once the uncontrolled support vibration, z_o, is determined, and the transfer matrix, T, is obtained using the off-line identification, the open-loop control scheme can be implemented following a block diagram shown in Fig. 3.

3.4 Closed-Loop Controller

If the disturbance force changes during the course of operation, an open-loop algorithm is in general no longer suitable and a closed-loop algorithm has to be employed instead. Using an approach adapted from Ref. 33, previously applied to the active vibration reduction problem, the closed-loop control scheme for the present semi-active disturbance-rejection problem is implemented in the discrete-time domain. The idea is to calculate adjustments in input, Δu_c, based on "current" support vibration levels, $z(k)$, such that vibration levels in the next time step are minimized. In such a case,

$$\Delta u_c(k) = \overline{T} z(k) \tag{12}$$

with $\overline{T}$ identical to that in Eq. 5, and the T matrix identified off-line, a priori. The total input to the controllable spring is then expressed as

$$u_c(k) = u_c(k-1) + \Delta u_c(k) \tag{13}$$

The block diagram corresponding to such a closed-loop control scheme is shown in Fig. 4. Updates to the inputs, $\Delta u_c(k)$, are carried out at intervals of every two disturbance cycles, based on calculated support vibration levels, $z(k)$, at these times.

3.5 Closed-Loop Adaptive Controller

In addition to basing control inputs on currently measured vibration levels to allow for variations in disturbance force, the system transfer matrix, T, would require identification and updating on-line if the system is undergoing changes (making it a closed-loop adaptive control scheme). However, the present semi-active system is non-linear (bi-linear), and the results in sections 4.3 and 4.4 will show that *on-line identification of the transfer matrix is required for effective disturbance rejection even when the system properties are not changing (and only the disturbance changes) during operation.* A detailed explanation of this phenomenon is provided in section 4.4. The closed-loop adaptive control algorithm is simply the closed-loop scheme described in the previous section with the controller gain, $\overline{T}$ (in Eqs. 5 and 12), updated using on-line identification of the transfer matrix, T, (as described earlier in Section 3.2.4). The block diagram for this closed-loop adaptive controller is shown in Fig. 5.

4. RESULTS AND DISCUSSION

4.1 Baseline System

Numerical simulations are carried out to evaluate the effectiveness of semi-active control (modulation in spring coefficient of the variable stiffness device) in rejecting harmonic disturbance inputs. The system parameters used in the simulations are given in Table 1. The baseline support force, F_s (z_o in the frequency domain), due to a disturbance force, $F_o \sin(\Omega t)$, is first calculated in the absence of any variations in stiffness (see Figs. 6). From Fig. 6b, the amplitude of the support force is seen to be 41% of the disturbance force, this attenuation being due to the passive isolation characteristics of K_o and C in parallel. In the following simulations, further reductions in the support vibrations due to semi-active modulation of stiffness are compared to this baseline vibration level (due to pure passive isolation).

4.2 Open-Loop Control Scheme

In this section, additional reductions in the disturbance transmitted to the support are examined when an open-loop control scheme is used. The inputs are calculated using Eq. 5, which specifies the stiffness modulation required to minimize the support vibration at frequency Ω. The first set of simulations used a penalty weighting of $W_1 = I$ (identity matrix), and $W_2 = 0$. Further, no "scaling down" of the control inputs was carried out (as described in section 3.2.2). For the mathematically optimal inputs determined directly from Eq. 5, $u^{2\Omega c} = 1.8186$ and $u^{2\Omega s} = -2.1271$, Fig. 7 shows the frequency content of the steady state vibrations transmitted to the support. Although the amplitude of the support force at the disturbance frequency, Ω, is seen to be reduced to 9% of the excitation force amplitude (compared to 41% in the absence of any stiffness modulations), a higher harmonic component at 3Ω □ (with an amplitude of 26% of the disturbance force) is now observed. This, of course, is expected due to the K_1xu term as discussed in section 3.2.3. From this perspective, some of the disturbance energy can effectively be thought of as being transferred to higher harmonics. This may be advantageous in certain conditions when it is important to avoid specific frequencies due to resonances, or to exploit the improved effectiveness of viscous and viscoelastic damping mechanisms at higher frequencies. However, it should be noted that the amplitude of the control input, $\left|u^{2\Omega}\right| = \sqrt{\left(u^{2\Omega c}\right)^2 + \left(u^{2\Omega s}\right)^2}$, exceeds unity, so the condition on the maximum permissible input, specified in Eq. 6, is violated. For the system considered, it is clear that the mathematically optimal control input, or stiffness variation, is not practically realizable. Since the stiffness variations required are larger than those that can be physically achieved by modulations of the variable stiffness device, it implies that to achieve the levels of disturbance rejection at Ω seen in Fig. 7, energy input would be actually required, and the system would no longer be semi-active.

In the next set of simulations, the control inputs were "scaled down" (as described in Section 3.2.2), so that the inputs ($u^{2\Omega c} = 0.6498$ and $u^{2\Omega s} = -0.7601$) never exceeded the maximum permissible values. In this case, Fig. 8 shows that the amplitude of the transmitted force at the disturbance frequency, Ω, is 28% of the disturbance force amplitude. Compared to a corresponding value of 41% in the absence of damping variation (recall Fig. 6b), this represents an additional 32% reduction in transmitted vibration over that achieved due to the pure passive isolation characteristics. The amplitude of the higher harmonic component at 3Ω is now 10% of the disturbance force amplitude.

Figure 9 shows the force transmitted to the support (both at the disturbance frequency, Ω, as well as the higher harmonic component at 3Ω), corresponding to different amplitudes of control input, u_c. Vibration levels corresponding to control inputs greater than unity represent only a mathematical solution not practically achievable by the variable stiffness device. In fact, for $\left|u_c\right| > 1.43$, the total stiffness would actually be *negative* over parts of the cycle. Since energy input would be required to realize the solutions corresponding to $\left|u_c\right| \geq 1$, this region has been marked as "active" on Fig. 9. Examining the support vibrations corresponding to different "semi-active" inputs, it can be observed that as the control input increases, the support force at the disturbance frequency decreases linearly (producing up to an additional 32% reduction over the passive isolation case for the present system), and the 3Ω component increases linearly. The control input levels were varied using two different methods – *(i)* the "scaling down" approach; and *(ii)* using different values of the input penalty weighting, W_2 (which produces an optimal solution). Since the results produced by both methods were identical, it is concluded that "scaling down" approach, also, essentially provides optimal inputs when considering physical limits in stiffness variation. Since scaling-down is simple and convenient, it is used in all subsequent simulations.

4.3 Closed-Loop Control Scheme

Benefits to using a closed-loop controller are expected when the disturbance changes during operation. In this section, the performance of both open-loop as well as closed-loop controllers are examined *when the disturbance phase changes during the simulation*. For the closed-loop controller, the control inputs are updated based on Eqs. 12 and 13 at intervals of every two disturbance cycles. The change in disturbance phase, ϕ, is introduced at t $= 10\pi$ s, as described below:

$$F(t) = \begin{cases} F_o \sin(\Omega t) & 0 \le t < 10\pi \\ F_o \sin(\Omega t + \phi) & t \ge 10\pi \end{cases} \qquad (14)$$

For a phase change of $\phi = 45°$, Fig. 10 shows the time history of the disturbance, as well as the force transmitted to the support, when the closed-loop controller is operational. It is seen that even after the change in disturbance phase occurs, the closed-loop controller is once again able to reduce the support vibration levels, in a short duration. Figure 11 shows the amplitude of the support force at the disturbance frequency, Ω, using both open- and closed-loop controllers. As expected the open-loop controller is no longer effective in disturbance rejection after the disturbance phase changes (since the control inputs, which are based only on the initial vibration levels, become non-optimal after the disturbance phase, and therefore the phase of the support vibrations, changes). However, with the closed-loop controller, after a transition period, the disturbance transmitted to the support is once again reduced. When a disturbance phase change of $\phi = 90°$ is introduced, Fig. 12 once again shows that the closed-loop controller performs better than the open-loop controller. For the 45° change in phase angle, the closed-loop control inputs are changed to $u^{2\Omega} = -0.3097$ and $u^{2\Omega} = -0.9508$, and the corresponding inputs for the 90° disturbance phase change are $u^{2\Omega} = -0.9265$ and $u^{2\Omega} = -0.3762$. It should be noted that although the closed-loop controller is more effective than the open-loop controller, the steady state disturbance levels transmitted to the support are not as low as those prior to change in disturbance phase (as would have been expected if an active force-generator type actuator had been used).

Figure 13 shows the steady-state support vibrations at the disturbance frequency, Ω, as a function of disturbance phase change (varying between -90° and 90°). It is observed that as the phase change increases, the effectiveness of the open-loop control scheme is degraded significantly, to the extent that the support vibration levels are larger than those for the uncontrolled case when the phase change exceeds ±40°. Performance degradation is also observed for the closed-loop scheme, but is milder. The reason that the closed-loop controller is not able to track phase changes perfectly (as would have been expected if an active force-generator type actuator) can be explained as follows: For a *pure active controller*, the control force can be written as Au (where A is some *constant* coefficient). When the disturbance phase changes during operation, the phase of the response and the support vibrations will correspondingly change, and since the input, u, is based on the current support vibrations, its phase is appropriately adjusted and comparable reductions in vibration are obtained. For the *semi-active controller*, the control force is $K_l xu$. Thus, when the closed-loop controller adjusts the input u proportional to the change in phase of the response (and the support vibration levels), the bi-linear semi-active force generated, $K_l xu$, is no longer simply proportional to this change in response phase. This suggests that an adaptive controller (recalculating the system transfer matrix, T, online) may be required for the semi-active narrow-band disturbance rejection if the disturbance is likely to change during operation, even when the system parameters are unchanged.

4.4 Closed-Loop Adaptive Control Scheme

The effectiveness of the closed-loop adaptive controller is evaluated in this section when the disturbance phase changes during operation. Online identification of the transfer matrix, T, using batch least square approach (Eq. 10) for initial estimates and recursive least square identification (Eq. 11), for updates is carried out, as described in section 3.2.4. For the present simulations, updates of the transfer matrix (in the recursive least squares approach) were carried out every four disturbance cycles. The disturbance phase changes as described in Eq. 14, except that it is introduced at t $= 32\pi$ s (instead of 10π s). The first 24π seconds is used for batch least square identification of the T matrix by inputting a sequence of small input signals, at the end of which period the controller is switched on. Figures 14 and 15, respectively, show the variation of the support force amplitude at the disturbance frequency, Ω, for disturbance phase changes of 45° and 90°. It is seen from both figures that when the adaptive controller is operational, after a transition period, the support vibration levels are reduced to those prior to the change in disturbance phase (unlike the "non-adaptive" closed-loop controller that did not retain its effectiveness; recall Figs. 11-13).

Figure 16 shows the steady-state support vibrations at the disturbance frequency, Ω, as a function of disturbance phase change (varying between -90° and 90°). It is observed that even as there is an increase in disturbance phase change, unlike the open- and closed-loop controllers, the closed-loop adaptive controller completely retains its effectiveness in reducing support vibrations. Thus, for a semi-active (bi-linear) system, a closed-loop adaptive controller (continuous on-line identification of system transfer matrix) is required even when there is only a disturbance change, and not a "direct" change in system properties. Although, it could be said that for a non-linear semi-active system, the "system", as such, is dependent on the excitation, so a change in the disturbance changes the system itself.

5. SUMMARY AND CONCLUDING REMARKS

The present study examines the potential of using a semi-active controllable stiffness device, whose spring coefficient can be modulated in real-time, for narrow-band disturbance rejection applications. For a harmonic disturbance, a frequency-domain optimal control algorithm is developed for determining the input (controllable stiffness variation) that minimizes the force transmitted to the support at the disturbance frequency. An input frequency of twice the disturbance frequency is used, so that the resulting bilinear semi-active force (proportional to xu) has a component at the disturbance frequency that cancels the force at the support. Such a scheme, however, will result in a higher-harmonic support force. The effectiveness of open-loop, closed-loop, and adaptive controllers in rejecting the transmitted disturbances are evaluated. Some of the key observations of the present study are presented next.

Mathematically optimal solutions often resulted in stiffness variations that were in excess of those practically achievable by controllable springs, and input scaling was used in these cases. The scaled inputs were found to be "optimal" since they produced results identical to those obtained by increasing the input weighting in the objective function (a true optimal solution). For physically achievable stiffness variations, the support force could be reduced by about an additional 30%, beyond the levels due to the passive isolation characteristics (no stiffness variation). When the disturbance phase changed during the simulation, the performance of the open-loop controller rapidly degraded, as the inputs were no longer optimal. While the closed-loop controller performed somewhat better (and generally reduced vibrations to levels lower than those due to pure passive isolation), there was still some performance degradation, and support vibration levels could not be reduced to those prior to change in disturbance phase, as would have been expected in the case of closed-loop control using an active force-generator-type actuator. However, when a closed-loop adaptive controller was used (with on-line system identification), the reduction in support vibrations was completely maintained even after the disturbance phase underwent change. Thus it is seen that for narrow-band disturbance rejection using a semi-active system (which is bi-linear), a closed-loop adaptive control scheme is required if the disturbance is likely to change during operation, even if the basic system parameters themselves are unchanged.

REFERENCES

1. Reichert, G., "Helicopter Vibration Control –Survey," *Vertica*, Vol. 5, No. 1, 1981, pp. 1-20.
2. Lowey, R. G., "Helicopter Vibrations: A Technological Perspecitve," *Journal of the American Helicopeter Society*, Vol. 29, No. 4, Oct. 1984, pp. 4-30.
3. Harris, C. M., "Shock and Vibration Handbook," McGraw-Hill, New York, 1988.
4. Staple, A. E., "An Evaluation of Active Control of Structural Response as a Means of Reducing Helicopter Vibration," *Proceedings of the 15th European Rotorcraft Forum,* Amsterdam, Netherlands, September 1989, pp. 51.1-51.18.
5. Rao, S. S., "Mechanical Vibrations," Addison-Wesley, Reading, Mass., 1990.
6. Scribner, K. B., Sievers, L. A., von Flotow, A. H., "Active Narrow-Band Vibration Isolation of Machinery Noise from Resonant Substructures," *Journal of Sound and Vibration*, v. 167, no. 1, Oct. 1993, pp. 17-40.
7. Jerkins, M. D., Nelson, R. J., Pinnington, R. J., and Elliott, S. J., " Active Isolation of Periodic Machinery Vibrations," *Journal of Sound and Vibration*, Vol. 166, No. 1, 1993, pp. 117-140.
8. Inman, D. J., "Engineering Vibration," Prentice Hall, Englewood Cliffs, New Jersey, 1994.
9. Welsh, W., Fredrickson, C., Rauch, C., and Lyndon, I., "Flight Test of An Active Vibration Control System on the UH-60 Black Hawk Helicopter," *Proceedings of the American Helicopter Society 51st Annual Forum,* May 1995, pp. 393-402.
10. Sun, J. Q., Jolly, M. R., Norris, M. A., "Passive, Adaptive and Active Tuned Vibration Absorbers - A Survey," *Journal of Mechanical Design*, Transactions Of the ASME, v. 117B, Jun. 1995, pp. 234-242.
11. Williams, Robert L. II, "Survey of Active Truss Modules," ASME, Design Engineering Division, DE, Vol. 82, no. 1, Sep. 1995, pp. 899-906.
12. Yang, J. N., Wu, J. C., and Li, Z., "Control of Seismic-excited Buildings using Active Variable Stiffness Systems," *Engineering Structures,* Vol. 19, No. 9, 1996, p. 589-596.
13. Patten, W. N., Mo, C., Kuehn, J., and Lee, J., "A Primer on Design of Semiactive Vibration Absorbers (SAVA)," *Journal of Engineering Mechanics,* Vol. 124, No. 1, January 1998, pp. 61-68.
14. Sadek, F. and Mohraz, B., " Semiactive Control Algorithms for Structures with Variable Dampers," *Journal of Engineering Mechanics,* Vol. 124, No. 9, September 1998, pp. 981-990.
15. Gavin, H. P. and Doke, N. S., "Resonance Suppression through Variable Stiffness and Damping Mechanisms," *Proceedings of the SPIE Smart Structures Conference*, SPIE Vol. 3671, March 1999, pp. 43-53.
16. Krasnicki, E. J., "Comparison of Analytical and Experimental Results for a Semi-Active Vibration Isolator," *The Shock and Vibration Bulletin,* The Shock and Vibration Information Center, Naval Research Laboratory, Vol. 50. 1980.
17. Krasnicki, E. J., " The Experimental Performance of an Off-Road Vehicle Utilizing a Semi-Active Suspension," *The Shock and Vibration Bulletin,* The Shock and Vibration Information Center, Naval Research Laboratory, 1984.
18. Hrovat, D., Margolis, D. L., and Hubbard, M., "An Approach Toward the Optimal Semi-Active Suspension," *Journal of Dynamic Systems, Measurement and Control,* Vol. 110, September 1988, pp. 288-296.
19. Karnopp, D., "Design Principles for Vibration Control Using Semi-Active Dampers," *Journal of Dynamic Systems, Measurement and Control,* Vol. 112, September 1990, pp. 448-455.
20. Symans, M. D. and Constantinou, M. C., "Semi-active Control System for Seismic Protection of Structures: A State-of-the-Art Review," *Engineering Structures*, Vol. 21, 1999, pp. 469-487.
21. Hac', A. and Youn, I., "Optimal Semi-Active Suspension with Preview Based on a Quarter Car Model," *Journal of Vibration and Acoustics*, Vol. 114, January 1992, pp. 84-92.

22. Yi, K., Wargelin, M., and Hedrick, K., "Dynamic Tire Force Control by Semi-Active Suspensions," *ASME, Dynamic Systems and Control Division*, DSC-Vol. 44, November 1992, pp. 299-310.

23. Symans, M. D. and Constantinou, M. C., "Seismic Testing of A Building Structure with A Semi-Active Fluid Damper Control System," *Earthquake Engineering and Structural Dynamics*, Vol. 26, 1997, pp. 759-777.

24. Dyke, S. J., and Spencer, B. F. Jr., Sain, M. K., and Carlson J. D., "An Experimental Study of MR Dampers for Seismic Protection," *Smart Materials and Structures*, Vol. 7, No. 5, October 1998, pp. 693-703.

25. Onoda, J., Endo, T., Tamaoki, H., and Watanabe, N., "Vibration Suppression by Variable-Stiffness Members," *AIAA Journal*, Vol. 29, No. 6, June 1991, pp. 977-983.

26. Fodor, M. and Redfield, R. C., "Resistance Control, Semi-Active Damping Performance," *ASME Advanced Automotive Technology*, DSC-Vol. 56/DE-Vol. 86, 1995, pp. 161-169.

27. McClamroch, N. H. and Gavin, H. P., "Closed Loop Structural Control Using Electrorheological Dampers," *Proceedings of the American Control Conference*, June 1995, pp. 4173-4177.

28. Kawashima, K. and Unjoh, S., "Seismic Response Control of Bridges by Variable Dampers," *Journal of Structural Engineering*, Vol. 120, No. 9, September 1994, pp. 2583-2601.

29. Symans, M. D. and Kelly, S. W., " Fuzzy Logic Control of Bridge Structures Using Intelligent Semi-Active Seismic Isolation Systems," *Earthquake Engineering and Structural Dynamics*, Vol. 28, 1999, pp. 37-60.

30. Johnson, W., "Self-tuning regulators for multicyclic control of helicopter vibration," NASA Technical Paper 1996, March 1982.

31. Astrom, K. J. and Wittenmark, B. "Adaptive Control," 2nd Ed., *Addison-Wesley Publishing Company Inc.*, Reading, MA, 1995.

32. Fortescue, T. R., Kershenbaum, L. S., and Ydstie, B. E., "Implementation of Self-Tuning Regulators with Variable Forgetting Factors," *Automatica*, Vol. 17, No. 6, 1981, pp. 831-835.

33. Hall, S. R. and Wereley, N. M., "Linear Control Issues in the Higher Harmonic Control of Helicopter Vibrations," *Proceedings of the American Helicopter Society 45th Annual Forum*, Boston, MA, May 1989.

Parameter	Numerical Value
m	1
C	0.4
K_o/m	1
K_1/K_o	0.7
F_o/m	1
Ω	2 rad/s

Table 1: Numerical values of system parameters

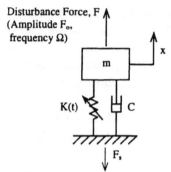

Figure 1: Schematic of single-degree-of-freedom system with semi-active actuator

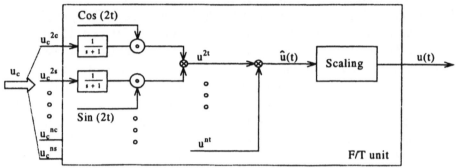

Figure 2: Frequency-to-Time domain conversion (F/T) unit

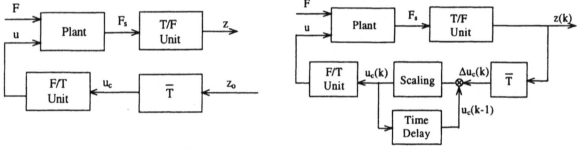

Figure 3: Block diagram of open-loop
control system

Figure 4: Block diagram of closed-loop
control system

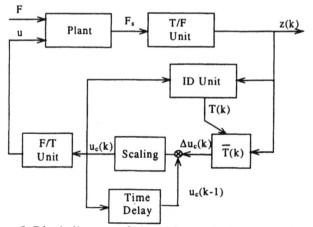

Figure 5: Block diagram of closed-loop adaptive control system

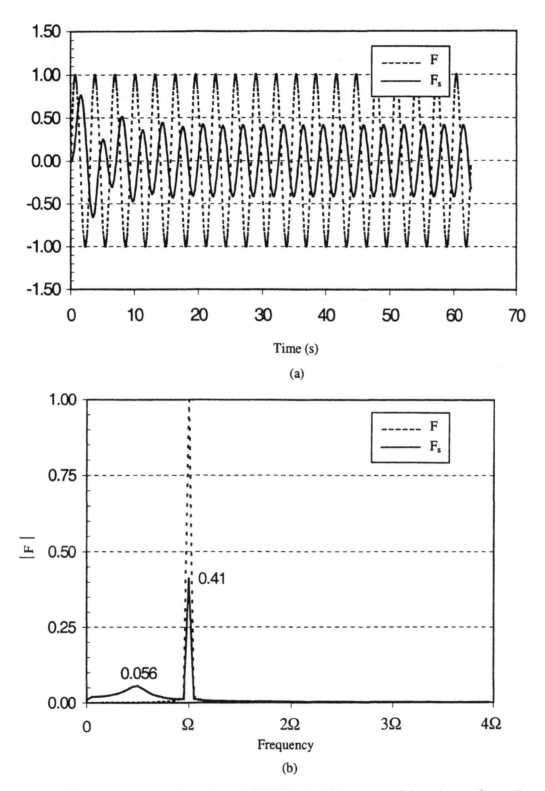

Figure 6: (a) Time history and (b) corresponding frequency content of disturbance force, F, and support force, F_s, of the baseline uncontrolled system

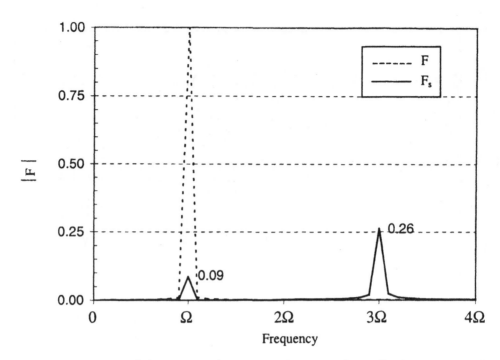

Figure 7: Frequency content of disturbance force, F, and support force, F_s, due to optimal control input (no input limits)

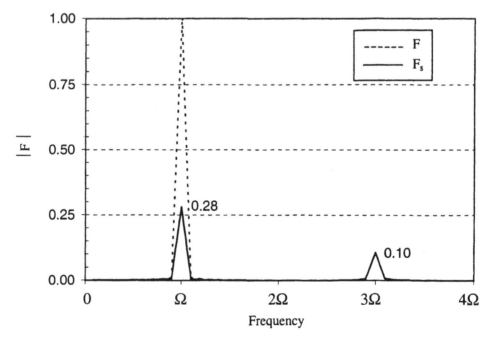

Figure 8: Frequency content of disturbance force, F, and support force, F_s, due to semi-active control (input "scaled-down")

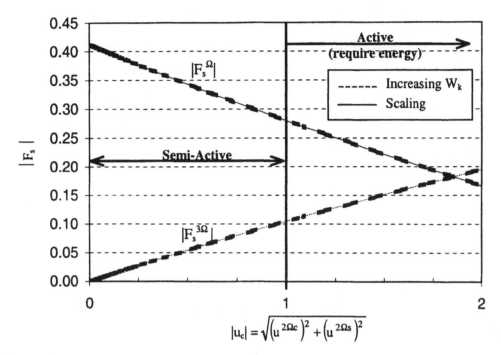

Figure 9: Frequency contents of support force, F_s, for increasing control input amplitudes.

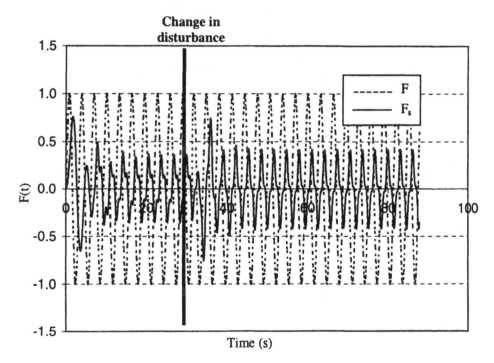

Figure 10: Time history of disturbance force, F, and support force, F_s, for closed-loop system, with a change in disturbance phase of 45°

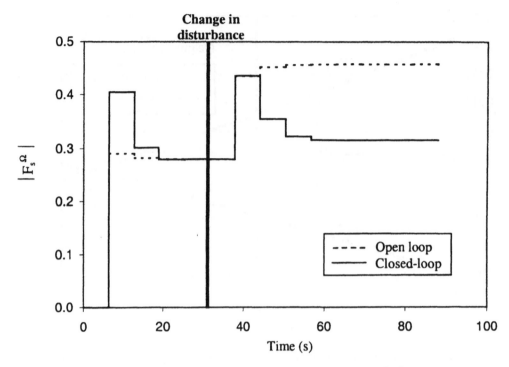

Figure 11: Amplitude of support force, F_s, at disturbance frequency, Ω, for open-loop and closed-loop systems, with a change in disturbance phase of 45°

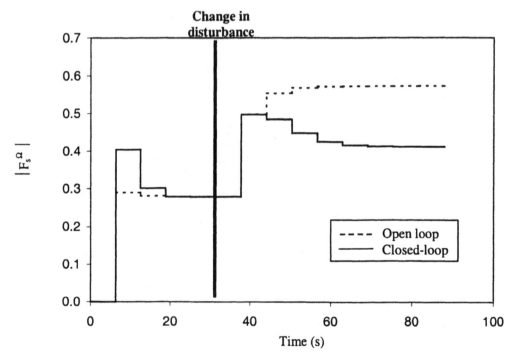

Figure 12: Amplitude of support force, F_s, at disturbance frequency, Ω, for open-loop and closed-loop systems, with a change in disturbance phase of 90°

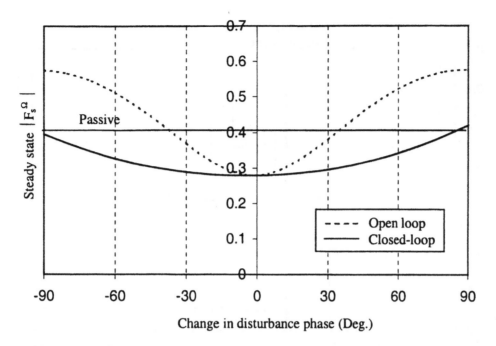

Figure 13: Variation in steady state support force, F_s, at Ω for closed-loop and open-loop systems as a function of change in disturbance phase angle

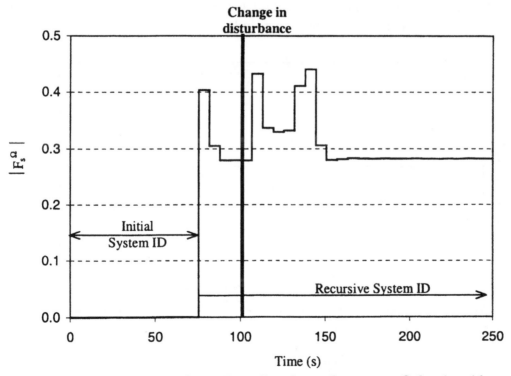

Figure 14: Amplitude of support force, F_s, at disturbance frequency, Ω, for closed-loop adaptive system, with a change in disturbance phase of 45°

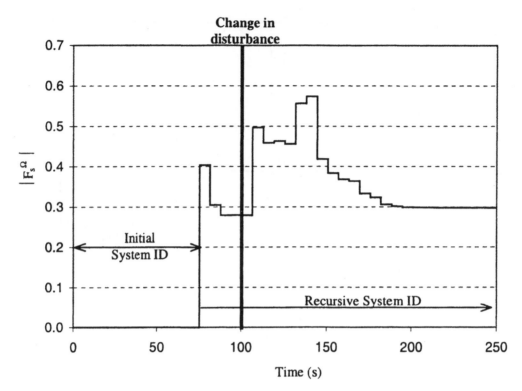

Figure 15: Amplitude of support force, F_s, at disturbance frequency, Ω, for closed-loop adaptive system, with a change in disturbance phase of 90°

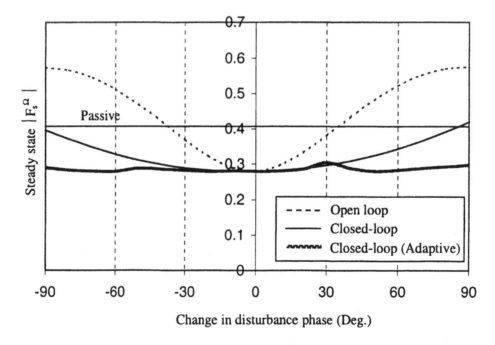

Figure 16: Variation in steady state support force, F_s, at Ω for closed-loop, closed-loop adaptive, and open-loop systems as a function of change in disturbance phase angle

Advanced Actuators

MECHANICALLY RATCHETING PIEZOELECTRIC MOTORS

George A. Lesieutre, Gary H. Koopmann, Eric M. Mockensturm, Jacob Loverich, and David Ericson

ABSTRACT

The main principle underlying the operation of mechanically-ratcheting piezoelectric motors is rectification and accumulation of small oscillatory displacements of piezoelectric drive elements using "mechanical diodes." An initial rotary version of a motor based on this concept used commercial roller clutches as mechanical diodes and inexpensive piezoelectric bimorphs as drive elements. Experiments with an initial 2-armed motor demonstrated a stall torque of about 0.04 N-m, a no-load speed of about 180 RPM, and peak power output of about 0.075 W. This output corresponds to a specific power of 1.1 W/kg.

A design optimization study indicated that, by increasing the length of the bimorph drive element, splitting the piezo elements at a strain node, and decreasing the end mass, output specific power could be substantially increased for a given roller clutch. Experiments confirmed a performance improvement of more than 10X in specific power. Other aspects of motor design and their effects on specific power were also considered, including the stiffness and backlash of mechanical diodes, the optimum drive frequency, and the role of resonance.

A different rotary motor was also developed, based on the use of a piezoelectric stack driver and larger roller clutches. Experiments with this motor demonstrated a stall torque of about 5.0 N-m, a no-load speed of about 350 RPM, and peak power output of about 40 W, corresponding to specific power of about 30 W/kg. Compact motors that rectify and accumulate the small, high frequency displacements generated by smart materials drive elements have the potential to achieve high specific power in combination with accurate positioning.

Penn State University, Center for Acoustics and Vibration, 157 Hammond, University Park, PA 16802

INTRODUCTION

The potential of high force piezoelectric actuators, both rotary and linear, as alternatives to electric or hydraulic actuators is evident in the attention dedicated to the topic in recent years. [1-4] The low strain output of piezoceramics has typically limited the output force of traditional inchworm-type piezoelectric actuators by limiting their dynamic clamping force, and friction-based clamping mechanisms have limited the life of such devices.

In the past several years, a variety of mechanically-ratcheting piezoelectric motors have been developed for use in large-displacement structural shape control applications. [5-8] The main principle underlying their operation is rectification and accumulation of small, perhaps resonant, displacements of piezoelectric drive elements using "mechanical diodes." *Passive* clamping mechanisms typically rely on the mechanical interference ("wedging") of a roller between two surfaces, as illustrated in Figure 1. On the driving half of each cycle, the forward motion of a drive element is converted to motion of a shaft when its output force exceeds that of the load. On the recovery half of each cycle, a second, fixed, mechanical diode prevents the load from back driving the shaft. This approach increases the output mechanical power relative to that obtained in other piezoelectric motors that use direct friction-based clamping mechanisms (such as inchworm or ultrasonic motors). [*c.f.,* 8-10]

The development of smart materials-based motors invites comparisons to competing approaches such as hydraulics and electric motors. Electric motors exhibit specific powers in the range of 100 W/kg, and some that exceed 1000 W/kg for intermittent operation. Electric motors and piezoelectric material-based motors have an advantage over hydraulic systems in that power may be transferred over long distances with relatively light wires. Piezoelectric material-based motors have advantages over typical electric motors in that they offer better potential to conform with geometric requirements associated with tightly-integrated adaptive structures, and in the potential for reduced electromagnetic field generation.

MOTOR CONCEPTS AND PERFORMANCE

Initial Bimorph-Driven Rotary Motor

Figure 2 shows an early version of a motor based on this concept that used commercial roller clutches as mechanical diodes and inexpensive piezoelectric bimorphs as drive elements. An additional feature of motors that use passive mechanical diodes is that only a single drive signal is required; thus, the drive electronics can be optimized for single-frequency operation. Experiments with this two-armed motor demonstrated a stall torque of about 0.075 N-m, a no-load speed of about 180 RPM, and peak power output of about 1.1W. [1]

Optimized Bimorph-Driven Rotary Motor

A design optimization study indicated that, by increasing the length of the bimorph drive element, from 30 to 51 mm, splitting the piezo element at about 40% of its length, and decreasing the size of the end mass, output power density could be maximized for a given commercial roller clutch. [1] Figure 3 shows a two-armed motor fabricated as a result of the optimization study. Experiments confirmed a significant performance improvement. [1, 11]

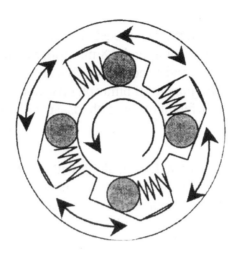

Figure 1. A roller clutch

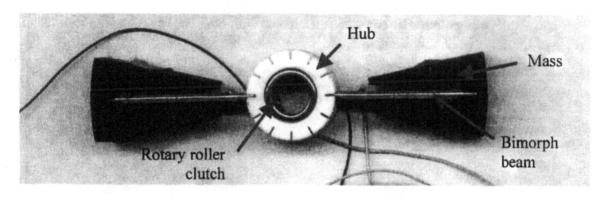

Figure 2: The initial bimorph-driven motor

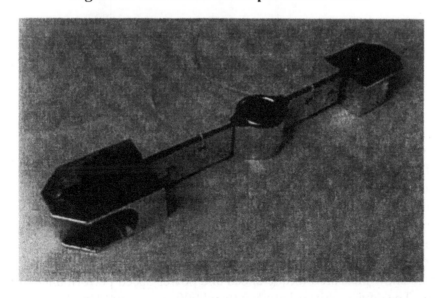

Figure 3: The optimized bimorph-driven motor

Despite this apparent success, performance models indicated that peak specific power for a bimorph-driven motor built around the commercial 8 mm roller clutch would be about 30 W/kg, even considering alternate elastic end conditions. This level is not competitive with existing technology and, consequently, pursuit of bimorph driven motors was suspended. Subsequent design work focused on different drive methods, specifically stack-driven devices.

Stack-Driven Rotary Motor

Figure 4 shows the initial stack-driven device, along with a solid geometric model of the device. Such concepts better exploit the inherent potential of the piezo material, although the challenge of generating adequate motion remains. The piezo stack is enclosed in a housing that provides a compressive preload and a heat conduction path. Linear stack motion acting though a stinger and an offset drive produces rotary motion of the output shaft. The output shaft meets two large diameter roller clutches through two couplers. These large roller clutches have lower angular backlash and compliance than smaller clutches. Four steel rods, a base, and a head provide the structural backbone of the motor, internally reacting the linear stack force with minimal deformation. The stack mass (with housing) is about 0.54 kg, while the mass of the entire assembly is about 1.3 kg.

Figure 4: The initial stack-driven motor

To operate this motor at frequencies above 300 Hz, the 35 mm commercial roller clutches were modified to increase the stiffness of the springs that force the rollers into the ramps. This enabled the drive frequency to be increased to above 600 Hz, albeit with an increase in over-running drag. Evidently, clutch performance was affected by an internal resonance, perhaps of the rollers oscillating on their compression springs.

Note that very high frequency operation associated with very small devices can also result in substantial heating, depending on the piezoelectric material used. Temperature excursions on the order of 20 C were observed when driving this device briefly at high power.

Figures 5 and 6 show measured torque-speed and power-speed performance of this initial stack-driven motor. The device, driven between 0 and 900 V_{pp}, at 600 Hz, developed a free angular velocity of 350 RPM, and a maximum torque of about 5.0 N-m.

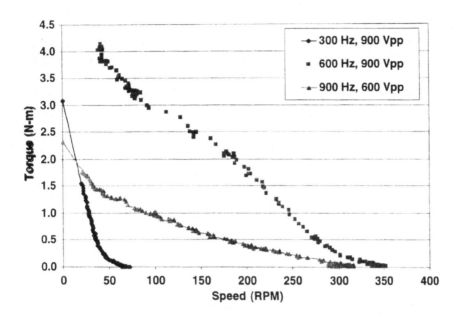

Figure 5: Torque-speed performance of the initial stack-driven motor

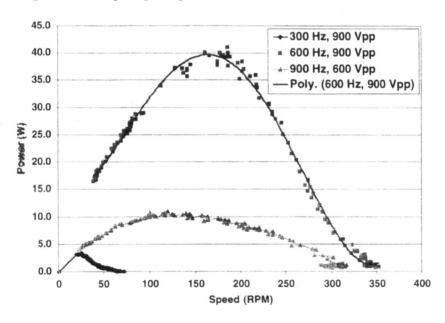

Figure 6: Power-speed performance of the initial stack-driven motor

OTHER DESIGN CONCERNS FOR MECHANICALLY-RATCHETING MOTORS

In real ratcheting motors , the holding elements are critical. Ideally, a clamping system behaves as a mechanical diode, permitting selective motion in one direction. Real holding mechanisms, however, exhibit backlash and/or low stiffness over some small (seating) range of motion. Because the step size must be somewhat larger than the backlash in the clamping system, a piezoelectric drive element cannot be so small that its free displacement is smaller than the backlash. An initial dynamic model was developed to address the roles of mechanical diode backlash and stiffness, and the geometric scale at which output power density is maximized.

This motor model was used to size components, as well as to estimate the performance of hypothetical motors. A key element of the model is a description of the angular compliance (including backlash) of the roller clutch used. At present, this information must be obtained experimentally for each roller clutch. This model of stack-driven rotary motors was validated by comparison of predictions to initial test results.

To enable accurate projections to arbitrary sizes, however, a validated *scalable* model of roller clutch compliance is essential. For example, larger clutches tend to have smaller angular backlash. For this purpose, four commercial roller clutches of various sizes were characterized to date. This data provides the basis for roller clutch scaling, as well as projections of motor performance.

Base-driven resonant motor concepts were also explored, shedding light on the reasons resonant drive systems can be so effective. A resonant system can act to amplify either displacement or force, depending on the drive frequency and the load. Small changes in frequency can result in large changes in drive impedance, so the frequency can be adjusted to best match the drive and load impedances. Drawbacks of this driving strategy include relatively low resonance frequencies, and considerable sensitivity of the optimal drive frequency to the load.

Base-driven resonant motors, however, offer the potential to maximize power output in the presence of stress limits for the driving material. Because tensile stresses must be avoided, there are limits on how hard stacks can be driven. Even in the presence of significant compressive preload, driving a stack at its own fundamental resonance frequency may be unacceptable. Base driving involves using a series mass-stiffness system, along with drive frequencies that are somewhat lower than the fundamental frequency of the stack itself.

A critical feature for many applications is 4-quadrant operation: *i.e.,* a useful motor should be able to deliver power to, or absorb power from a load while moving in either direction. A reversible mechanical diode is one approach to enabling this kind of operation in future ratcheting piezomotors. Several initial concepts were built and evaluated. The reversible diodes require, however, additional active elements to change direction, reminiscent of the active clamps of inchworm motors, but without relying on direct friction for holding.

SUMMARY

Mechanically-ratcheting piezoelectric motors rectify and accumulate small oscillatory displacements of piezoelectric drive elements using "mechanical diodes." An initial piezoelectric stack-driven rotary motor achieved higher perforrnance than a second-generation bimorph-driven motor. Measured performance included a stall torque of about 5.0 N-m, a no-load speed of about 350 RPM, and peak power output of about 40 W, corresponding to specific power of about 30 W/kg.

With continued development, mechanically ratcheting motors have the potential to displace established technologies in some applications. Issues that remain to be addressed include backlash/compliance, 4-quadrant operation, and scaling to small sizes to enable very high frequency operation and the accompanying high specific powers.

ACKNOWLEDGMENTS

This research was supported by the Defense Advanced Research Projects Agency (DARPA) under the Mechanical Diode Resonant Rectifying Actuator (MEDIRRA) project of the Compact Hybrid Actuator program (CHAP). Dr. Ephrahim Garcia and Dr. Gary Anderson were the Technical Monitors.

REFERENCES

1.   Lesieutre, G.A., G.H. Koopmann, E.M. Mockensturm, J.E. Frank, and W. Chen, "Mechanical Diode Based, High-Torque Piezoelectric Rotary Motor," AIAA Adaptive Structures Forum, Seattle, WA, April, 2001.

2.   Regelbrugge, M.E., and E. Anderson, "Piezoelectric Pump Actuators: Concepts Models, Capabilities and Limitations," International Conference on Adaptive Structures Technology, College Park, MD, October 15, 2001.

3.   Teter, J.P., Musoke H. Sendaula, John Vranish, E.J. Crawford, "Magnetostrictive linear motor development," IEEE Transactions on Magnetics, v 34, n 4, pt 1, July, 1998, pp. 2081-2083.

4.   Mauck, Lisa, Christopher S. Lynch, "Piezoelectric hydraulic pump," Proceedings of SPIE Smart Structures Conference, v 3668, n II, March 1-4, 1999, pp. 844-852.

5.   Frank, J.E., E.M. Mockensturm, W. Chen, G.H. Koopmann, and G.A. Lesieutre, "Roller Wedgeworm: A Piezoelectrically-Driven Rotary Motor," International Conference on Adaptive Structures Technology, Paris, France, October 11-14, 1999.

6.   Chen, Quanfang, Da-Jeng Yao, Chang-Jin Kim, Greg P. Carman, "Development of mesoscale actuator device with micro interlocking mechanism," Journal of Intelligent Material Systems and Structures, v 9, n 6, 1997, pp. 449-457.

7.   Zhang, Q.M., "A d15-based rotary motor," at the DARPA Smart Structures Technology Interchange Meeting, NASA Langley Research Center, Hampton, VA, USA, June 17, 1999.

8.   Frank, J.E., G.H. Koopmann, W. Chen, G.A. Lesieutre, "Design and performance of a high force piezoelectric inchworm motor," Proceedings of SPIE Smart Structures Conference, v 3668 n II, March 1-March 4, 1999, pp. 717-723.

9.   Uchino, Kenji, "Piezoelectric ultrasonic motors: Overview," Smart Materials and Structures, v 7, n 3, June, 1998, pp. 273-285.

10.   Glenn, T.S., and N.W. Hagood, "Development of a two-sided piezoelectric rotary ultrasonic motor for high torque," Proceedings of SPIE Smart Structures Conference, v 3041, March 3-6, 1997, pp. 326-338.

11.   Lesieutre, G.A., et al., "Mechanical Diode Resonant Rectifying Actuator," DARPA Technical Interchange Meeting, WPAFB, Dayton, OH, October 5-7. 2001.

SOLID-FLUID HYBRID ACTUATION: CONCEPTS, MODELS, CAPABILITIES AND LIMITATIONS

Marc Regelbrugge and Eric Anderson

ABSTRACT

This paper addresses concepts, models and fundamental performance limitations of hybrid solid-fluid actuators. The actuators discussed incorporate solid-state driving elements such as electrostrictive, piezoelectric or magnetostrictive materials to pressurize a working fluid which, in turn, is directed to a conventional hydraulic piston-type actuator for mechanical applications. First-principles models of hybrid solid-fluid actuators are derived and employed in simulation to estimate ranges of performance likely achievable from such devices in practice. The influences of various design and physical parameters are explored in the context of these models. The results of simulation studies indicate that achieving mechanical output power levels up to 75% of that theoretically obtainable based on physical properties of the solid-state driving elements should be possible with well-engineered, hybrid actuators.

INTRODUCTION

Materials possessing significant piezoelectric, electrostrictive, magnetostrictive and shape-memory properties offer high specific energies for actuation in mechanical systems. However, limitations inherent in the solid state often render the actuation properties of these materials unsuitable for a wide variety of mechanical applications. For example, field-induced strains in electrostrictive, magnetostrictive and piezoelectric materials are limited to a range below 1%, although field-induced stresses can range as high as 80-100 MPa. Transformation-induced strains of shape-memory materials can range as high as 6-8% with transformation stresses in the range of 40-60 MPa, but actuation bandwidth of thermally activated shape-memory materials is low; typically below a few Hertz. New materials, such as single-crystal piezoelectrics and magnetically activated shape-memory alloys have relaxed these limitations somewhat, but generally at the cost of reduced actuation energy density.

These natural limitations have forced application developers to search for ways to match the natural actuation capabilities of these "active" materials to their specific applications. Approaches taken include development of electrical or mechanical transformers and, recently, development of repetitive-motion devices such as motors and pumps. All of these techniques attempt to harness the naturally high energy densities of active materials in a form suitable for end-use application.

This paper discusses aspects of a particular approach to matching the actuation capabilities of electrostrictive and piezoelectric ceramics to generic applications demanding motions of several centimeters, forces of several hundred Newtons, and bandwidths in the range up to 50 Hz. The approach involves coupling the solid-state actuator to a working fluid, which is used to transmit power to a mechanical load.

Marc E. Regelbrugge, Rhombus Consultants Group, 2565 Leghorn St., Mountian View, CA 94043.
Eric H. Anderson, CSA Engineering, Inc. , 2565 Leghorn Street, Mountian View, CA 94043.

Two principles are involved. First, the actuation energy is rectified in the working fluid, allowing more energy to be extracted from the driven, active material element. Second, the rectification allows operation of the active material at a frequency that differs from the actuation frequency of the load. These two factors, when taken together, allow the solid-state actuator to deliver power to the load at suitable levels of force and speed. A development program to demonstrate workable actuators based on this approach is now underway.

While theoretically attractive, practical limitations arise that limit the efficacy of this solid-fluid hybrid actuation concept. In particular, real-fluid effects of viscosity and compressibility combine with loss mechanisms inherent in the active material to limit the power bandwidth of the driving actuator. Also, great care must be taken in design to match the characteristics of the driving actuator to the fluid transmission and output actuator if maximum power is to be available to drive the mechanical load. While prior work has described efforts to build and/or model such systems [1-6], it has largely addressed development of specific devices rather than investigation of the bounds and limitations of the approach.

This paper explores the issues arising in solid-fluid hybrid actuation using simple, generic models of the active material, working fluid, and hydraulic components of such an actuator. The models are derived, key features and limitations of the models are discussed, and simulation results are examined in some depth to draw observations and develop insight about aspects influencing design and operation of such hybrid actuation systems.

HYBRID SOLID-FLUID ACTUATOR

Figure 1 illustrates the generic class of devices considered by the present development. As this figure shows, the device considered here comprises several elements: a solid-state element of stiffness k driving a piston of area A_1 to pressurize the working fluid by induced-strain actuation; and fluid passages connecting the pressurization chamber with an hydraulic output cylinder and accumulator volume through four valves.

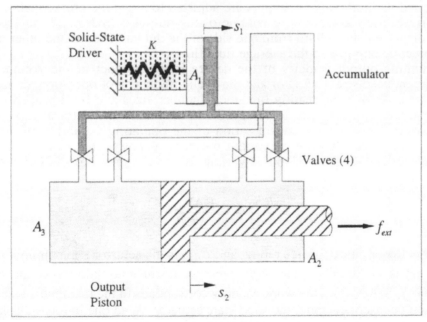

Figure 1. Hybrid Solid-Fluid Actuator Concept

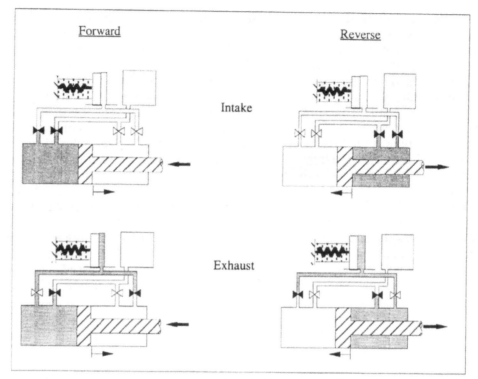

Figure 2. Sequences of Hybrid Actuator Operation

Figure 2 shows the hybrid actuator's sequence of operation. Valve openings are sequenced to allow pressurized fluid into one of the output cylinder's chambers. During this stroke, the alternate output-piston chamber communicates directly with the accumulator volume to allow the output piston to displace different volumes in each chamber. Once the pressurization stroke has reached its limit, valve openings are adjusted to allow the pressurization chamber to take fluid in from the low-pressure volume of the output cylinder and the accumulator volume. This displacement of fluid from one side of the output piston to the other moves the piston in the direction opposite to the average fluid flow.

The free-running output velocity of the device is the product of the volume of fluid displaced by the induced-strain actuator and the cyclic frequency of operation, divided by the output piston area. As one can readily imagine, chamber sizes and fluid properties may be adjusted to achieve a wide range of force-speed output characteristics for any given driving element. However, elementary consideration of the operating capability of the induced strain actuator will reveal that the maximum mechanical power that can theoretically be delivered by the solid-state actuator to the working fluid per cycle of operation at frequency $f = 1/T$ is

$$W_{MAX} = \frac{F_b \delta_{MAX}}{4T} = \frac{P_S \Delta V_{MAX}}{4TA_1^2} \tag{1}$$

where F_b is the actuator blocked-force rating, and δ_{MAX} is the actuator's maximum free induced stroke (P_S is the "stall" pressure – the pressure at which no fluid can be moved by the actuator, and $\Delta V_{MAX} = A_1 \delta_{MAX}$). This work quantity corresponds to the maximum-area rectangle that can be inscribed under the solid-state actuator's load line in force-displacement, or, equivalently, pressure-volume space, as illustrated in Figure 3 (with inspiration from [2]). Consideration of compressibility of the working fluid dictates that the pressurization chamber acted upon by the solid-state actuator presents a finite fluid stiffness to the actuator. Using β as the fluid compressibility ($\beta = dV/VdP$), the stiffness of the pressurization chamber is $A_1^2/\beta V_1$. The chamber fluid stiffness presents a loading to the actuator, as illustrated in Figure 3, which reduces the maximum achievable chamber pressure and power output, respectively, to

$$P_{MAX} = \frac{A_1^2 P_S}{\beta V_1 K + A_1^2} \quad \text{and} \quad W_A = \frac{P_S A_1}{4T}\left[\delta_{MAX} - \beta P_S \frac{V_1}{A_1}\right].$$

The available output of the integrated device is further reduced by the necessary pressure drop incurred to move fluid from the pressurization chamber to the output cylinder:

$$W_O = \frac{(P_S - \Delta P_V)A_1}{4T}\left[\delta_{MAX} - \beta P_S \frac{V_1}{A_1}\right] \qquad (2)$$

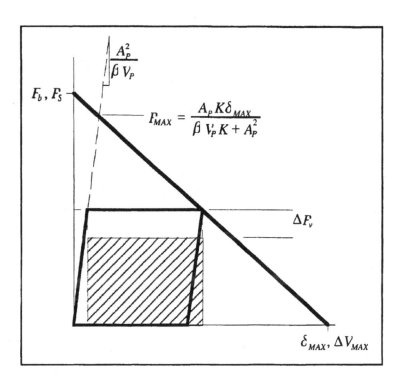

Figure 3. Explanation of Hybrid-Actuator Work showing Loss Elements

Depending on specifics of the hybrid device's design and working fluid properties, this level of power output may be substantially less than that predicted theoretically by (1). However, clear direction can be recovered from (2) regarding maximization of output power for a given actuator. In particular, the valve pressure drop, fluid compressibility, and the effective height of the pressurization space (V_1/A_1) must be minimized. These goals are far from straightforward to accomplish, owing to the necessity of fluid passages, losses associated with viscous flow through these passages (included in ΔP_V), and the variable nature of fluid properties with temperature, entrained gas, etc.

To investigate specific choices of design parameters on the power output of these hybrid devices, we have resorted to simulations in which parameters can be varied and the operation of the device can be simulated directly in time. These models are based on a continuity formulation which expresses the evolution of fluid mass in each of four volumes. These volumes are separated by, and can communicate through the valves. Pressure drop through each valve is approximated by

$$\Delta P_V \cong \frac{\rho_2}{2} U_2^2 = \frac{\dot{m}_v^2}{2A_2^2}$$

where A_2 is the effective area of the fluid jet downstream of the valve as approximated by the empirical relation

$$A_2 = \frac{2ah}{2+\pi}$$

Here, h is the valve opening displacement, and ah is the valve opening area. Mass flow through a valve connecting the ith and j^{th} chambers is simply estimated [] by

$$\dot{m}_{ij} \cong \frac{2a_{ij}h_{ij}(t)}{2+\pi}\mathrm{sgn}\left(P_i - P_j\right)\sqrt{2\rho_j\left|P_i - P_j\right|} \tag{3}$$

using the convention that fluid flows from the ith chamber into the jth chamber when $P_i > P_j$ (i.e. when $\dot{m}_{12} > 0$). Here, we note that the following relation holds due to compressibility of the fluid:

$$\rho(P_i) \equiv \rho_i = \bar{\rho}e^{\beta\left(P_i - \bar{P}\right)} \tag{4}$$

where $\bar{\rho}$ and $\bar{P}$ represent fluid density and pressure at initial pre-charge conditions. The quantity $\bar{P}$ is referred to as the "charge pressure."

For the sequences depicted in Figure 2, and using chamber numbering from Figure 1, we have the following four continuity conditions depending on the phase of the pressurization cycle (intake and exhaust) and the direction of travel (forward is taken as positive s_2 in Fig. 1).

TABLE I. MASS FLOW CONSTRAINTS IN EACH OPERATIONAL PHASE

Intake		Exhaust	
Forward	Reverse	Forward	Reverse
$\dot{m}_1 = \dot{m}_{12}$	$\dot{m}_1 = \dot{m}_{13}$	$\dot{m}_1 = -\dot{m}_{13}$	$\dot{m}_1 = -\dot{m}_{12}$
$\dot{m}_2 = -\dot{m}_{12} - \dot{m}_{2A}$	$\dot{m}_2 = 0$	$\dot{m}_2 = -\dot{m}_{2A}$	$\dot{m}_2 = \dot{m}_{12}$
$\dot{m}_3 = 0$	$\dot{m}_3 = \dot{m}_{3A} - \dot{m}_{13}$	$\dot{m}_3 = \dot{m}_{13}$	$\dot{m}_3 = -\dot{m}_{3A}$
$\dot{m}_A = \dot{m}_{2A}$	$\dot{m}_A = -\dot{m}_{3A}$	$\dot{m}_A = \dot{m}_{2A}$	$\dot{m}_A = \dot{m}_{3A}$

The table above defines a constraint space for each phase of operation of the hybrid actuator. Note that mass-flow between chambers identically satisfies the overall condition of continuity, namely $\sum_i \dot{m}_i = 0$.

The simulation model is completed by expressing pressure rate in each in each chamber of the device as a function of mass flux and mechanical quantities, as shown in Equations (5), where δ and $\dot{\delta}$ represent the commanded solid-state actuator extension and extension-rate, respectively, and the quantities V_i represent respective chamber initial volumes at pre-charge conditions. Note that Eqs. (5) have also incorporated the relation $\dot{\rho}_i = \rho_i \beta \dot{P}_i$.

$$\dot{P}_1 = \frac{\dot{m}_1/\rho_1 + A_1\dot{\delta}}{\beta(V_1 - A_1\delta) + (1 + \beta P_1)A_1^2/K}$$

$$\dot{P}_2 = \frac{\dot{m}_2/\rho_2 - A_2\dot{s}_2}{\beta(V_2 + A_2 s_2)}$$

$$\dot{P}_3 = \frac{\dot{m}_3/\rho_3 + A_3\dot{s}_3}{\beta(V_3 - A_3 s_2)}$$

$$\dot{P}_A = \frac{\dot{m}_A/\rho_A}{\beta V_A + \left[1 + \beta(P_A - \overline{P})\right]A_A^2/k_A}$$

(5)

Simulation proceeds by explicit integration in time using mass fluxes computed from current-step pressures and valve openings, and computing current-step pressure rates from these updated mass fluxes and current-step mechanical quantities. Pressures, chamber fluid mass allocations and mechanical quantities are then updated for the next time step.

SIMULATION RESULTS

Figure 4 shows results obtained from the simulation model described above in terms of device pressures (P_1, P_2 and P_3) and actuator output motion s_2 for a candidate hybrid actuator configuration working against a 7500 N static load and operating at 700 Hz. Charge pressure is 5 MPa. Other key parameters are listed in the table below.

TABLE II. PARAMETERS FOR SIMULATION STUDIES

Parameter	Assumed Value
Actuator Stiffness (K)	250 kN/mm
Actuator Peak Stroke (δ_{MAX})	120 μm
Max. Work/cycle per Eq. (1)	0.9 J
Fluid Compressibility (β)	450 μm^2/N
Fluid Density (ρ)	1.0 gm/cm^3
Areas: A_1, A_2, A_3	628 mm^2, 370 mm^2, 416 mm^2
Valve Area, Opening (ah_{MAX}, h_{MAX})	9 mm^2, 0.4 mm

Examination of Figure 4 reveals that hybrid actuator output velocity has an oscillatory (i.e. non-constant) character, and that positive and negative pressure spikes occur while valves are opening. The former characteristic indicates that caution should be exercised when applying actuators of this type to mechanisms exhibiting dynamic responses in the range of actuator operating frequencies. The latter characteristic implies that device operation will be limited by fluid cavitation when the low-pressure spikes associated with the intake of fluid into the pressurization chamber fall below the vapor pressure of the fluid.

Limitations on device performance caused by cavitation are quite distinct from those discussed earlier, which relate more specifically to capability of the piezoelectric driving element and inherent properties of the working fluid. In particular, maximum operating frequency and output force (and hence the maximum mechanical output power available from the actuator) are limited chiefly by charge pressure level and intake-valve pressure drops, and not by the capability of the active-material driver.

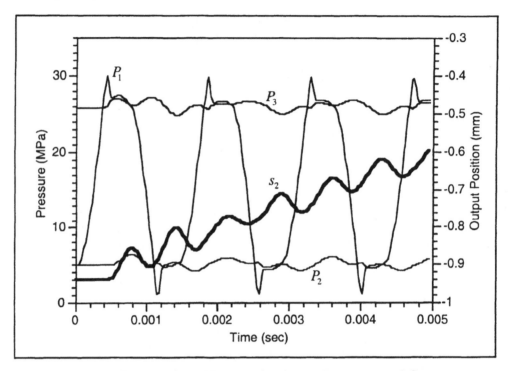

Figure 4. Typical Time History of Actuator Pressures and Output

Given the inherent material and fluid-dynamic limitations of such hybrid actuator devices, the observations made above beg the questions, "how much useful mechanical power could such an actuator produce, and how should such a device be built for maximum power output?" The answers to these questions are explored here by examining simulation results for a variety of candidate hybrid actuator configurations. The target actuator is sized using the relation of Eq. (2) for 750 W mechanical output (10 cm/sec output velocity at 7500 N output force). Figure 5 contains plots of estimated output power versus operating frequency for various permutations of simulated-device parameters, as predicted by the model described above.

Figure 5(a) shows average power obtained driving a static, 7500 N load for cases where charge pressure is set to 2 MPa and 5 MPa. Note that the maximum power available from this case is predicted by the simulation model at 150 W for 2 MPa charge pressure, and at 400 W for the 5 MPa charge pressure. Also, note the relatively linear increase in output power with frequency up to the maximum. Above the maximum power level for each case, the power output of the device is limited by predicted cavitation of the working fluid on the intake stroke. Hence, output power available from the device is reduced to keep inlet valve flow velocities small enough to avoid cavitation. The power is modulated by adjusting the maximum commanded stroke of the solid driving element (δ_{MAX}) to a level that avoids inlet cavitation.

Figure 5(b) shows the influence of valve opening area on maximum available power. In this case, 5 MPa charge pressure is assumed, and the maximum valve opening areas are adjusted from the baseline 9 mm^2 to 15 mm^2. Maximum output power of the hybrid actuator increases from roughly 500 W to 600 W before inlet cavitation occurs. Note that valve dynamics have also been improved in these cases. First resonance of the passive valves is set at 2.8 kHz for the cases of Fig. 5(b), and at 1.4 kHz for the cases of Fig. 5(a) (the authors note that frequencies in these ranges have been observed in the laboratory for carefully engineered passive valves).

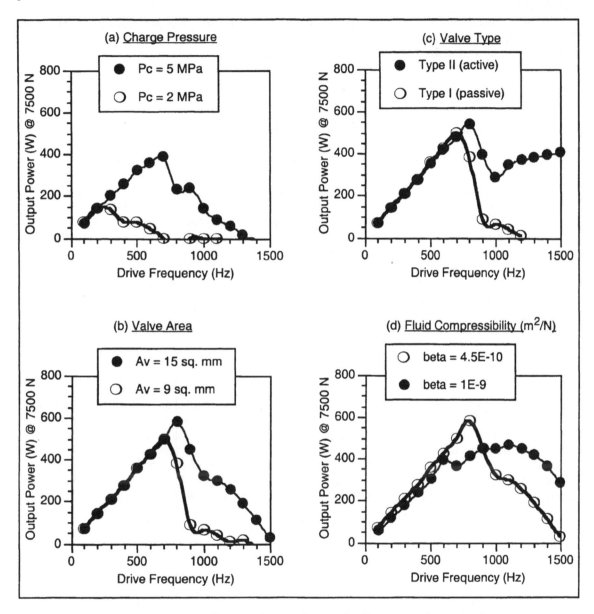

Figure 5. Simulated Device Output Power for Parameter Perturbations

Referring back to the typical characteristics visible in the device pressure history (Fig. 4), one should note that cavitation occurs first at the lowest pressure excursion of the pressurization chamber. If no flow restriction was present between the pressurization chamber and the low-pressure chamber of the output piston, it would be impossible to lower the pressure below the initial charge pressure. While realization of lossless flow paths is not possible, careful attention to the timing of valve opening displacements may allow the severity of the negative pressure spike to be mitigated and cavitation to be avoided. This effect is shown in Figure 5(c), which compares operation of a passive valve (2.8 kHz first resonance) and an active valve (2.8 kHz activation rate). The active valve opening area is 3 mm² versus 9 mm² for the passive valve, yet the faster, externally actuated opening of the active valve reduces inlet pressure drop sufficiently to increase the maximum power of the hybrid actuator to nearly 600 W.

Finally, Figure 5(d) graphically shows the effect of real-fluid compressibility. As shown in Fig. 5(d), increasing the fluid compressibility to 1000 μm²/N reduces maximum output power of the hybrid actuator to the range around 400 W.

CONCLUSION

The most clear implication of the present work is that hybrid solid-fluid actuators are possible, and the approach of rectifying energy produced at a frequency considerably above that at which it is applied to an external load appears to merit further development. However, fundamental limitations on the maximum actuation power available from a solid-fluid hybrid actuator include not only the inherent physical limitations of the solid driving element (e.g., stiffness, stroke, stable operating frequency), but also limitations imposed by the working fluid (e.g., compressibility, vapor pressure). Indeed, the likelihood of fluid cavitation is noted here as the principal factor limiting the increase of output power with increasing operating frequency of the solid-state driving element.

This latter observation has not yet been well treated in the published literature [1-6] despite the fact that it is a first-order effect which can be illuminated by even the simplest valve-flow models. Indeed, from (3), one may develop a simple expression to estimate the operating frequency at which maximum output power can be achieved:

$$f_{MAX} \approx \frac{A_{1i}}{A_1 \pi \delta_{MAX}(2+\pi)} \sqrt{\frac{2\overline{P}}{\rho_i}} \tag{6}$$

The value predicted by Eq. (6) for the parameters listed in Table II and 5 MPa charge pressure is 740 Hz, which compares well with the maximum power deduced by simulation and presented in Fig. 5(a). While such good correspondence is encouraging, one should note that cavitation-limited maximum output power of such devices also depends inherently on compressibility of the working fluid, and the external loading applied to the actuator. Nevertheless, Equation (6) provides general guidance regarding selection of inlet valve opening areas (A_{1i}) with respect to piston area (A_1). Equation (6) predicts maximum-power frequency to be directly proportional to this ratio.

Finally, the present study has illuminated the need for well-designed valves to rectify the high-frequency pressurization of the working fluid. Valves must operate at high frequency and, in particular, they must operate with sufficient speed and opening area to cause low pressure drop in the pressurization chamber's inlet passages. For these reasons, active valves may be quite desirable, despite the added complexity and power they would require to be realized in practice. The key to successful valve design is without surprise to allow maximum opening (to maximize the ratio A_{1i}/A_1) at speeds sufficiently faster than f_{MAX} if drive frequency f_{MAX} is to be achieved.

In sum, realization of practical actuators exploiting the promising energy-density properties of solid-state actuators will require careful attention to the engineering of the entire actuation device, not just the solid-state driver components or the pressure-rectifying valves. The systemic influences of variable external loads and real-world fluid properties must be considered to be integral parts of the actuation system if design details are to be chosen to maximize net output power of the device. Simple models, like those presented here, can serve as the groundwork for this further development, but a level of understanding needed to achieve the best hybrid actuator designs will also require extensive experimentation and development.

REFERENCES

1. Gerver, M.D. et al., "Magnetostrictive Water Pump," *Proc. SPIE Conf. on Smart Structures and Integrated Systems*, **3329**, M. E. Regelbrugge, ed., 1998, pp. 694-705.
2. Mauck, L.D. and C. S. Lynch, "Piezoelectric Hydraulic Pump," *Proc. SPIE Conf. on Smart Structures and Integrated Systems*, **3668**, N. Wereley, ed., 1999, pp. 844-852.
3. Mauck, L.D. and C. S. Lynch, "Piezoelectric Hydraulic Pump Development," *Proc. SPIE Conf. on Smart Structures and Integrated Systems*, **3991**, N. Wereley, ed., 2000, pp. 729-739.
4. Nasser, K., D. J. Leo and H. Cudney, "Compact Piezohydraulic Actuation System," *Proc. SPIE Conf. on Industrial and Commercial Applications of Smart Structures Technologies*, **4327**, J. H. Jacobs, ed., 2000, pp. 312-322.
5. Hagood, N. et al., "Development of Micro-Hydraulic Transducer Technology", *Proc. Tenth Int'l Conf. on Adaptive Structures and Technologies*, R. Ohayon and M. Bernadou, eds., Technomic, Lancaster, PA, 2000, pp. 71-81.
6. Sirohi, J. and I. Chopra, "Development of a Compact Piezoelectric-Hydraulic Hybrid Actuator," *Proc. SPIE Conf. on Smart Structures and Integrated Systems*, **4327**, P. L. Davis, ed., 2001, pp. 401-412.
7. Merritt, H.E., *Hydraulic Control Systems,* John Wiley & Sons, New York, 1967.

COMPACT PIEZO-HYDRAULIC HYBRID ACTUATOR

Jayant Sirohi and Inderjit Chopra

ABSTRACT

This paper discusses the preliminary design, fabrication and testing of a compact piezo-hydraulic hybrid actuator. The device is envisaged as a potential actuator for a trailing edge flap on a full scale smart rotor system. Preliminary sizing was carried out to meet the requirements of a trailing edge flap on a section of a full scale MD900 Explorer helicopter rotor blade. The device consists of two parts. The first part is a pump driven by piezoceramic stacks and the second part is an output stage comprising a hydraulic cylinder and a directional valve. The overall size of the pump is 4"x1.5"x1.5". The present design is focused towards a high pumping frequency in order to obtain a compact device with a large power density. Experiments were performed to measure the unidirectional output characteristics of the device as a function of system parameters such as bias pressure, pumping frequency and check valve design. A no-load velocity of 0.65 in/sec and a blocked force of 22 lbs were measured. The maximum pumping frequency was 300 Hz.

INTRODUCTION

Several actuation mechanisms based on piezoceramic stacks have been investigated for full scale smart rotor applications [1–8]. In these applications, the major challenge is to increase the small stroke of the piezoceramic stack actuators. Several stroke amplification mechanisms have been designed and implemented to address this issue. Amplification mechanisms in general involve many moving parts that contribute to actuation losses and degrade rapidly under high loading, especially the large centrifugal loads encountered on a helicopter rotor blade. In addition, the maximum practical amplification ratio is on the order of 15. In order to overcome the stroke limitations of piezoceramic stacks without mechanical amplification mechanisms, a device based on the piezo-hydraulic concept is proposed as a potential trailing edge flap actuator on a full scale rotor blade.

[1] Jayant Sirohi, Graduate Research Assistant, Alfred Gessow Rotorcraft Center, Department of Aerospace Engineering, University of Maryland, College Park, MD 20742.
[2] Inderjit Chopra, Alfred Gessow Professor and Director, Alfred Gessow Rotorcraft Center, Department of Aerospace Engineering, University of Maryland, College Park, MD 20742

The piezo-hydraulic hybrid actuator concept is based on frequency rectification of small piezostack displacements. This concept combines the advantages of the high energy density of piezostacks with the versatility of hydraulics as a power transmitting medium. The basic frequency rectification principle involves trading off the high frequency, low displacement stroke capability of piezoceramic stacks into a lower frequency, higher displacement output stroke. Several other actuation concepts based on the same fundamental principle have been extensively investigated, for example, inchworm motors and ultrasonic motors. However, these concepts rely on frictional forces in order to achieve rectification of piezostack displacements and also require close tolerances and precise machining. As a result, they are very susceptible to wear and have a limited lifetime. In comparison, the piezo-hydraulic concept does not rely on frictional forces, and is more suited to moderately large force outputs.

Several researchers have investigated piezo-hydraulic hybrid devices in the past. Mauck and Lynch [9,10] investigated a system consisting of a pump driven by a high voltage piezostack of length 10.2 cm and cross-sectional area 1.9 cm × 1.9 cm. The final device achieved a blocked force of 61 lbs and a no-load output velocity of 7 cm/sec. The large current requirements and heating of the piezostack limited the pumping frequency of the system to 60 Hz. Nasser et. al. [11] presented a piezohydraulic actuation system that makes use of the compressibility of the working fluid in order to eliminate accumulators and 4-way valves. Solenoid valves were used as active check valves, that could achieve both rectification of the flow as well as control the direction of the output actuator. The system was run at 10 Hz at an input voltage amplitude of 150 V, and demonstrated an overall amplification factor of 1.42. The solenoid valves were the main barrier to increase the pumping frequency. Konishi et. al. [12,13] developed a piezoelectric hydraulic hybrid actuator using a piezostack of length 55.5mm and diameter 22mm at an operating voltage ranging from -100V to +500V. This piezoelectric pump was excited at 300 Hz and delivered an output power of approximately 34 W.

A common feature of the previous research efforts on piezo-hydraulic hybrid actuators is that the required flow rates were achieved primarily by using large piezostacks, operating at a relatively low pumping frequency. Compactness was not a major design driver in these configurations, and the high frequency capability of piezostacks was not effectively utilized. However, in the present work, compactness was considered an important factor due to the severe volumetric constraints in a helicopter rotor blade. The present work describes the preliminary sizing and development of a compact piezo-hydraulic hybrid actuator, that comprises a piezoelectric pump coupled to an output hydraulic actuator. The focus is on operation at a high pumping frequency in order to achieve a high power density.

PRELIMINARY DESIGN AND SIZING

Typical requirements for the output of the device are shown in Figure 1. As the present device is intended to be the next generation actuator for a typical smart rotor system, the output design goals are along the lines of a recently developed trailing edge flap actuator driven by piezoelectric stacks [3]. The above mentioned actuator was designed for a full

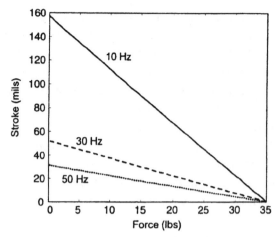

Figure 1. Nominal output design goals

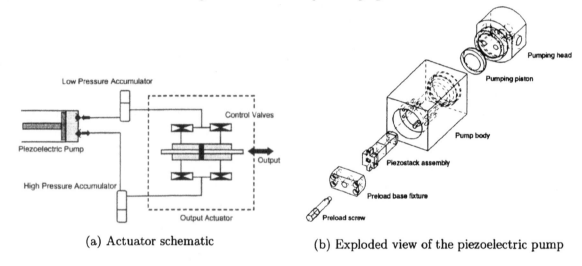

(a) Actuator schematic   (b) Exploded view of the piezoelectric pump

Figure 2. Piezoelectric hydraulic hybrid actuator

scale MD-900 Explorer rotor blade section, with a force requirement of 18 lbs peak at 43 mils displacement. A schematic of the present device is shown in Figure 2(a). The device broadly consists of two parts: a pump driven by piezoceramic stacks, and an output hydraulic cylinder and directional valve arrangement. The basic design, fabrication and testing of a proof of concept piezoelectric pump (Figure 2(b)) was described in a previous paper [14]. The fluid is pumped by means of a piston actuated by piezoceramic stacks, and rectification of the flow is achieved by means of check valves installed in the pumping head.

Quasi-static pumping assumptions

The performance of the device is highly sensitive to the dimensions of both the pumping chamber and the output hydraulic cylinder. The viscosity of the fluid, diameter of the tubing, and the modulus of elasticity of the tubing material also play important roles in the frictional losses that occur in the device as well as its frequency response. However, for an initial design, it is convenient to neglect any frequency dynamics of the system, and assume a constant response at all pumping frequencies. As a result of this quasi-static pumping assumption, the volumetric flow rate of the pump, Q, at any given pumping frequency is equal to the product of the volumetric displacement of the pump per cycle, Δ_{pump} and the pumping frequency f_{pump}.

$$Q = \Delta_{pump} f_{pump} \tag{1}$$

The large flexibility in trading off force and stroke afforded by the use of hydraulics results in many possible combinations of pumping chamber and output actuator dimensions. The final design is therefore very dependent on the output requirements of the device. Keeping in mind the ultimate goal of maximizing output power density, for a device of known external dimensions, the output power is considered to be the primary performance metric.

The major variables on which the power output depends are:

1. Pumping chamber geometry: diameter d_{cham} and height Δ_{gap}

2. Output actuator diameter, d_{out}

3. Piezostack characteristics : blocked force F_{block} and free displacement δ_{free}

4. Fluid compressibility, β

A piezostack which is known to have a high energy density [3], PI-804.10 [15] is chosen to drive the pump, and hence the piezostack characterisitics are assumed to be constant for this trade-off study. Similarly, the fluid compressibility is also assumed constant, because a low viscosity standard hydraulic oil, MIL-H-5606F was chosen as the working fluid.

Work output per cycle: Load-line analysis

The power output of the device is given by the area under the load-force curve (area OABCO, Figure 3). As a result of the quasi-static pumping assumption (Equation (1)), to maximize the output power, it is sufficient to maximize the work done by the device per pumping cycle.

The equation of the piezostack load line is given by

$$\delta = \delta_{free} \left(1 - \frac{F}{F_{block}} \right) \tag{2}$$

and the fluid stiffness in the pumping chamber is given by

$$K_f = \beta \frac{A_p}{\Delta_{gap}} \tag{3}$$

where A_p is the area of the pumping chamber. Figure 4 shows a schematic of the fluid column between the pumping chamber and the output actuator when one of the check

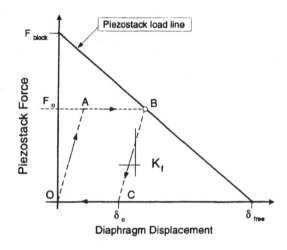

Figure 3. Load-displacement variation

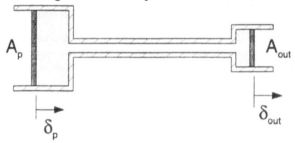

Figure 4. Fluid line between pumping chamber and output actuator

valves is open. The pumping chamber has a displacement δ_p, while the output actuator has a cross-sectional area A_{out} and a displacement δ_{out}. The area ratio is given by

$$A_R = \frac{A_{out}}{A_p} \qquad (4)$$

Solving the equations of the fluid stiffness in the pumping chamber and the load line of the piezostack, the work done per cycle can be derived in terms of the force and stroke of the output actuator as

$$\Delta W_{cyc} = \frac{F_{out}}{A_R}\left[\delta_{free} - \frac{F_{out}}{A_R}\left(\frac{1}{K_p} + \frac{\Delta_{gap}}{\beta A_p}\right)\right] \qquad (5)$$

where F_{out} is the force at the output hydraulic cylinder.

Maximum output work

The dimensions of the output actuator are fixed by overall geometric constraints; a commercially available 0.5" bore hydraulic cylinder was chosen for the present device. This fixes the value of A_{out} and the only parameters that remain to be selected are the diameter and height of the pumping chamber.

It should be noted here that although maximum work output is the primary goal of the device in order to achieve maximum possible power density, the application specifications

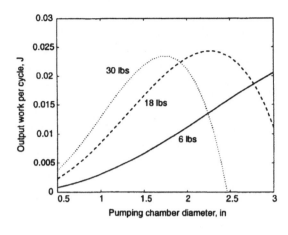

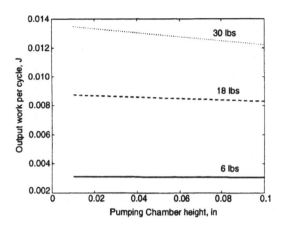

(a) Work output as a function of pumping chamber diameter, for a chamber height of 0.05"

(b) Work output as a function of pumping chamber height, for a chamber diameter of 1"

Figure 5. Dependence of work output per cycle on pumping chamber geometry

may dictate a certain displacement requirement at a certain bandwidth (Figure 1). This requirement directly translates into the flow rate of the piezoelectric pump. A given output power, being the product of force and velocity, can be achieved with a high force and low velocity or a low force and a high velocity. Therefore, in addition to sizing the parameters of the device for maximum output work per cycle, the constraint of achieving the required output displacement must also be included in the design process.

Plots of the variation of output work with A_{out} and Δ_{gap} are shown in Figure 5. In Figure 5(a), the work output per cycle is plotted as a function of pumping chamber diameter for various output loads, with a pumping chamber height of 0.05". The maximum work output per cycle is achieved at a certain value of pumping chamber diameter, and this maximum decreases and takes place at a lower diameter with higher output load. Figure 5(b) shows the work output per cycle as a function of pumping chamber height, for a pumping chamber diameter of 1". The maximum work output increases monotonically (at a slower rate) with decreasing pumping chamber height. This is to be expected as a smaller pumping chamber height increases the stiffness of the fluid and essentially provides a direct energy transfer to the output load. As this function has no extremum, a value of pumping chamber height between 0.02" and 0.05" can be chosen depending on other factors such as machinability and clearances.

Based on the above discussion, taking into consideration any geometric constraints on the overall size of the device, an optimum pump geometry can be arrived at for a given output load.

For the present proof of concept experimental device, the diameter of the pumping chamber was set at 1" and the height of the pumping chamber was set at 0.05". A list of all the important parameters of the current design is shown in Table. 1. The piezostack data correspond to each piezostack [3].

Piezostack – Model P-804.10			Hydraulic Fluid – MIL-H-5606F		
Number of piezostacks	2		Density	859	gm/cc
Length	0.3937	in	Kinematic Viscosity	15	centistokes
Width	0.3937	in	Reference Bulk Modulus β_{ref}	260,000	psi
Height	0.7087	in	Pumping Chamber		
Blocked Force	1133	lbs	Diameter	1	inch
Free strain	1035	$\mu\epsilon$	Height	0.050	inch
Free displacement	≈ 26	μm	Output Actuator - Double Rod		
Maximum voltage	120	V	Bore diameter	0.5	inch
Minimum Voltage	-24	V	Shaft diameter	0.25	inch
Capacitance	≈ 7	μF	Stroke	2	inch

Table 1. Prototype device parameters

EXPERIMENTAL SETUP AND RESULTS

Experimental Setup

Experiments were performed to measure the output power of the device as a function of pumping frequency and other operating parameters such as system bias pressure. The experimental setup is shown in Figure 6. In order to make the final device as compact as possible, it is envisaged that the piezoelectric pump will drive the output hydraulic cylinder directly, without any accumulators in the circuit. Therefore, in the experimental setup, valves V1 and V2 were closed, cutting off the accumulators from the rest of the hydraulic circuit. A 10μm line filter was included upstream of the pump inlet in order to prevent any contamination in the working fluid from affecting the operation of the pump. Deadweights were hung off the end of the output hydraulic actuator and the position of the output hydraulic cylinder, h, was measured using a linear potentiometer. The pressures in the system were measured using pressure transducers and the displacement of the piezostacks was measured using strain gages bonded on the piezostacks. All data acquisition was performed using a Windows NT based PC with a National Instruments PCI-6031E 16-bit DAQ card, operated by a virtual instrument application programmed in LabVIEW™ 5.1.

The piezostacks were actuated by a sinusoidal waveform from 0 to 100 V, at frequencies of 50-300 Hz. A conventional linear amplifier, a HERO precision power amplifier (Model PA9810), custom designed by Rohrer Meß-& Systemechanik [16], was used to actuate the stacks. The model PA9810 amplifier is a 700 Watt amplifier with a ±500 V, ±1.5 A output, and the maximum pumping frequency was limited by the current limit of the amplifier.

It should be noted that the setup as shown in Figure 6 only enables uni-directional actuation of the output hydraulic actuator. While this may be sufficient for the present measurements, bi-directional actuation will require the installation of a 4-way valve in the hydraulic circuit.

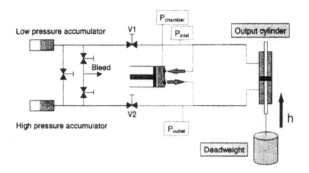

Figure 6. Experimental setup for the evaluation of device performance

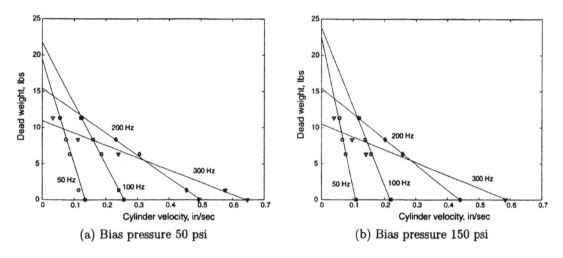

(a) Bias pressure 50 psi   (b) Bias pressure 150 psi

Figure 7. Ouput force-velocity characteristics

Experimental Results

The output force-velocity characteristics of the device are shown in Figure 7. The output was measured at two different values of system bias pressure, 50 psi and 150 psi. It can be seen that the performance is approximately the same at both the bias pressures. At the higher bias pressure of 150 psi, the no load cylinder velocity is less than that for the case of 50 psi bias pressure. This is believed to be due to seizing of the piston in its bore due to possibly higher off-axis loads. A tighter tolerance may solve this problem. For the time being, 50 psi is chosen as an acceptable working bias pressure for the device.

Figure 8 shows the measured pressure traces in the pumping chamber of the piezoelectric pump, converted into the force acting on the piezostack. Each force-velocity trace is measured at a particular value of output load. Also shown for comparison is the load line of the piezostack. The effect of fluid compressibility in the pumping chamber is clearly visible from this plot. It can be seen that a large part of the force capability of the piezostack is not utilized as a result of the finite compressibility of the fluid.

The fluid compressibility, which is a key parameter on which the power output of the device depends, can be very different from the reference value specified in the data sheet.

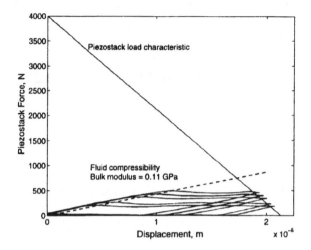

Figure 8. Measured pumping chamber force-displacement curves, for 50 Hz pumping frequency, and 50 psi bias

The actual compressibility depends on several factors such as the system bias pressure and the amount of air entrained, and can be as low as 10% of the reference value [17]. This can have a significant effect on the output performance. It should be noted that the reference fluid bulk modulus is specified at 2500 psi bias pressure , with no entrained air. The actual fluid bulk modulus is given by Reference [18] as

$$\beta_{\text{fluid}} = E_{\text{p}} \times E_{\text{a}} \times \beta_{\text{ref}} \tag{6}$$

where E_{p} and E_{a} are the correction factors for bias pressure and entrained air, and β_{ref} is the reference fluid bulk modulus, which for the current fluid is given as 260,000 psi (Table 1). For the current system bias pressure of 50 psi, $E_{\text{p}} = 0.9$. The fraction of entrained air is assumed to be 1%, which is not an uncommon value for hydraulic systems. For this amount of entrained air, $E_{\text{a}} = 0.1$. It can be seen that the entrained air has a strong influence on the effective fluid bulk modulus, however, dissolved air is known to have no effect on the effective bulk modulus [17]. Substituting these values into Equation (6), the effective fluid bulk modulus is obtained as 23400 psi (0.16 GPa). This value is in good agreement with the measured value of 0.11 GPa, from Figure 8.

Another important parameter on which the power output of the device depends is the flow rate of the hydraulic fluid. The flow rate can be theoretically increased by increasing the pumping frequency of the piezoelectric pump. However, as the pumping frequency is increased, the dynamics of the check valves becomes a significant factor affecting the flow rate. The performance of the device was measured using two different types of check valves in the pumping head. The first type, called the Type I valves, are commercially available ball type check valves [18]. The second type check valves, called the Type II valves, are designed and fabricated in-house, to have a natural frequency around 1.5 kHz. Both valves were installed in the pumping head and the performance of the device was measured at pumping frequencies of upto 300 Hz. The results are shown in Figure 9. It can be seen that larger flow rates, and hence, larger power outputs are achievable with Type II valves, that have a higher natural frequency .

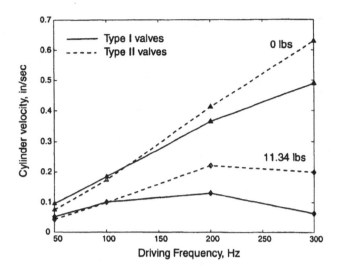

Figure 9. Comparison of output performance, Type I and Type II check valves

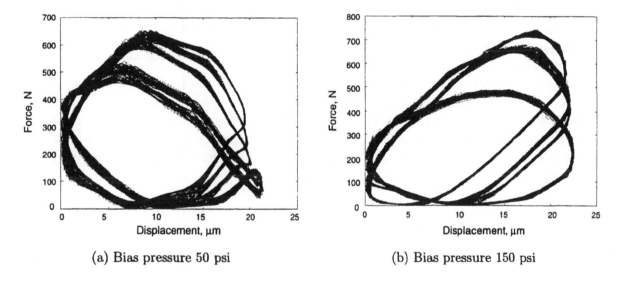

(a) Bias pressure 50 psi   (b) Bias pressure 150 psi

Figure 10. Ouput force-velocity characteristics

Figure 10 shows a comparison of the pumping chamber pressure traces at an excitation frequency of 300 Hz, for the Type I valves and the Type II valves. It can be seen that at this frequency, the shape of the pressure curves for the case of the Type I valves are far from the ideal parallelogram shapes. It can be concluded that at high frequencies, the dynamics of the Type I valves becomes a significant factor affecting the pressure history in the pumping chamber. In comparison, the pressure traces for the case of the Type II valves (Figure 10(b)) are much closer to the ideal shape. It can be concluded that the Type II valves allow larger flow rates because they do not affect the dynamics of the fluid in the pumping chamber as much as the Type I valves.

SUMMARY AND CONCLUSIONS

The performance of the piezo-hydraulic hybrid actuator is very dependent on the output mechanical load. For a given output load and a given set of operating parameters such as piezostack characteristics, the dimensions of the piezoelectric pump can be calculated in order to obtain maximum power density.

The power output of the device depends on its blocked force and its no-load velocity. The blocked force of the device is directly dependent on the blocked force of the piezostacks, as well as the dimensions of the pump. However, the no-load velocity, which is a measure of the flow rate of the pump, is dependent on piezostack characteristics, pump dimensions, and pumping frequency. Neglecting any frequency dynamics in the pump, the flow rate is linearly dependent on the pumping frequency. Hence, in order to achieve high power densities, the pumping frequency must be as high as possible. It was observed that the commercially available Type I check valves, as a result of their low natural frequency, caused a roll-off of the flow rate with frequency. Type II valves with a high natural frequency enabled an increase in the flow rates of approximately 25% over Type I valves. With the Type II valves, at a pumping frequency of 300 Hz, a no-load velocity of 0.65 in/sec was measured. The blocked force of the device was measured to be approximately 22 lbs. The pumping frequency was constrained by the current limit of the amplifier.

Future work would involve exploring means to increase the pumping frequency of the device. The first barrier to higher frequency pumping is the current requirements of the piezostacks. Another major barrier is the temperature rise in the piezostacks caused by self-heating. In addition to increasing the pumping frequency, future work would involve incorporating a 4-way valve in the hydraulic circuit and characterizing the bi-directional performance of the device.

ACKNOWLEDGEMENTS

The authors would like to acknowledge CSA Engineering Inc., and DARPA for their support. The authors would also like to thank Dr. Eric Anderson and Jason Lindler for their useful comments.

REFERENCES

1. Inderjit Chopra. Status of application of smart structures technology to rotorcraft systems. *Journal of the American Helicopter Society*, 45(4):228–252, October 2000.

2. Taeoh Lee and Inderjit Chopra. Development and validation of a refined piezostack-actuated trailing-edge flap actuator for a helicopter rotor. In *SPIE symposium on smart structures and materials, conference on smart structures and integrated systems*, Newport Beach, CA, March 1999.

3. Taeoh Lee. *Design of a High Displacement Smart Trailing-Edge Flap Actuator Incorporating Dual-Stage Mechanical Stroke Amplifier for Rotors*. PhD thesis, University of Maryland, College Park, MD 20742, USA, 2000.

4. P. Jänker, V. Klöppel, F. Hermle, T. Lorkowski, S. Storm, M. Christmann, and M. Wettemann.

Development and evaluation of a hybrid piezoelectric actuator for advanced flap control technology. In *25th European Rotorcraft Forum*, Rome, Italy, September 1999. Paper G-21.

5. B.G. van der Wall R. Kube, A. Büter, U. Ehlert, W. Geissler, M. Raffel, and G. Schewe. A multi concept approach for development of adaptive rotor systems. In *8th Army Research Office (ARO) workshop on the aeroelasticity of rotorcraft systems*, State College, PA, October 1999.

6. T. Lorkowski, P. Jänker, F. Hermle, , S. Storm, M. Christmann, and M. Wettemann. Development of a piezoelectrically actuated leading edge flap for dynamic stall delay. In *25th European Rotorcraft Forum*, Rome, Italy, September 1999. Paper G-20.

7. Friedrich K. Straub. Development of a full scale smart rotor system. In *8th Army Research Office (ARO) workshop on the aeroelasticity of rotorcraft systems*, State College, PA, October 1999.

8. Friedrich K. Straub. Smart rotor technology developments - challenges and payoffs. In *The 12th International Conference on Adaptive Structures Technology*, University of Maryland, College Park, MD, October 2001.

9. Lisa D. Mauck and Christopher S. Lynch. Piezoelectric hydraulic pump development. *Journal of Intelligent Material Systems and Structures*, 11:758–764, October 2000.

10. Lisa D. Mauck, William Sumner Oates, and Christopher S. Lynch. Piezoelectric hydraulic pump performance. *Proceedings of the 8th SPIE Conference on Smart Structures and Materials: Industrial and Commercial Applications of Smart Structures Technologies, Newport Beach, CA*, 4332:246–253, March 2001.

11. Khalil Nasser, Donald J. Leo, and Harley H. Cudney. Compact piezohydraulic actuation system. *Proceedings of the 7th SPIE Conference on Smart Structures and Integrated Systems, Newport Beach, CA*, 3991:312–322, March 2000.

12. Katsunobu Konishi, Toshio Yoshimura, Kyoji Hashimoto, and Nobuhiko Yamamoto. Hydraulic actuators driven by piezoelectric elements(1st report, trial piezoelectric pump and its maximum power). *Journal of Japanese Society of Mechanical Engineering (C)*, 59(564):213–220, 1993.

13. Katsunobu Konishi, Toshio Yoshimura, Kyoji Hashimoto, Takahiko Hamada, and Taro Tamura. Hydraulic actuators driven by piezoelectric elements(2nd report, enlargement of piezoelectric pumps output power using hydraulic resonance). *Journal of Japanese Society of Mechanical Engineering (C)*, 60(571):228–235, 1994.

14. Jayant Sirohi and Inderjit Chopra. Development of a compact piezoelectric-hydraulic hybrid actuator. *Proceedings of the 8th SPIE Conference on Smart Structures and Integrated Systems, Newport Beach, CA*, 4327:401–412, March 2001.

15. Physik Instrumente (PI). *Products for Micropositioning, US-edition*, 1997.

16. HERO precision power amplifier, model pa9810. ROHRER Meß und Systemtechnik, email: rohrer@t-online.de, Munich, Germany, 1999. Technical manual.

17. D.McCloy and H.R. Martin. *Control of Fluid Power: Analysis and Design*. Ellis Horwood Limited, Chichester, England, second(revised) edition, 1980.

18. The Lee Company. *Technical Hydraulic Handbook*, 1997.

PERFORMANCE MODELING OF A PIEZO-HYDRAULIC ACTUATOR

Christopher Cadou and Bing Zhang

ABSTRACT

A simple quasi-static model has been developed as an engineering tool for improving the performance of a piezo-hydraulic actuator's fluid system. The model's predictions compare reasonably well with experimental data at low frequencies (< 150 Hz) as trends in the dependence of actuation speed on driving frequency and applied load are captured appropriately. The model indicates that there is an optimum load and driving frequency that corresponds to maximum power output but that the operating conditions for optimum power output and efficiency are different. The efficiency of the fluid system is maximum (100%) at 0 Hz independent of load but decreases with increasing frequency at a rate that depends on the load. Viscous losses through the valves and tubing are negligible compared with the inertial losses associated with accelerating and decelerating the load. This acceleration and deceleration process occurs twice per piezo cycle because of the configuration of the fluid system used to rectify the oscillatory motion of the piezo stack. Accordingly, the inertia of the load dominates the behavior of the device at high frequencies. The performance of the fluid system is most sensitive to the stiffness of the fluid in the pumping chamber, which should be maximized for maximum power output.

INTRODUCTION

The next generations of aircraft and rotorcraft will increasingly rely on a network of small aerodynamic actuators coupled with sensors and distributed across aerodynamic surfaces to alter and control, among other things, the vehicle's stall and acoustic characteristics. These actuators must be compact enough to fit inside wings, control flaps and helicopter rotor blades, while still being capable of delivering significant amounts of power [1]. Since some applications like active control of separation-induced vibration [2] on helicopter rotors require the actuator to be located on moving aerodynamic elements that are subject to high inertial loads, traditional hydraulic actuators that require high-pressure lines for routing fluid are not practical. In applications like these, it is simpler to route power electrically. What is required, however, is a high power density actuator to make the electrical to mechanical power

Christopher P. Cadou, University of Maryland Department of Aerospace Engineering, 3179 D Glenn L. Martin Hall, College Park, MD
Bing Zhang, University of Maryland Department of Aerospace Engineering, 3179 Glenn L. Martin Hall, College Park, MD

conversion. Piezoelectric materials offer a promising means of achieving this conversion in a low-mass device (high power density) because they can produce very high forces and be driven at high frequencies. The difficulty is that the displacement of piezoelectric materials is relatively small because maximum voltage-induced strains are ~0.1%. As a result, some sort of rectification is required to convert the high-frequency, low amplitude motion of the piezoelectric material to lower frequency but higher amplitude motion suitable for moving realistic control surfaces.

Various strategies for rectifying the piezo-motion have been developed for a range of applications. Inchworm motors [3] use a pair of piezoelectric elements to alternately drive and clamp a bar or rod 'inching' it forward or backward. These are now in routine use in optical instruments like interferometers that require very precise motion control but their longevity at high force levels is somewhat limited by wear on the clamps and bar. A variant of this approach that is capable of producing either linear or rotary motion is the micro-pulse actuator [4]. Ultrasonic motors excite an elastic wave in a ring pressed against a disk pressed against it causing the disk to rotate. While these are efficient because they operate at resonance, they are relatively slow and small [5]. Mechanically coupling the piezoelectric stack to a lever or other mechanical linkage can produce amplitude amplification but at the expense of actuation force [6,7]. Moreover, the additional components increase the complexity and inertia of the system, which complicates operation at the high frequencies required for achieving high power density [8].

The focus of this work is a particularly promising approach that achieves amplitude rectification without sacrificing actuation force by using a hybrid piezo-fluid system. In such a system, pictured in figure 1, a piezoceramic element pressurizes a fluid-filled cavity equipped with check valves to create a compact fluid pump. The pressurized fluid drives a co-located hydraulic cylinder creating a high power density hybrid actuation device. A detailed discussion of the operation of the system follows later. In brief, a time-varying voltage applied to the piezoceramic stack causes it to expand and contract alternately pressurizing and de-pressurizing the fluid in the pumping chamber. This time-varying pressure coupled with the pair of oppositely oriented check valves result in a net displacement of fluid through the system and motion of the actuator piston. The direction of actuation could be reversed by inverting each check valve (or by using an active 4-way valve).

The concept of piezo-hydraulic pumping has been around for at least a decade and was first recognized in the context of micro-fluidic applications. Accordingly, there has been at least one attempt to use very simple fluid models to predict the performance of single and multiple chamber pumps [9]. Recently, other researchers have created devices that work in both directions as either high power density actuators or 'energy harvesting devices' that could, for example, transmit the mechanical deformation undergone by the heel of a boot during walking into strain on a piezoelectric stack to produce electrical power [10,11]. However, this work is focused primarily on the development of the device and does not address the dynamics of the fluid during operation. The application of piezo-hydraulic pumping to large-scale actuators for aerospace applications is more recent and researchers in this area have begun to address the coupling between the piezo actuator and the fluid in the pumping chamber [12]. Some basic principles governing efficiency of the transfer of power from the piezo to the fluid in the pumping chamber have been presented, but a realistic model of the entire fluid system that is useful for predicting the performance of an actuator has not been presented. Extensive work on the overall efficiency of the conversion of electrical to mechanical power in piezo-driven structures [13] and in hybrid actuators using lumped parameter modeling [14] has been

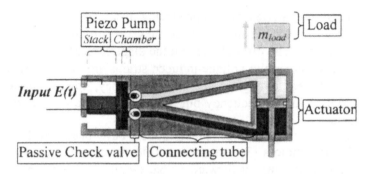

Figure 1 Schematic Diagram of Hybrid Actuator System

performed but there is no work to date that integrates realistic models of the fluid behavior with a model of the overall fluid system.

Accordingly, the work presented here is a preliminary attempt to fill this gap by focusing on the design of the fluid system and incorporating more realistic fluid models. The behavior of the fluid in the actuator is described and is used to predict and explain the behavior of a prototype actuator system. The efficiency of the conversion of mechanical power in the pumping chamber to mechanical power at the actuator is addressed and some suggestions are made for improving the performance of these devices.

QUASI-STATIC HYBRID ACTUATOR MODEL

Figure 2 shows the simplified model used to develop the mathematical representation of the actuator. We assume that the actuator operates in a quasi-static mode in which the pressures in the various fluid components and the positions of the diaphragm and actuator piston are determined by static force balances across the components. This is justified because the speed of an acoustic wave propagating through the fluid is much larger than the actuation speed or the velocity of the fluid through the system. Basic models for the piezo stack and the fluid components illustrated in figure 2 are developed in the following two sections. These sub-models are then assembled to create a system of coupled equations that are used to predict the performance of the actuator. The dimensions of the principal components are listed in Table 1.

Piezo Stack

Previous work [15] models the stress-strain relationship in a piezoelectric material as follows

$$S_{ij} = s_{ijkl}^E T_{kl} + d_{kij} E_k$$

where S_{ij} are the strain components, s_{ijkl}^E are the compliance components at fixed electric field, T_{ij} are the stress components, d_{kij} are the piezoelectric components, and E_k are the components of the electric field. Figure 3 shows the orientation of the piezoceramic stack being considered here. Since the electric field is being applied along axis 3, we restrict our attention

Table 1 Dimensions of fluid components

Component		Known				Unknown		
		$A_{x\text{-section}}$		Length		Length	Pressure	Mass
Description	Index		in^2		in			
Piezo stack	p					x_p	p_p	
Pump chamber	2	A_d	0.7854			x_2	p_2	m_2
Tube A	3	A_{tA}	0.0123	x_3	22		p_3	m_3
Act. A	4	A_a	0.1473			x_4	p_4	m_4
Act. B	5	A_a	0.1473			x_5	p_5	m_5
Tube B	6	A_{tB}	0.0123	x_6	22		p_6	m_6
Pump Body	p&2	A_d	0.7854	L_o	0.95			
Diaphragm	p&2			δ_d	1.0			
Actuator Body	4&5			L_a	6.0			
Actuator Piston	4&5			δ_a	0.25			

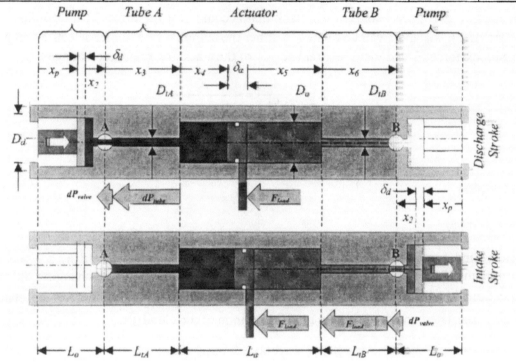

Figure 2 Simplified representation of actuator used to develop mathematical model

Figure 3 Piezo Stack Orientation

Figure 4 Fluid Element

to stresses and strains along that axis. Equation 1 reduces to:

$$S_{33} = s_{33}^E T_{33} + d_{333} E_3$$

which can be re-written using simpler notation as

$$\varepsilon = \frac{1}{k}\sigma + \frac{1}{\beta}E$$

where ε is the strain, σ is the stress, E is the voltage applied to the stack along the 3 axis, k is an elastic modulus and β is an electric modulus. Recognizing that the stress is simply the stack force divided by the cross-sectional area $\sigma = F/A_p$, that the strain $\varepsilon = (x - l_o)/l_o$ where l_o is the unstressed, unenergized length of the stack, and rearranging leads to a simplified expression for the force applied by the piezo stack:

$$F = k_{eff}\left[x - \zeta(E)l_{op}\right] \tag{1}$$

Note that k_{eff} is an 'effective' spring constant of the stack and ζ is a multiplying factor on the un-loaded, unenergized length of the stack that accounts for the effect of the applied voltage (electric field). k_{eff} and ζ can be related to either the compliance and piezoelectric components or the block-force F_b and the free-strain $\varepsilon_{free,E_{ref}}$ of the stack at a reference voltage E_{ref}:

$$k_{eff} = \frac{A_p}{S_{33}^E l_{op}} = -\frac{F_b}{l_{op}\varepsilon_{free,E_{ref}}} \tag{2}$$

$$\zeta = 1 + d_{333}E = 1 + \varepsilon_{free,E_{ref}}\frac{E}{E_{ref}} \tag{3}$$

Fluid Model

Compressiblity

Each fluid component is modeled as a compressible element with stiffness given by the bulk modulus B, which describes the change in pressure associated with changing the volume of the fluid ie. $B = \partial P/(\partial\rho/\rho)$. Integrating gives a general expression for the change in pressure associated with a change in mass or volume of the element:

$$p = p_o + B\left[\ln\left(\frac{m}{m_o}\right) - \ln\left(\frac{V}{V_o}\right)\right]$$

Assuming that the bulk modulus and the cross-sectional area of the element are constants enables us to write the pressure of the fluid in the element p_e in terms of the mass of fluid in the element m_e, the length of the element x_e, and a reference pressure p_{oe} associated with m_{oe} and x_{oe} (see figure 4).

$$p_e = p_{oe} + B\left[\ln\left(\frac{m_e}{m_{oe}}\right) - \ln\left(\frac{x_e}{x_{oe}}\right)\right]$$

This, in turn, can be linearized for small (<1%) changes in x_e and m_e to yield a first order expression for the pressure of the fluid in an element e:

$$p_e \approx p_{oe} + \frac{B}{m_{oe}}(m_e - m_{oe}) - \frac{B}{x_{oe}}(x_e - x_{oe}) \tag{4}$$

Note that the lengths of the fluid elements corresponding to tubes A and B are fixed which means that the pressure in the tubes is only a function of the initial pressure and the mass of fluid added or removed from the tube. The dimensions of each fluid element along with those of other system components are presented in table 1.

Viscous Losses

Viscous losses are neglected in all of the components except the valves and the tube connecting the open valve with the actuator. The pressure drop across the valve is estimated using the standard assumption that the mass flow rate through the passive check valves is proportional to the square root of the pressure drop through the valve. This leads to

$$\dot{m} = C_v \sqrt{\Delta P_{valve}}$$ (5)

where C_v is the flow coefficient for the valve. Note that the valve has a small cracking pressure and inertia but these effects are not included in the analysis.

The fluid in the connecting tube is assumed to be accelerated by a pressure gradient that is applied impulsively as the check valve opens. Assuming for the purposes of this calculation that the fluid is incompressible and that the velocity profile develops uniformly along the length of the tube, we use Szymansky's [16] closed-form expression for the fluid response to an impulsively applied pressure gradient:

$$u(x, y) = \frac{h}{4}\left[x^2 - 1 - 8\sum_{n=1}^{\infty}\left(\frac{J_1(a_n x)}{a_n^3 J_1'(a_n)} e^{-a_n^2 y} \right) \right]$$ (6)

Here, u is the fluid velocity with h and the non-dimensional parameters x, y, defined as follows:

$$x = \frac{r}{R}$$

$$y = \frac{\mu t}{\rho R^2}$$

$$h = \frac{R^2}{\mu}\frac{\Delta P_{tube}}{L} = 4U_{max}$$ (7)

In these expressions, R is the tube radius, μ and ρ are respectively the fluid viscosity and density, ΔP_{visc} is the fluid pressure and L is the length of the tube. x and y are, respectively, non-dimensional radial position, and time. Note that the first two terms in equation 10 correspond to steady Poiseuille flow while the unsteady component of the velocity field is represented by the third term: a series expansion in first order Bessel functions $J(x)$ with coefficients a_n chosen such that the solution converges to the steady Poiseuille flow solution as $t \to \infty$ (ie. the third term goes to zero at large time). Szymansky showed that these coefficients can be approximated adequately as follows:

$$a_n = n\pi - \frac{\pi}{4} + \frac{1}{8\pi n} + O\left(\frac{1}{n^2} \right)$$

Integrating the product of density and velocity over the tube x-sectional area and the valve opening time t_{open} gives the mass moved per piezo stroke while the mass moved per stroke times the frequency gives the overall mass flow rate of the system:

$$\dot{m} = f \int\limits_{0}^{t_{open}} \left[\int\limits_{0}^{R} \rho u(r,t) 2\pi r dr \right] dt \qquad (8)$$

Load

Weights of different sizes were attached to the output rod of the actuator to simulate load conditions. The static component of the applied load is simply the mass of the weight times the acceleration due to gravity:

$$F_{load,static} = m_{weight} g \qquad (9)$$

The dynamic or inertial component of the applied load is simply the mass of the weight times the acceleration of the actuator rod.

$$F_{load,inertial} = m_{weight} a_{act} \qquad (10)$$

If the acceleration is assumed to be constant during actuation, it can be written in terms of the displacement of the actuator and the valve opening time as follows:

$$a_{act} = \frac{2\delta_{act}}{t_{open}^2} \qquad (11)$$

System Model

Since table 1 shows that a total of 15 unknown quantities characterize this system, 15 equations relating these quantities to each other are required in order to solve it. Equations 1 through 4 yield six relations for the pressure associated with each component in terms of the component's length and the mass of the fluid contained with it. An additional five relations come from balancing pressures at the interface between each component. For the forward (discharge) stroke of the piezo stack they are:

$$P_p - p_2 = 0$$
$$p_2 - p_3 - dP_{vA} - dP_{tA} = 0$$
$$p_3 - p_4 = 0 \qquad (12a\text{-}e)$$
$$p_4 - p_5 - dP_{load,s} - dP_{load,i} = 0$$
$$p_5 - p_6 = 0$$

where P_p is the pressure exerted by the piezo stack on the fluid through the diaphragm, dP_{vA} is the pressure loss through valve A, dP_{tA} is the pressure loss through tube A that arises from fluid motion through the tube, $dP_{load,s}$ is the pressure drop across the actuator piston associated with the static component of the load (F_{Load}/A_a) while dP_{load_i} is the pressure drop across the actuator piston associated with the inertial component of the load. We will discuss how dP_{vA}, dP_{tA}, and dP_{load_i} are estimated during the discussion of the solution procedure. For the reverse (intake) stroke of the piezo stack:

$$P_p - P_2 = 0$$
$$P_2 - P_6 + dP_{vB} + dP_{tB} = 0$$
$$P_6 - P_5 = 0 \qquad \text{(13a-e)}$$
$$P_5 - P_4 - dP_{load,s} - dP_{load,i} = 0$$
$$P_4 - P_3 = 0$$

Two equations arise from the requirement that the fluid mass in the system be conserved. The first is that the total mass of fluid in the system is constant

$$m_2 + m_3 + m_4 + m_5 + m_6 = M \qquad \text{(14)}$$

where M is the total mass of fluid in the system. The second is that since only one valve is open at a time, the fluid mass in each half of the actuator is fixed during forward (discharge) and reverse (intake) piezo strokes. During the forward piezo stroke when valve A is open, the following equations hold:

$$m_2 + m_3 + m_4 = M_A$$
$$m_5 + m_6 = M_B \qquad \text{(15a,b)}$$

During the reverse piezo stroke (intake) when valve B is open, the equations become:

$$m_3 + m_4 = M_A$$
$$m_5 + m_6 + m_2 = M_B \qquad \text{(16a,b)}$$

Note that since $M_A + M_B = M$ holds all the time, only two of the three conservation equations for the forward and reverse piezo strokes bay be used because only two are linearly independent.

Finally, two equations arise from the requirement that the overall length of the piezo pump and the actuator are fixed. For the piezo pump, this means that:

$$x_p + \delta_d + x_2 = L_o \qquad \text{(17)}$$

where L_o is the overall length of the pump and δ_d is the thickness of the pump diaphragm. For the actuator, this results in:

$$x_4 + \delta_a + x_5 = L_a \qquad \text{(18)}$$

where L_a is the overall length of the actuator bore and δ_a is the thickness of the actuator piston.

Equations 1 through 9 can be written in matrix form as follows:

$$Ax + B = 0 \qquad \text{(19)}$$

Where A is a modified stiffness matrix, x is the vector describing the state of the system, and B are the boundary conditions. The matrices A and B have different forms depending on the state of the valves.

Valves A and B closed:

$$\begin{bmatrix}
-(k_p+k_d) & -BA_d/m_{02} & BA_d/L_v & 0 & 0 & 0 & 0 & 0 & 0 & 0 & 0 & 0 \\
0 & 1 & 0 & 1 & 0 & 1 & 0 & 1 & 0 & 1 & 0 \\
1 & 0 & 1 & 0 & 0 & 0 & 0 & 0 & 0 & 0 & 0 \\
0 & 0 & 0 & -B/m_{01} & B/L_{vA} & 0 & 0 & 0 & 0 & 0 & 0 \\
0 & 0 & 0 & B/m_{03} & -B/L_{vA} & -B/m_{04} & B/x_{04} & 0 & 0 & 0 & 0 \\
0 & 0 & 0 & 0 & 0 & B/m_{04} & -B/x_{04} & -B/m_{05} & B/x_{05} & 0 & 0 \\
0 & 0 & 0 & 0 & 0 & 0 & 0 & B/m_{05} & -B/x_{05} & -B/m_{06} & B/L_{vB} \\
0 & 0 & 0 & 0 & 1 & 0 & 0 & 0 & 0 & 0 & 0 \\
0 & 0 & 0 & 0 & 0 & 0 & 0 & 0 & 0 & 0 & 1 \\
0 & 0 & 0 & 0 & 1 & 0 & 1 & 0 & 0 & 0 & 0 \\
0 & 1 & 0 & 1 & 0 & 1 & 0 & 0 & 0 & 0 & 0
\end{bmatrix}\begin{bmatrix} x_p \\ m_2 \\ x_2 \\ m_3 \\ x_3 \\ m_4 \\ x_4 \\ m_5 \\ x_5 \\ m_6 \\ x_6 \end{bmatrix}+\begin{bmatrix}
k_p l_{sp}\zeta(E)+k_d l_{sd}-P_{02}A_d \\
-M \\
\delta_d-L_o \\
P_{02}-P_{03}-dP_{valve}-dP_{tube} \\
P_{03}-P_{04} \\
P_{04}-P_{05}-dP_{load,s}-dP_{load,s} \\
P_{03}-P_{06} \\
-L_{vA} \\
-L_{vB} \\
\delta_a-L_o \\
-M_A
\end{bmatrix}=0$$

(20a)

Valve A open, valve B closed:

$$\begin{bmatrix}
-(k_p+k_d) & -BA_d/m_{02} & BA_d/x_{02} & 0 & 0 & 0 & 0 & 0 & 0 & 0 & 0 & 0 \\
0 & B/m_{02} & -B/x_{02} & -B/m_{05} & B/x_{05} & 0 & 0 & 0 & 0 & 0 & 0 \\
0 & 0 & 0 & B/m_{03} & -B/x_{03} & -B/m_{04} & B/x_{04} & 0 & 0 & 0 & 0 \\
0 & 0 & 0 & 0 & 0 & B/m_{04} & -B/x_{04} & -B/m_{05} & B/x_{05} & 0 & 0 \\
0 & 0 & 0 & 0 & 0 & 0 & 0 & B/m_{04} & -B/x_{04} & -B/m_{06} & B/L_{vB} \\
0 & 1 & 0 & 1 & 0 & 1 & 0 & 1 & 0 & 1 & 0 \\
1 & 0 & 1 & 0 & 1 & 0 & 1 & 0 & 1 & 0 & 1 \\
1 & 0 & 1 & 0 & 0 & 0 & 0 & 0 & 0 & 0 & 0 \\
0 & 0 & 0 & 0 & 1 & 0 & 0 & 0 & 0 & 0 & 0 \\
0 & 0 & 0 & 0 & 0 & 0 & 0 & 0 & 0 & 0 & 1 \\
0 & 1 & 0 & 1 & 0 & 1 & 0 & 0 & 0 & 0 & 0
\end{bmatrix}\begin{bmatrix} x_p \\ m_2 \\ x_2 \\ m_3 \\ x_3 \\ m_4 \\ x_4 \\ m_5 \\ x_5 \\ m_6 \\ x_6 \end{bmatrix}+\begin{bmatrix}
k_p l_{sp}\zeta(E)+k_d l_{sd}-P_{02}A_d \\
P_{02}-P_{03}-dP_{valve}-dP_{tube} \\
P_{03}-P_{04} \\
P_{04}-P_{05}-dP_{load,s}-dP_{load,s} \\
P_{05}-P_{06} \\
-M \\
-(L_a+L_{vA}+L_a+L_{vB}-\delta_a-\delta_d) \\
\delta_d-L_o \\
-L_{vA} \\
-L_{vB} \\
-(m_{02}+m_{03}+m_{04})
\end{bmatrix}=0$$

(20b)

Valve A closed, valve B open:

$$\begin{bmatrix}
-(k_p+k_d) & -BA_d/m_{02} & BA_d/x_{02} & 0 & 0 & 0 & 0 & 0 & 0 & 0 & 0 & 0 \\
0 & B/m_{03} & -B/x_{02} & 0 & 0 & 0 & 0 & 0 & 0 & -B/m_{06} & B/x_{06} \\
0 & 0 & 0 & B/m_{03} & -B/x_{03} & -B/m_{04} & B/x_{04} & 0 & 0 & 0 & 0 \\
0 & 0 & 0 & 0 & 0 & B/m_{04} & -B/x_{04} & -B/m_{05} & B/x_{05} & 0 & 0 \\
0 & 0 & 0 & 0 & 0 & 0 & 0 & B/m_{05} & -B/x_{05} & -B/m_{06} & B/x_{06} \\
0 & 1 & 0 & 1 & 0 & 1 & 0 & 1 & 0 & 1 & 0 \\
1 & 0 & 1 & 0 & 1 & 0 & 1 & 0 & 1 & 0 & 1 \\
1 & 0 & 1 & 0 & 0 & 0 & 0 & 0 & 0 & 0 & 0 \\
0 & 0 & 0 & 1 & 0 & 0 & 0 & 0 & 0 & 0 & 0 \\
0 & 0 & 0 & 0 & 0 & 0 & 0 & 0 & 0 & 0 & 1 \\
0 & 0 & 0 & 1 & 0 & 1 & 0 & 0 & 0 & 0 & 0
\end{bmatrix}\begin{bmatrix} x_p \\ m_2 \\ x_2 \\ m_3 \\ x_3 \\ m_4 \\ x_4 \\ m_5 \\ x_5 \\ m_6 \\ x_6 \end{bmatrix}+\begin{bmatrix}
k_p l_{sp}\zeta(E)+k_d l_{sd}-P_{02}A_d \\
P_{02}-P_{06}+dP_{valve}+dP_{tube} \\
P_{03}-P_{04} \\
P_{04}-P_{05}-dP_{load,s}-dP_{load,s} \\
P_{05}-P_{06} \\
-M \\
-(L_a+L_{vA}+L_a+L_{vB}-\delta_a-\delta_d) \\
\delta_d-L_o \\
-L_{vA} \\
-L_{vB} \\
-(m_{03}+m_{04})
\end{bmatrix}=0$$

(20c)

Note that the system is driven by the time varying voltage $E(t)$ applied to the piezo stack that enters through $\zeta(E)$ in the first element of matrix B.

Solution Procedure

An iterative method is used in which MatLab's matrix inversion routines are used to solve equations 20a through 20c for x as the piezo voltage is varied through at least two cycles. On the first iteration, dP_{vA}, dP_{vB}, dP_{vB}, dP_{tB}, and $dP_{load,i}$ and dP_{vA} are set equal to zero. The time-evolution of the solution is used to determine the mass flow rate of fluid through the valves and tubing, and the displacement and acceleration of the actuator. This, in turn, is used to estimate dP_{vA}, dP_{vB}, dP_{vB}, dP_{tB}, and $dP_{load,i}$ using equations 5 through 11. Equations 20a through 20c are solved again and dP_{vA}, dP_{vB}, dP_{vB}, dP_{tB}, and $dP_{load,i}$ are estimated again until the system converges. Typically only a few iterations are required to get convergence to better than 1%.

A few comments about the solution of equations 20a through 20c during a single step of the iteration are also in order. First, note that only one equation is solved per time step. Which one is solved depends on the state of the valves, which, in turn, is determined by the pressures computed at the previous time step. Second, all of the forces acting on the system (the static component of the load, the inertial component of the load and the viscous forces in the valve and tube) are taken to be constant during this part of the calculation. While this calculation resembles an integration over time, this quasi-steady assumption means that it is not a true one. Third, the time step is kept small so that m/m_o and V/V_o are small (<0.01) and equation 4 remains a reasonable approximation to equation 3. Fourth, the state of the system (represented by vector x) is saved at each time step so that the time evolution of any state variable can be observed.

RESULTS

Actuation Velocity

Figure 5 compares the predictions of the simple quasi-static model to measurements of actuation velocity in a system with dimensions equivalent to those presented in table 1. A complete description of the apparatus and methods used to collect this data appears elsewhere [17,18]. In the figure, the markers connected by dotted lines represent the experimental data while the solid lines represent the model predictions. Data is presented for two simulated loads consisting of 6 lb. and 11 lb. weights attached to the actuator output rod. While not in perfect agreement with the data, the model does capture several important features. At low frequencies the model and the experimental data are in reasonable agreement as they both show an initially linear increase in velocity with frequency and a decrease in velocity with increasing actuator load. The model does over-predict velocity at some high frequencies, but the correspondence is much better at low frequencies where the unsteady effects not accounted for in this model are expected to be especially weak. Finally, the model predicts that velocity peaks at a particular actuation frequency and falls to zero as frequency is increased further. While the peak velocity and the cutoff frequency are not predicted exactly, the correspondence is reasonable considering the simplicity of the model and the fact that it does not account for any of the valve dynamics.

Explanation of Hybrid Actuator Operation

The more detailed presentation of a single pumping/actuation cycle in figure 7 explains the key features of the velocity results. The left-hand plot shows pressure in the pumping chamber as a function of the change in piezo-stack length (pressure-displacement characteristics) while the right-hand plot shows pressures on both sides of the actuator piston as a function of the change in actuator position. Before any load is applied, the entire system is at the same pre-load pressure (state 0). Applying a load to the actuator increases the pressure in the lower section of the actuator, decreases the pressure in the upper section, and causes the actuator to deflect downward slightly (state 1). Since the check-valves are assumed to remain closed, the pressure in the pumping chamber remains unaffected. Note also that the pressure difference times the area of the actuator piston equals the applied load at any point in the cycle.

These plots can be used to visualize the actuation cycle. As the piezo-stack is energized, the pressure-displacement characteristic of the stack sweeps upward along the pressure-displacement characteristic associated with the fluid in the pumping chamber. Valve A opens (state 2) when the pressure in the pumping chamber is greater than the pressure in the lower

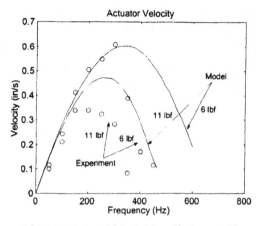

Figure 5 Comparison of Model Predictions with Experimental Data

Figure 6 Pressure Drops in Various Components of Actuator System

section of the actuator. Fluid is now allowed to flow from the pumping chamber into the actuator cylinder. As the actuator cylinder translates to accommodate the fluid being added to section A, the fluid in section B is compressed and the pressure in both sections rises as the piezo-stack characteristic sweeps along the slightly inclined pressure-displacement characteristic of the actuator. The pressure peaks when the piezo-stack voltage peaks (3). When the stack voltage and hence the pumping chamber pressure starts to decrease, check valve A closes and the pressure in the pumping chamber decreases. Meanwhile, the pressure in the actuator remains constant because the check valves are closed. The pressure in the pumping chamber continues to decrease until it is lower than the pressure on the upper side of the actuator piston. At this point, valve B opens (4), fluid flows back into the pumping chamber, and the actuator piston translates again. Translation stops when the piezo stack voltage and hence the pressure in the pumping chamber reaches a minimum. At this point

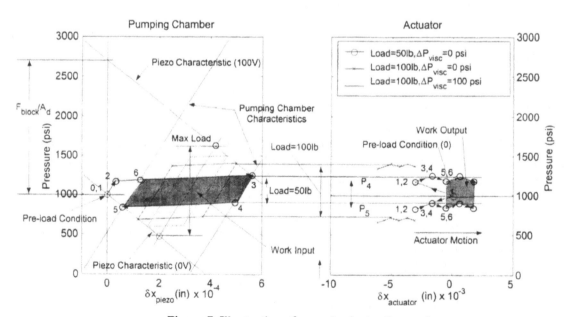

Figure 7 Illustration of pumping/actuation cycle

valve B closes (5), the pumping chamber is full of fluid again, and a new pressurization/actuation cycle begins.

There are several important features of the pumping/actuation cycle illustrated in figure 7 that are worth noting. The first is that actuation occurs in two steps during one pumping cycle: the first is associated with pump discharge while the second is associated with pump intake. The load comes to rest between actuation steps. This motivated the inclusion of the inertial force in the quasi-static model to account for the fact that the load must be accelerated twice during each pumping cycle. Second, the figure shows that increasing the load decreases the length of the piezo stroke and hence the actuation stroke. This explains the experimental observation that increasing the load decreases actuation velocity for fixed operating frequency. Viscous and inertial forces increase the 'effective' load on the piezo stack further reducing the displacement per stroke and hence the actuation velocity. This explains the tendency of the actuation velocity to 'roll off' at high frequencies since viscous and inertial forces increase rapidly (and non-linearly) with operating frequency. Figure 6 shows that of the forces that increase non-linearly with frequency, the pressure drop associated with accelerating the load is far more important than the fluid losses in the valves and connecting tubing. Hence, in this model it is the inertial effects associated with accelerating the load that govern the velocity 'roll off'. Third, figure 7 shows that the compressibility of the fluid elements limits the maximum force that the piezo-stack can deliver to the load to something less than that corresponding to the block force. In fact, if the fluid in the pumping chamber is too compliant (ie. its pressure-displacement characteristics are too shallow), no pumping action will occur at all.

Efficiency and Power Output

While the metrics used for measuring efficiency are the subject of considerable debate [19] (some use the ratio of the real power dissipated in the load to the apparent power [20] while others use the ratio of the stored energy in the structure to the stored energy in the actuator [21]), the measure used here focuses exclusively on the fluid system. Accordingly, the efficiency is defined as the ratio of the work output of the actuator $W_{act\ out}$ to the work input to the fluid in the pumping chamber $W_{pump\ in}$:

$$\eta_{fluid\ system} = \frac{W_{act\ out}}{W_{pump\ in}}$$

The work input is the area of the shaded region enclosed by the pressure-displacement trajectory of the fluid in the pumping chamber. Similarly, the work output is the area of the shaded region between the pressure-displacement trajectories of the fluid in the upper and lower sections of the actuator. The work output is also the actuation force times the velocity.

Figure 8 shows the work input, work output, and efficiency *in a single cycle* as a function of the operating frequency. The work output *per stroke* is maximum at 0 Hz because viscous and inertial losses are zero. As frequency increases, the work output decreases because viscous and inertial losses increase the 'effective' load on the piezo stack and decrease the displacement of the actuator. Interestingly, however, work input increases initially as the increase in the 'effective' load felt by the piezo-stack more than offsets the decrease in the piezo-stack displacement. At some point, however, this trend reverses and the work input begins to decrease with frequency. Figure 9 shows the power input (work input per stroke times frequency) and output (static load force times actuator velocity) as a function of frequency. While the trends are basically the same, the power input and output start at zero and peak at a higher frequency than the work input per stroke. The maximum power output for the 11 lb load is 0.5 W and occurs at approximately 275 Hz. The efficiency at that point is about 12%. Note that there is no compelling reason to operate the device at higher frequencies since adding more power to drive at higher frequency actually decreases the output power.

Finally, figures 10 and 11 are contour plots of the power output and mechanical efficiency of the hybrid actuator system. Note that while peak efficiency is independent of the load and always occurs at 0 Hz, the load does affect the rate at which efficiency decreases with increasing frequency. Figure 10 indicates that the efficiency is least sensitive to the driving frequency for loads around 30 lbf. However, figure 11 shows that the peak power output depends on both the operating frequency and the load and is maximized when the load is ~40 lb and the driving frequency is ~ 100 Hz

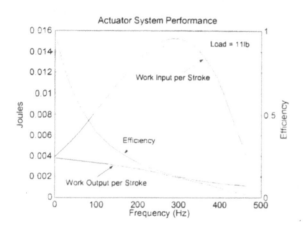

Figure 8 Actuator Work Input, Output, and Efficiency

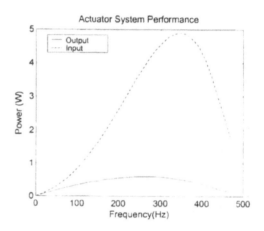

Figure 9 Actuator Power Input and Output

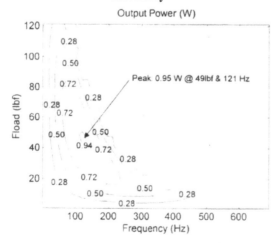

Figure 10 Actuator Output Power

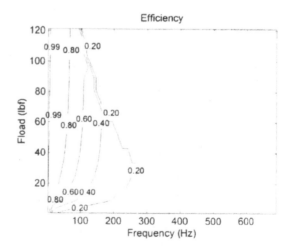

Figure 11 Actuator Efficiency

CONCLUSIONS

A quasi-static model has been developed to predict the performance of a piezo-hydraulic actuator's fluid system. In spite of the fact that it does not formally address the dynamic behavior of the fluid system, the quasi-static approximation seems to do a reasonably good job of predicting actuator performance at frequencies below 150 Hz. At higher frequencies, the agreement is not as good, but still shows the proper qualitative behavior that the actuation velocity and power output decrease to rather rapidly with increasing frequencies after their respective peaks. While viscous losses in the valve and tubing play a role at low frequencies, the peak power output and high frequency cutoff is dominated by the inertia of the load. As a result, practical devices that expect to achieve high power density by driving the piezo-stack at high frequency will need to find a way to minimize the acceleration and deceleration of the load during the piezo actuation cycle. The description of the actuation cycle indicates that maximizing the stiffness of the fluid in the pumping chamber is critical to maximizing the performance of the fluid system. This can be achieved by using a fluid with a large bulk modulus and by reducing the length of the pumping chamber. Other ways to improve performance without operating at high frequency are to increase the piezo-stack driving voltage and to increase the compliance of the rest of the fluid system besides the pumping chamber.

FUTURE WORK

Future efforts will focus on improving the fidelity of the model in two areas. The first has to do with the treatment of the fluid bulk modulus that is assumed to be fixed in the current fluid model. While this makes the model simpler, in certain situations it can lead to non-physical results like negative pressures in the pumping chamber that can only be avoided by increasing the pre-load pressure. In practice, however, the bulk modulus of the fluid depends on the pressure because of the small fraction of gas that is always dissolved in the fluid. This fact means that the fluid in the pumping chamber becomes ever more compliant as its volume is increased so that the pressure goes smoothly to zero as the volume of the pumping chamber goes to infinity. The second area has to do with the dynamics of the valves, which are not considered in this work. Since recent experiments with different valves indicate that high frequency performance is quite sensitive to valve selection, this issue must be addressed.

ACKNOWLEDGEMENTS

This work was conducted in collaboration with CSA Inc. and was supported under the DARPA Compact Hybrid Actuator Program (CHAP) F33615-00-C-3026, Ephraim Garcia Program manager. Special thanks go to Dr. Jayant Sirohi, Dr. Interjit Chopra, and Dr. Norman Wereley for the experimental data against which our model's predictions were compared. Also, thanks to Dr. Eric Anderson at CSA inc. for his assistance/support.

REFERENCES

1.   Straub, F. K. and Merkley, D. J., 'Design of Smart Material Actuator for Rotor Control', *Smart Structures and Materials 1995*, Proc SPIE 2443, pp. 89, 1995

2.   Giurgiutiu, Z. A., Chaudry, Z. A., Rodgers, C. A., 'Engineering Feasibility of Induced Strain Actuators for Rotor Blade Active Vibration Control', *Smart Structures and Materials 1995*, Proc SPIE 2443, 1995

3.   Galante, T. F., Frank, T., Bernard, J., Chen, W., Lesieutre, G. A., and Koopman, G. H., 'Design, Modeling, and Performance of a High Force Piezoelectric Inchworm Motor', *Journal of Intelligent Material Systems and Structures*, 10(12), pp. 962-972, 2000

4.   Micro Pulse Systems, Inc. 3950 Carol Ave., Santa Barbara, CA 93110

5.   Walaschek, J., 'Piezoelectric Ultrasonic Motors', *Journal of Intelligent Material Systems and Structures*, 6., pp. 71-83, 1995

6.   Newnham, R. E., Xu, Q. C., and Yoshikawa, S., U.S. Patent No. 4,999,819

7.   Robbins, W. P., Polla, D., Glumac, D. E., 'High-Displacement Piezoelectric Actuator Utilizing a Meandering-Line Geometry – Part I: Experimental Consideration', *IEEE Trans. Ultrason., Ferroelectr., Freq. Control*, 39 [5] pp. 454-460, 1991

8.   Samak, D. K. and Chopra, I., 'Design of High Force, High Displacement Actuators for Helicopter Rotors', *SPIE North American Conference on Smart Materials and Structures*, Orlando FL, Feb. 1994

9.   Ullman, A., 'The Piezoelectric Valve-less pump – Performance Enhancement Analysis', *Sensors and Actuators A*, 69, pp. 97-105, 1998

10. Hagood, N. W., Roberts, D. C., Saggere, L., Breuer, K. S., Chen, K., Carretero, J. A., Mlcak, R., Pulitzer, S., Schmidt, M. A., Spearing, S. M., Su, Y., 'Micro Hydraulic Transducer Technology for Actuation and Power Generation', *Smart Structures and Materials 2000: Smart Structures and Integrated Systems*, Proceedings of SPIE Vol. 3985, pp. 680-688, 2000.

11. Roberts, D. C., Hagood, N. W., Su, Y., Li, H., Carretero, J. A. 'Design of a Piezoelectrically-Driven Hydraulic Amplification Microvalve for High Pressure, High Frequency Applications', , *Smart Structures and Materials 2000: Smart Structures and Integrated Systems*, Proceedings of SPIE Vol. 3985, pp. 616-628, 2000.

12. Mauck, L. D., and Lynch, C. S., 'Piezoelectric Hydraulic Pump Development', Journal of Intelligent Material Systems and Structures, Vol 11, pp. 758-764, October 2000

13. Leo, D. J., 'Energy Analysis of Piezoelectric-Actuated Structures Driven by Linear Amplifiers', Journal of Intelligent Material Systems and Stuctures, Vol. 10, pp. 36-45, January 1999

14. Nasser, K., Leo, D. J., and Cudney, H. H., 'Compact Piezohydraulic System', *Smart Structures and Materials 2000: Industrial and Commercial Applications of Smart Structures Technologies*, Proceedings of SPIE Vol. 3991, pp. 312-322, 2000

15. Lines, M. E., and Glass, A. M., Principles and Applications of Ferroelectrics and Related Materials, Oxford: Clarendon Press, 1977

16. Szymansky, Piotr *L'Hydronamique Du Fluide Visqueux*, J. Math. Pure Appl., ser.9, vol11, p. 67

17. Sirohi, J. and Chopra, I., 'Compact Piezo-hydraulic Hybrid Actuator', Twelfth International Conference on Adaptive Structures and Technologies, University of Maryland, October 15-17, 2001

18. Sirohi, J., 'Piezoelectric Hydraulic Hybrid Actuator for a Potential Smart Rotor Application', PhD. Thesis, University of Maryland, December 2002.

19. Leo, D. J., 'Energy Analysis of Piezoelectric-Actuated Structures Driven by Linear Amplifiers', Journal of Intelligent Material Systems and Stuctures, Vol. 10, pp. 36-45, January 1999

20. Determination of Actuator Power Consumption and System Energy Transfer', Journal of Intelligent Material Systems and Structures, 5 (1), pp. 12-20, 1994

21. Giurgiutiu, V., Rogers, C. A., 'Dynamic Power and Energy Capabilities of Commercially-Available Electroactive Induced-strain Actuators', Journal of Intelligent Material Systems and Structures, 7 (6), pp. 656-667, 1996

EXPERIMENTAL STUDIES OF ZERO SPILLOVER SCHEME FOR ACTIVE STRUCTURAL ACOUSTIC CONTROL SYSTEMS

Moustafa Al-Bassyiouni and Balakumar Balachandran

ABSTRACT

In designing a controller, one way to avoid energy spillover is to use what is called the *zero spillover control scheme*. Here, the feasibility of this scheme is studied experimentally for actively controlling sound fields inside an enclosure with a flexible boundary. Noise is transmitted into the enclosure through the flexible boundary, and piezoceramic patches, which are mounted on the flexible boundary, are used as actuators. Polyvinylidene fluoride sensors are used on the flexible boundary and microphone sensors are used inside and outside the enclosure. For narrowband disturbances, attenuation ranging up to 17.0 dB has been realized. Controller stability and robustness of zero spillover controller are discussed along with other issues.

KEYWORDS: Active structural acoustic control (ASAC), piezoelectric actuators, enclosed sound field, spillover.

INTRODUCTION

Control of noise and vibration is important for many civil, industrial, and defense applications. Passive control techniques, which are typically effective at high frequencies, require expensive system redesign and this redesign does not always result in a significant improvement. Active control methods, which can be grouped under Active Noise Control (ANC) and Active Vibration Control (AVC) categories, are effective solutions for low frequency applications. Since the 1930s, when Paul Leug [1] proposed the use of a feedforward control scheme to globally attenuate noise propagating inside a tube, feedforward approaches have been used for problems ranging from spatially one-dimensional systems to spatially three-dimensional systems. From the 1950s, spurred by the work of Olsen and May [2] who proposed a feedback control scheme to locally attenuate the three-dimensional sound

M. Al-Bassyiouni and B. Balachandran, Department of Mechanical Engineering, University of Maryland, College Park, MD 20742

field around a head seat, feedback approaches have been used extensively in many problems. The various advances in active control methodologies and applications made since these times are well documented in the literature [3,4,5,6].

Active Structural Acoustical Control (ASAC) which can be considered a modified version of ANC, takes advantage of vibrating structural elements as secondary noise sources to cancel the sound fields generated by a primary noise source (e.g., Sampath and Balachandran [7], Balachandran and Zhao [8], and Al-Bassyiouni and Balachandran [9]). It appears that ASAC requires much less dimensionality than ANC in order to achieve widely distributed spatial noise reduction. However, active research is still being pursued to address issues such as actuator nonlinearities [8], actuator power levels, sensors, controller architecture, etc. In the previous work of the authors [9], which was conducted with the objective of developing an ASAC feedforward controller for broadband and narrowband attenuation of sound fields, it was shown as to how a zero spillover control scheme can be applied towards the control of three-dimensional sound fields with the aid of simulation results. Here, as a continuation of this work, the zero spillover scheme has been experimentally implemented with the control actuators being piezoelectric patches bonded to a flexible side of the three dimensional enclosure. Spatially local attenuation of sound fields in the frequency range of 100 Hz to 200 Hz is studied here.

The rest of this article is organized as follows. In the next section, the controller development and construction is briefly addressed (additional details can be found in reference [9]), and in the following section, the experimental arrangement is described along with identification of path transfer functions and controller implementation. Robustness and performance analysis is presented next, and subsequently, simulation and experimental results are provided and discussed.

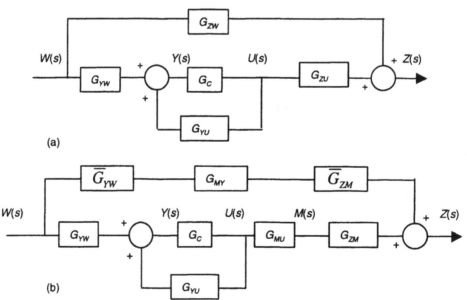

Figure 1. Block diagrams of (a) typical ANC system, and (b) reconstructed zero spillover controller for ASAC system.

CONTROLLER DEVELOPMENT

A typical block diagram for a single-input, single-output feedforward ANC system is shown in Figure 1(a), where $W(s)$ is the transform of the external noise signal, $Y(s)$ is the transform of the noise signal at the reference microphone, $U(s)$ is the transform of the control signal, $Z(s)$ is the transform of the noise signal at the error microphone, G_{ZW} is the primary noise path transfer function, G_{ZU} is the secondary noise path transfer function, G_{YW} is the noise path transfer function from the external noise source to the reference microphone, G_{YU} is the inherent feedback transfer function from the actuator to the reference microphone, and G_C is the control transfer function [3,10].

In the previous work of the authors [9], a near collocated microphone is installed at actuator's location in order that $U(s)$ can be mapped to the corresponding acoustic signal transform $M(s)$ through the transfer function from $U(s)$ to $M(s)$. Now, in the context of active structural acoustic control (ASAC) systems, the quantity $U(s)$ corresponds to the vibrations experienced by the structure as a result of actuators (secondary sources) bonded directly to that structure. The spatial separation between the actuator and the corresponding collocated microphone can be made "small enough" in order to minimize the propagating phase delay between them. In such case, the transfer function between them is most likely minimum phase; however, one has to ensure that the microphone does not spatially interfere with the structure's vibrations. Assuming that the influence of the system nonlinearities and the effects of three-dimensional wave propagation can be neglected, the primary path transfer function now can be constructed as a cascade given by

$$G_{ZY}(s) \cong \overline{G}_{ZM}(s) G_{MY}(s) \tag{1}$$

Even though the transfer function cascade (1) may be practically realizable, it may be obtained analytically by equating G_{ZM} and $\overline{G}_{ZM}$, and solving for G_{MY}.

Now, let us consider the system described by the block diagram shown in Figure 1(b). It is simply a reconstruction of the ANC block diagram of Figure 1(a) with the addition of the collocated microphone effect. In this reconstruction, the transfer function G_{ZW} is composed in the form

$$G_{ZW}(s) = \overline{G}_{ZM}(s) G_{MY}(s) \overline{G}_{YW}(s) \tag{2}$$

and the following assumptions have been made:
- Magnitude of G_{YU} is "small" so that

$$\frac{U(s)}{Y(s)} \cong G_C(s) \tag{3}$$

-
$$\overline{G}_{YW}(s) \cong G_{YW}(s) \tag{4}$$

-
$$\overline{G}_{ZM}(s) \cong G_{ZM}(s) \tag{5}$$

With the above-mentioned assumptions, a possible zero spillover controller for the system can take the following form

$$G_{ZSC}(s) = -\frac{G_{MY}(s)}{G_{MU}(s)} \tag{6}$$

Throughout the rest of this article, the transfer function G_{ZW} is called the *primary direct path*, and the cascade of equation (2) is called the *primary indirect path*. In the next section, the experimental part of this work is presented.

EXPERIMENTAL ARRANGEMENT AND SYSTEM REALIZATION

In Figure 2, the experimental arrangement is illustrated. All details on the experimental components can be found in our earlier studies (e.g., [7], [8] and [9]). For the results reported in this study, the middle PZT actuator pair B2 is used, the microphone M-B2 is used as a near-collocated sensor, and microphone Mic.1 is used as error sensor (The other actuator pairs, PVDF sensors, and microphones are being used in ongoing experiments).

An experimental approach is used to develop the system model [9]. This approach can be useful for implementing the controller on full-scale systems such as helicopter cabins. In earlier work, it has been shown how the locations and orientations of collocated and error microphones can be adjusted to satisfy the assumptions provided in equations (3) to (5). Accordingly, it was concluded that vertical orientation of the reference microphone minimizes the feedback effect due to the piezoceramic actuation, and hence this effect can be, to some extent, neglected at the reference microphone by comparison to the effect of loudspeaker. Furthermore, the distance between the PZT patch pair B2 and the collocated microphone M-B2 was adjusted to 10 mm to insure that G_{MU} is minimum phase. It has been also noted that in the selected frequency range, the assumption stated in equation (4) is met. In addition to that, condition (5) was ensured by analytically determining G_{MY} as explained in the previous section. These results indicate that the earlier made assumptions are valid.

ROBUSTNESS AND PERFORMANCE

In the previous section, it was discussed as to how the sensor locations could be chosen to satisfy the assumptions made during the controller development. From an analytical standpoint, the controller defined in equation (6) is supposed to give perfect cancellation of noise under the stated assumptions.

However, since the identified transfer functions do not always capture all of the system characteristics, perfect cancellation may not be realizable in an experimental setting.

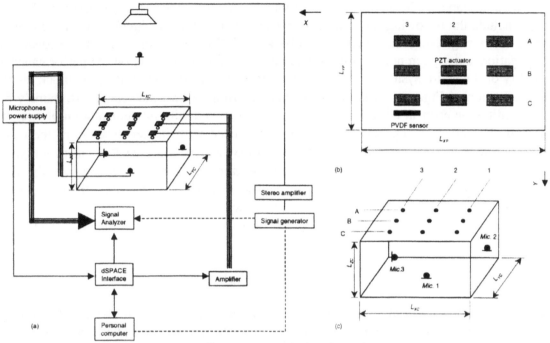

Figure 2. Experimental setup. (a) overall arrangement, (b) locations of actuators and sensors on the flexible panel, and (c) locations of microphones inside enclosure.

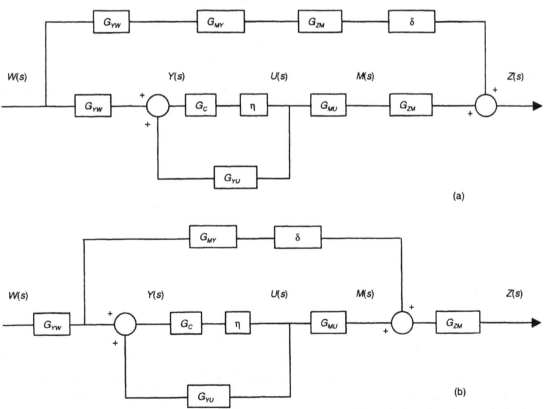

Figure 3. Reconstructed zero spillover controller for ASAC system for robustness analysis.

There are three different sources of errors and effects that affect the experimental results; these include the following: (1) errors from experimental transfer function identification, (2) errors from neglecting the inherent feedback, and (3) effects of three-dimensional wave propagation and system nonlinearities. The impact of different errors on the performance and stability of the entire system is investigated in this section. In Figure 3(a), the block diagram representation of Figure 1(b) is modified with insertion of two new functions, namely $\delta(s)$, representing the error in the primary direct path associated with implementation of equations (4) and (5), and $\eta(s)$, representing the error associated with implementation of the control transfer function (6). In the nominal case, both $\delta(s)$ and $\eta(s)$ are each equal to unity. It is noted that the three-dimensional effects can be directly related to $\delta(s)$ while the errors associated with experimental identification appear in both $\delta(s)$ and $\eta(s)$. In Figure 3(b), a possible simplification for the block diagram is shown.

Now, the primary noise path, or the *uncontrolled* sound response at the error microphone, is given by

$$Z = G_{ZM} G_{MY} \delta G_{YW} W \tag{7}$$

where the argument s of the functions is not shown explicitly as before. After substituting for G_C from equation (6), the *controlled* sound response at the error microphone becomes

$$Z_C = G_{ZM} G_{MY} \delta \left[\frac{(1-\frac{\eta}{\delta}) + \eta G_{MY} \dfrac{G_{YU}}{G_{MU}}}{1 + \eta G_{MY} \dfrac{G_{YU}}{G_{MU}}} \right] G_{YW} \tag{8}$$

where the subscript "c" denotes the controlled response. From equations (7) and (8), the following attenuation ratio is determined:

$$\frac{Z_C}{Z} = \frac{\left[(1-\frac{\eta}{\delta}) + \eta G_{MY} \dfrac{G_{YU}}{G_{MU}} \right]}{\left[1 + \eta G_{MY} \dfrac{G_{YU}}{G_{MU}} \right]} = 1 - \frac{\dfrac{\eta}{\delta}}{1 + \eta G_{MY} \dfrac{G_{YU}}{G_{MU}}} \tag{9}$$

From the expression (9), it is clear that the terms $[\, 1 + \eta\, G_{MY}\, G_{YU}\, /\, G_{MU} \,]$ and $[\, \eta\, /\, \delta \,]$ are very significant. The first of these terms, which does not depend on δ, is important for the stability of the control system (here it is called the *critical characteristic polynomial*) while the second term is called as *attenuation index*. Furthermore, the term $G_{YU}\, /\, G_{MU}$ is expected to be always less than unity, since the collocated microphone is much closer to the actuator than the reference microphone. The second term represents a single input, single output (SISO) feedback system, thus stability bounds on η can be obtained from any classical SISO algorithm that tests the associated open loop transfer function $[\, \eta\, G_{MY}\, G_{YU}\, /\, G_{MU} \,]$ (such as root locus or Bodé plot). Since all experimental and identified transfer functions are stable, η and δ must also be stable. However, δ should not have any nonminimum phase zeros that can cause system instability.

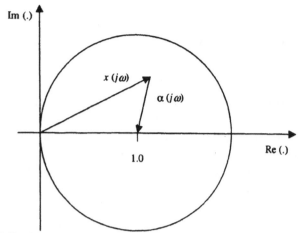

Figure 4. Performance test: Graphical representation of equation (9).

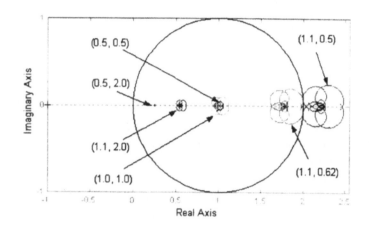

Figure 5. Performance test for different values of η (first argument) and δ (second argument).

For the purpose of this article, both δ and η will be assumed to be real values around +1.0. This is consistent with the results of the experiments, since $\overline{G}_{ZM}$ has been obtained analytically to exactly match G_{ZM}, and the phase in all identified transfer functions matches the experimental phase [9]. Although the above analysis is still valid for complex values, it is expected that this will be more involved. Equation (9) is represented graphically in Figure 4 at a single frequency ω, where x is the second term of equation (9) and α is the attenuation achieved by the controller. For $|Z_c|$ to be less than $|Z|$, the vector value of x must lie in a unit circle centered at +1. Thus, the robustness and performance analysis for the system can be summarized in terms of the following steps:

i. The open loop transfer function $\left[\eta \, G_{MY} \dfrac{G_{YU}}{G_{MU}} \right]$ is plotted on Nyquist plot and both gain margin and phase margin are noted for the maximum bounding value of η. This step tests the robustness (stability) of the system to identification error, and hence, it is noted as the "*robustness test*".

ii. The closed loop transfer function $\left[\dfrac{\dfrac{\eta}{\delta}}{1 + \eta G_{MY} \dfrac{G_{YU}}{G_{MU}}} \right]$ is plotted on Nyquist plot and it is

checked to lie within the (+1) unit circle for different values of δ and η. Therefore, this step is noted as the *"performance test"*.

When applying the robustness test to the experimentally identified system transfer functions, it turns out that the system is stable as long as $\eta \leq 1.1$. In Figure 5, it is supposed that η varies between [0.5,1.1] and δ varies between [0.5,2.0]. The performance test is run for the four extreme corners plus the pair (1.0,1.0), at which the attenuation index is unity, and (1.1,0.62), at which the attenuation index is 1.77. It is clear that as the attenuation index reaches unity, the attenuation in the noise signal increases significantly.

It should be mentioned here that if the inherent feedback was zero (e.g., if a non-acoustic reference sensor was used), the stability of the system would depend only on δ, and the attenuation achieved by the controller would be

$$\frac{Z_C}{Z} = 1 - \frac{\eta}{\delta} \qquad (10)$$

Now, applying the stability and performance tests to our experimental data and identified model, it has been found that $\eta = [0.8,1.4]$ and $\delta = [0.6, 1.0]$. For this range of η, it has been found that further improvement in the model identification is necessary.

However, if a non-acoustic sensor is used, then the system is still stable, and the controlled response varies from 14% to 133% of the uncontrolled one (i.e., it is also expected that spillover at some frequencies within the frequency band of interest will occur). These numbers are calculated using the bounds given above for η and δ and equation (10). In decibels, the controller is expected to produce from 17 dB reduction to 2.5dB increase in sound level. For example, at frequency $f = 124$ Hz, which is close to plate's (3,1) modal frequency $f_{31} = 126$ Hz, $\eta = -1.0$ dB while $\delta = -3.0$ dB; thus the attenuation index = 1.26 and then 11.7 dB attenuation is expected. Also, at $f = 174$ Hz, which is close to plate's (3,2) modal frequency $f_{32} = 172$ Hz, $\eta = 2.0$ dB while $\delta = -2.5$ dB; thus the attenuation index = 1.68 and then the expected attenuation is about 3.4dB.

RESULTS AND DISCUSSION

In this section, off-line simulation results obtained by using the identified transfer functions are provided. For this purpose, MatLab / SimuLink software is used to construct the mathematical model that represents the system. In Figure 6, the noise (in SPL) of the uncontrolled system is compared with that of the controlled system at the error microphone Mic.1, while in Figure 7, the differences between pressure values of the uncontrolled and controlled cases at Mic.1 are shown. It is seen that the simulated zero-spillover ASAC controller is effective for the considered panel-enclosure system. In addition, the effect of the consideration of inherent feedback is also illustrated, and as expected, this feedback is negligible when the sensor location is properly chosen. However, when applying the controller to the experimental set up, instability was encountered at several tones making it difficult to record any results. As mentioned before, instability was expected because of the 'small' error in identifying G_{MU}. The authors are considering this problem now.

Experimental results have also been obtained for the case of a non-acoustic sensor. In this particular case, a voltage signal proportional to the noise signal was fed directly into the controller. In Figure 8, the uncontrolled and controlled responses at the error microphone Mic. 1 are shown, and in Figure 9, the corresponding differences are shown. To construct these figures, the experiment was run twice for the same tone, once with the controller off, and then with the controller on. With a step size of 2.0 Hz and a total of 102 runs, the results have been generated. The experimental results are in agreement with the performance analysis presented in the previous section, where attenuation up to 17 dB was achieved, and spillover up to 7.9 dB arose at some points. Furthermore, the experimental result agrees quantitatively with the simulation result. It is noted that there is a shift in the high frequency spikes to the right and the low frequency spikes to the left. This is attributed to a band-pass filter, which was used to filter out the control signal in the experiments but not included in the system model.

In this study, a "modified and simplified" zero-spillover feedforward control scheme has been tested for ASAC and studied for tonal reduction of the noise transmitted into a rectangular enclosure. Robustness analysis has shown that a 'slightly' incorrect model can introduce system instability. A new strategy that tests system robustness and performance has been demonstrated. Ongoing studies at this juncture are focused on the use of this strategy to improve the controller performance, in addition to other issues such as broadband attenuation and multi-dimensional ASAC systems. Also, the experiments with the controller were terminated at tones of 100 Hz and 144 Hz, as control voltage values were higher than 200 V, which was used as the limiting value for the experiments (the required voltage values were 205.1 V and 204.8 V, respectively).

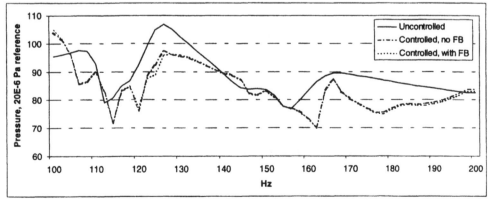

Figure 6. Pressure response at the error microphone Mic.1 obtained through simulation in uncontrolled and controlled cases. In case (a), inherent feedback is neglected, and in case (b), this feedback is considered.

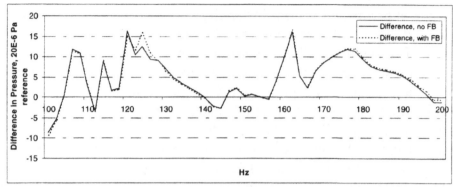

Figure 7. Simulation results: difference in pressure responses between the uncontrolled and controlled cases, at the error microphone location.

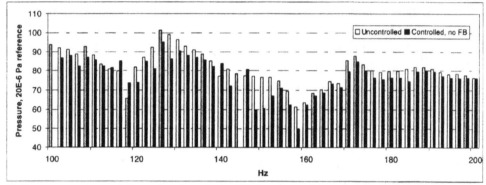

Figure 8. Pressure response at the error microphone Mic.1 measured experimentally in uncontrolled and controlled cases. Non-acoustic sensor was used as reference sensor.

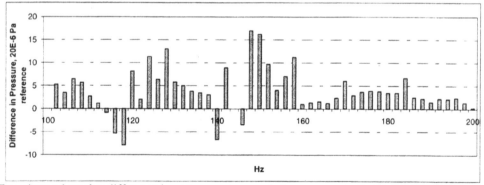

Figure 9. Experimental results: difference in pressure responses between uncontrolled and controlled cases at the error microphone location.

ACKNOWLEDGEMENT

Partial support received for this work from the U.S. Army Research Office through Contract No. DAAH 049610334 is gratefully acknowledged. Drs. Tom Doligalski and Gary Anderson are the technical monitors for this contract.

REFERENCES

1. P. Leug, "Process of silencing sound oscillations", US. Patent No. 2,043,416, 1936.
2. H. F. Olson and E. G. May, "Electronic sound absorber", *Journal of Acoustical Society of America*, Vol. 25, pp. 1130-1136, 1953.
3. A. Roure, "Self-adaptive broadband active sound control systems", *Journal of Sound and Vibration*, Vol. 101, pp. 429-441, 1985.
4. P. A. Nelson and S. J. Elliott, *Active control of sound*, Academic Press, London, 1992.
5. C. R. Fuller and A. H. Von Flotow, "Active control of sound and vibration", *IEEE Control systems*, pp. 9-19, Dec. 1995.
6. S. J. Elliott, "Down with noise: practical control systems for combatting audible noise show up in aerospace, general aviation, and military roles", *IEEE spectrum*, pp. 54-61, June1999
7. A. Sampath and B. Balachandran, "Active control of multiple tones transmitted in an enclosure", *Journal of Acoustical Society of America*, Vol. 106, No. 1, pp. 211-225, July 1999.
8. B. Balachandran and M.- Z. Zhao, "Actuator nonlinearities in interior acoustics control", *Proceedings of the SPIE's 7th Annual International Symposium on Smart Structures and Materials*, Newport Beach, CA, March 5-9, Vol. 3984, Paper No. 3984-13, 2000.
9. M. Al-Bassyiouni and B. Balachandran, "Zero spillover control of enclosed sound fields", *SPIE's 8th Annual International Symposium on Smart Structures and Materials*, Newport Beach, CA., March 4-8, Vol. 4362, Paper No. 4326-7, 2001.
10. J. Hong and D. S. Bernstein, "Bode integral constraints, colocation, and spillover in active noise and vibration control", *IEEE Transaction on Control Systems Technology*, Vol. 6, No. 1, January 1998.

Smart Damping

AN ADAPTABLE ACTIVE-PASSIVE PIEZOELECTRIC ABSORBER FOR NONSTATIONARY DISTURBANCE REJECTION — THEORY AND IMPLEMENTATION

Ronald A. Morgan and Kong Well Wang

ABSTRACT

It is well known that piezoelectric materials can be used as passive electromechanical vibration absorbers by shunting them with electrical networks. To suppress harmonic excitations with varying frequency, semi-active piezoelectric absorbers have been proposed in the past. However, these semi-active devices have limitations that restrict their applications. The authors have developed a high performance active-passive alternative to the semi-active absorber that uses a combination of a passive electrical circuit and active control actions. The active control consists of three parts: an *adaptive inductor tuning action*, a *negative resistance action*, and a *coupling enhancement action*. In this study, the performance of the new absorber design is experimentally verified. The new test stand consists of a piezoelectric stack integrated into an active-passive isolation mount. The difficulty of obtaining an accurate system model is addressed by proposing an experimental tuning procedure. Using this method, it is demonstrated that the active-passive absorber can provide effective isolation of harmonic excitations with varying frequency.

INTRODUCTION

Piezoelectric materials have been shown to be useful for passive vibration damping when they are shunted with an electrical network. For example, an electro-mechanical vibration absorber can be created by coupling the piezoelectric material with a circuit containing resistive and inductive elements [1]. These absorbers can be tuned to suppress a harmonic excitation at a given frequency (tonal tuning). The principal drawback of tonally tuned absorbers is that they require the absorber damping to be very low to achieve good performance, which causes the effective bandwidth to be quite small [2]. For this reason, passive absorbers are not generally useful in off-resonance situations, especially when the excitation frequency is unsteady or varying.

In recent years, semi-active piezoelectric absorber concepts have been proposed for suppressing harmonic excitations with time-varying frequency. The implementation of these semi-active absorbers requires either a variable inductor or a variable capacitor element, and both of these methods have some inherent limitations. For instance, the variable capacitor method [3] limits the tuning of the piezoelectric absorber to a relatively small frequency range. The variable inductor approach [4], which is accomplished using a synthetic inductance circuit, can add a significant parasitic resistance to the circuit, which is generally

---

[1] Ronald A. Morgan: Northrop Grumman Oceanic Systems, P.O. Box 1488, MS 9105, Annapolis, MD 21404
[2] K. W. Wang: The Pennsylvania State University, 157E Hammond Building, University Park, PA 16802

undesirable for narrow-band applications. In either case, the variable passive elements can be difficult to tune rapidly and accurately. In a previous study, the authors have introduced a high performance adaptive active-passive piezoelectric absorber as an alternative to these semi-active absorbers [5]. The fundamental concept behind this new absorber design is the integration of a fixed passive inductance with an active variable inductance, which allows fast and accurate frequency tuning while minimizing the required control power. In addition, negative resistance and active coupling enhancement actions were added to the control law to significantly improve the performance and robustness of the treatment, compared to a passive or semi-active piezoelectric absorber. The vibration control capabilities of the adaptive absorber have been successfully demonstrated using a cantilever beam test stand [5] and expanded to address cases with multiple frequency excitations [6].

THEORY

A general schematic of the system under consideration is illustrated in Figure 1, which consists of a piezoelectric actuator and sensor integrated into a mechanical structure. The sensor provides feedback to an external voltage source that is in series with an inductive shunt. It is assumed that the system model can be obtained, either analytically or experimentally, in the form shown in Eq. (1).

$$M\ddot{\underline{q}} + C\dot{\underline{q}} + K^D \underline{q} + K_c Q = \hat{\underline{F}} \cdot f(t)$$
$$L_p \ddot{Q} + R_p \dot{Q} + \frac{1}{C_p^S} Q + K_c^T \underline{q} = V_C \tag{1}$$

The matrices M, C, and K^D are the mass, damping, and open-circuit stiffness matrices of the system, q is a vector of the generalized coordinates of the structure, and Q is the charge on the piezoelectric. The vector K_c represents the coupling between the mechanical and electrical systems, $f(t)$ is the excitation force, and the vector $\hat{F}$ contains the weightings of the excitation force on the individual structural coordinates. In the circuit equation, L_p and R_p are the passive inductance and inherent resistance of the shunt circuit, C_p^S is the capacitance of the piezoelectric under constant strain, and V_c is the control voltage.

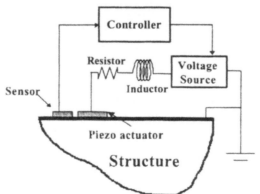

Figure 1. General System Configuration

The passive inductance L_p is used to tune the circuit to the nominal or steady-state excitation frequency. An active controller is then developed, which consists of three actions. The first part of the control law is designed to imitate a variable inductor so that the absorber can be adaptively tuned to the correct frequency. The advantages of the active inductor include fast and accurate adjustment, no parasitic resistance, and easier implementation compared to a semi-active inductor. No passive resistance is intentionally added to the shunt circuit, however the passive inductor will contain some internal resistance, which is denoted R_p. If this internal resistance is excessively large, the narrow-band disturbance rejection performance of the absorber is degraded. In this case, an active negative resistance action is used to reduce the effective resistance in the absorber circuit. Finally, the effective coupling in the circuit equation is enhanced using the third part of the control law, which is the active coupling enhancement action. This action significantly improves the performance and robustness of the absorber. The total control law for the single frequency adaptive piezoelectric absorber is given in Eq. (2), where L_a is the active variable inductance, R_a is the negative resistance, and G_{ac} is the active coupling gain. The closed-loop circuit equation, which is shown in Eq. (3), shows that the effective inductance, resistance, and electromechanical coupling can be changed.

$$V_C = -L_a(t) \cdot \ddot{Q} + R_a \dot{Q} - (G_{ac} - 1) K_C^T q \tag{2}$$

$$(L_p + L_a(t)) \ddot{Q} + (R_p - R_a) \dot{Q} + \frac{1}{C_p^S} Q + G_{ac} K_C^T \underline{q} = 0 \tag{3}$$

To ensure that the active inductance is properly tuned, an expression for the optimal tuning on a general multiple degree of freedom (MDOF) structure is derived. The tuning ratio δ is defined in Eq. (4), where ω_e is the estimated excitation frequency.

$$\delta = \frac{\omega_a}{\omega_e} \quad where \quad \omega_a = \frac{1}{\sqrt{(L_p + L_a) C_p^S}} \tag{4}$$

The frequency estimate, ω_e, is obtained on-line using an algorithm based on the recursive least squares method [7]. The first step in the derivation of the optimal tuning law is to transform the system model into modal space using the transformation $q = U\eta$, where U is a matrix containing the eigenvectors of the structural model. The displacement of the structure at a desired point is chosen as the objective function, which is then expressed as a weighted sum of the modal responses, as shown in Eq. (5), where W is the modal weighting vector. The optimal tuning ratio for the frequency ω_e is given in terms of the system modal parameters by Eq. (6), where ω_n is a vector of the modal frequencies of the structure. The ~ notation denotes the system parameters after transformation to the modal space, and the subscripts i and j denote the ith and j^{th} elements of a vector.

$$w_d(t) = W^T \eta(t) \tag{5}$$

$$\delta_{opt} = \left[1 + G_{ac} C_p^S \frac{\sum_{j=1}^{n} W_j M_j \left(a_j - \tilde{F}_j b \right)}{\sum_{j=1}^{n} W_j M_j \tilde{F}_j} \right]^{-\frac{1}{2}} \tag{6}$$

$$where \quad M_j = \frac{1}{\omega_{n_j}^2 - \omega_e^2} \quad a_j = \tilde{K}_{c_j} \sum_{i=1}^{n} \frac{\tilde{K}_{c_i} \tilde{F}_i}{\omega_{n_i}^2 - \omega_e^2} \quad b = \sum_{i=1}^{n} \frac{\tilde{K}_{c_i}^2}{\omega_{n_i}^2 - \omega_e^2}$$

The key assumptions used in this derivation are that the excitation frequency is changing relatively slowly compared to the system dynamics (quasi steady-state) and that the structural and circuit damping can be neglected.

EXPERIMENTAL VALIDATION

Experimental Setup Description

The test stand used in this study, as shown in Figure 2, is an active-passive mount incorporating a piezoelectric stack actuator. In addition to vibration control, a vibration isolation application is also explored to demonstrate the versatility of the adaptive active-passive absorber.

Figure 2. Active-Passive Mount Test Stand

An inertial shaker (Wilcoxon Model F4) is used to provide external disturbance. At the bottom of the test stand is a large spring-mass system simulating foundation dynamics. The active-passive mount, which is contained by two aluminum plates, is sandwiched between the vibrating source and the foundation. Between the plates is a piezoelectric stack actuator and four nylon struts, which are used to support the plate under moment and shear loads. While the bottom end of the actuator is rigidly bolted, the top end is a spherical ball connection that maintains a point contact with the mounting plate. This arrangement prevents shear, moment, and tensile load from being transmitted to the actuator and potentially damaging it. The nylon struts act as adjustable passive stiffness elements that can be used to preload the actuator and to maintain point contact with the mounting plate under dynamic loads. The entire test stand is mounted on an optical isolation table.

The schematic of the complete experimental setup is illustrated in Figure 3. The shaker contains an impedance head, which provides the force and acceleration at the disturbance input point. The piezoelectric stack actuator used here also contains integrated strain gauges that provide a collocated displacement measurement. The disturbance signal is generated by the Dspace DSP board, amplified using an audio amplifier, and applied to the inertial shaker. The output of the impedance head load cell is used as an input to the digital controller for estimating the frequency of the excitation. The other control inputs are the displacement measurement from the actuator's integrated strain gauges (used for the active coupling control action) and the measured voltage across the circuit inductor (used for the active inductance tuning and negative resistance actions). The control voltage output is then amplified using a piezoelectric power amplifier and applied to the active-passive circuit. The inductive element in the active-passive circuit is realized using a synthetic inductance circuit [8]. The synthetic inductance circuit was used in place of a passive inductor because of its ease of adjustment, which made it more convenient for the purpose of experimental demonstration. The accelerometer on the foundation is monitored and recorded using a Signal Analyzer. More details on the construction of the test stand are given in Table I.

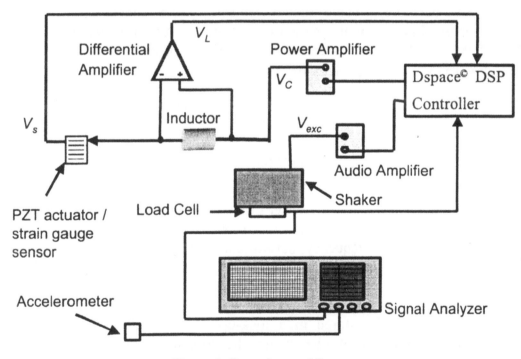

Figure 3. Experimental Setup

TABLE I. TEST STAND SPECIFICATIONS

Actuator stiffness (constant field)	12e6 N/m
Actuator stroke	15e-6 m @ 100 Volts
Actuator length	28 mm
Actuator Capacitance (constant stress)	340e-9 F
Mounting Plate dimensions	10" x 8" x ½ thick
Foundation Plate dimensions	10" x 10" x 1" thick
Foundation Stiffness (each)	3.2e4 N/m
Nylon strut stiffness (each)	1.5e6 N/m

Experimental Tuning

One of the purposes of this study is to demonstrate the application of the adaptive absorber design on a complex structure for which an analytical model is not available. For this reason, the test stand presented in the previous section was designed and constructed somewhat arbitrarily, and was not analytically modeled. Without an analytical model in the form of Eq. (1), the optimal tuning law (Eq. (6)) cannot directly be utilized. One alternative is to directly determine the optimal tuning ratios from experimental tests.

The approach used here is to construct an optimal tuning curve using experimental observation and a relatively straightforward trial and error approach. The first step is to tune the active-passive absorber to some known frequency ω_a, as shown in Eq. (7). Next, the frequency response function of the closed-loop system is obtained experimentally, and the frequency of the absorber notch is found from this experimental data. The optimal tuning ratio at this frequency ω_{notch} is given by Eq. (8).

$$\omega_a = \left[\left(L_p + L_a \right) C_p^s \right]^{-.5} \tag{7}$$

$$\delta_{opt} \left(\omega_{notch} \right) = \frac{\omega_a}{\omega_{notch}} \tag{8}$$

There is some trial and error involved because the optimal tuning ratio obtained with each test corresponds to the notch frequency, not the absorber frequency. This means that to obtain the optimal tuning at a given frequency, several attempts might be necessary. In reality, the optimal tuning ratio is not needed at exact frequencies because it is usually only slightly dependent on frequency, which was also shown in analysis results in [5]. Therefore, several points within the bandwidth of interest are usually sufficient. The frequency and optimal tuning ratio data can then be stored in a look-up table. This can be very conveniently implemented by a DSpace controller using the Simulink look-up table block, which automatically performs linear interpolation for frequencies between the entries of the look-up table.

Experimental Results

In this section, the application of the adaptive piezoelectric absorber design for vibration isolation purposes is explored. The optimal tuning ratios for the isolation experiments are determined using the experimental tuning method. The resulting tuning ratio is relatively constant over the bandwidth of interest and is approximately equal to $\delta = 1.22$. The effect of the active coupling gain is shown in Figure 4. This plot shows that an active coupling gain of $G_{ac} = 3$ significantly increases the notch depth, but can only increase the notch width slightly. Larger active coupling gains could not be used in this experiment due to the relatively poor signal-to-noise ratio of the actuator's integrated strain gauge sensor. Because the actuator is quite stiff, the displacement for a small force is barely large enough to be detected by the strain gauge sensor. On the other hand, the force input cannot be excessively increased because this will cause the voltage levels in the active-passive circuit to increase and exceed the hardware limit. The combination of these voltage constraints and the resolution of the strain gauge sensor limited the maximum usable active coupling gain.

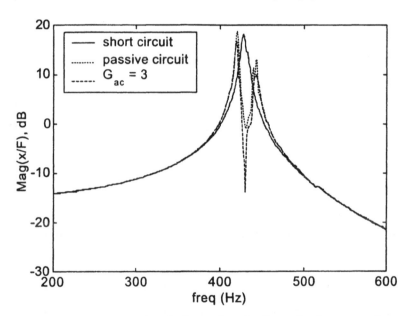

Figure 4. Effect of Active Coupling Gain on Steady-State Performance ($\omega_o = 430\,Hz$)

Next, we consider the transient performance of the closed-loop adaptive absorber. The excitation selected is a linear chirp signal, which is a sinusoid with linearly increasing frequency. The excitation and passive circuit parameters are given in Table II. Note that the frequency rates of change are expressed relative to the nominal excitation frequency. The physical meaning of $\dot\omega = .10\omega_o$, for example, is a 10% increase per second, relative to the nominal frequency.

Table II. Vibration Isolation Experimental Parameters

$\omega_o = 430$ Hz	$G_{ac} = 3$
$L_p = .221$ H	$\Delta = .025$
$R_p = 6\,\Omega$	$t_s = 1$ sec
$R_a = 0\,\Omega$	$\dot{\omega}/\omega_o = .005, .02, .05, .10$ [rad/s²]

The passive baseline used here is an optimally-damped passive piezoelectric absorber. A typical transient response envelope comparison is shown in Figure 5. The performance and control effort results for each excitation case are summarized in Table III.

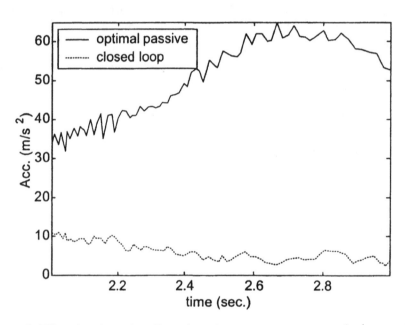

Figure 5. Vibration Isolation Transient Response Envelopes ($\dot{\omega}/\omega_o = .02$)

Table III. Summary of Vibration Isolation Test Results

$\dot{\omega}/\omega_o$	% Reduction RMS Response	% Reduction Peak Response	Control Power Watts, RMS
0.005	83.7	80.1	3.55e-6
0.020	83.6	65.1	3.02e-6
0.050	86.5	82.8	3.68e-6
0.100	82.7	51.2	3.96e-6

These results confirm that the adaptive active-passive absorber can be very effective for vibration absorption and isolation, despite the limitations of the present test stand instrumentation hardware. In other words, the adaptive absorber can outperform the optimal passive treatment significantly, for both high and low speed variations of the disturbance frequency.

SUMMARY

The goal of this paper is to outline the theory of the active-passive piezoelectric absorber concept and evaluate the feasibility of implementing such a device on an unmodeled complex structure. An experimental tuning method is derived and applied to an active-passive mount test stand. Despite the limitation of the instrumentation and sensor resolution in the current test set up, it is shown that the adaptive absorber can produce a transient RMS response attenuation of greater than 80% (compared to an optimal passive absorber), even for relatively rapid changes in excitation frequency.

REFERENCES

1. Hagood, N. W., and von Flotow, A. H., 1991 , "Damping of Structural Vibrations with Piezoelectric Materials and Passive Electrical Networks", *Journal of Sound and Vibration,* 146(2): 243-268.

2. von Flotow, A. H., Beard, A., and Bailey, D., 1994, "Adaptive Tuned Vibration Absorbers: Tuning Laws, Tracking Agility, Sizing, and Physical Implementations", *Proceedings of NOISE-CON 94 (Ft. Lauderdale, FL),* 1: 437-455.

3. Davis, C. L., Lesieutre, G. A., and Dosch, J., 1997, "A Tunable Electrically Shunted Piezoceramic Vibration Absorber", *Proceedings of the SPIE,* 3045: 51-59.

4. Hollkamp, J. H., and Starchville, Jr., T. F., 1994, "A Self-Tuning Piezoelectric Vibration Absorber", *Journal of Intelligent Material Systems and Structures,* 5: 559-566.

5. Morgan, R. A. and Wang, K.W, 2000, "An Active-Passive Piezoelectric Vibration Absorber for Structural Control under Harmonic Excitations with Time-Varying Frequency", *Proceedings of the Adaptive Structures and Material Systems Symposium: 2000 ASME IMECE,* AD-60: 285-298.

6. Morgan, R. A. and Wang, K.W, 2001, "A Multi-Frequency Piezoelectric Vibration Absorber for Variable Frequency Harmonic Excitations," *Proc. SPIE Conf. on Smart Structures and Materials,* 4331: 130-140.

7. Handel, P., and Tichavsky, P., 1994, "Adaptive Estimation for Periodic Signal Enhancement and Tracking", *Intl. Journal of Adaptive Control and Signal Processing,* 8: 447-456.

8. Chen, C. T., 1984, *Linear System Theory and Design,* Holt, Rinehart and Winston, Inc., NY, pp. 327.

HYBRID PIEZOELECTRIC DAMPING SYSTEM FOR FLEXIBLE THREE-STORY STRUCTURE WITH BASE EXCITATION

Kazuhiko Adachi, Yoshitsugu Kitamura, and Takuzo Iwatsubo

ABSTRACT

This paper aims at proposing an integrated design method of the active / passive hybrid type of the piezoelectric damping system for reducing the dynamic response of the multi-story structures due to external dynamic loads. The integrated design method is based on the numerical optimization technique; whose objective function is the active control power requirement of the hybrid piezoelectric damping system. Availability of the proposed design method is numerically demonstrated. The hybrid piezoelectric damping system for a three-story structure is successfully designed. The numerical simulation result indicates that the hybrid piezoelectric damping system is effective comparing with the pure active piezoelectric damping system from the viewpoint of the active control power requirement under the equal vibration suppression performance condition.

INTRODUCTION

A great deal of research is currently in progress on designing the active and passive types of damping system such as active mass damper and oil damper for the multi-story structures in order to reduce the dynamic response due to external dynamic excitations: earthquake and wind (e.g., as [1], [2] and [3]). Passive damping utilizes the response of the structures to generate the control force without external power supply and then guarantees the stability of the whole system. On the other hand, active damping achieves excellent performance of the vibration suppression and necessarily requires large amount of external power supply. The active / passive hybrid type of damping system (we call it "hybrid damping system") combines the advantages of both the passive and active damping systems. The authors proposed the optimum integrated design method of the hybrid damping system for the n-story structure, in which the hybrid damping system

Kazuhiko Adachi, CIMSS, Dept. of Mech. Eng., Virginia Tech, Blacksburg, VA 24061-0261, U.S.A.

Yoshitsugu Kitamura, Dept. of Mech. Eng., Kobe Univ., Rokkodai-cho 1-1, Nada-ku, Kobe 657-8501, Japan.

Takuzo Iwatsubo, Dept. of Mech. Eng., Kobe Univ., Rokkodai-cho 1-1, Nada-ku, Kobe 657-8501, Japan.

was constructed by an active mass damper installed on the top of the structure as the active damping and n oil dampers installed in every story of the structure as the passive damping [4].

This study aims at proposing an integrated design method of the active / passive hybrid type of the piezoelectric damping system for the n-story structure. In the present paper, the hybrid piezoelectric damping system is installed on the structure instead of the active mass damper and the oil dampers. The availability of the hybrid piezoelectric damping system from the viewpoint of the active control power requirement under the equal vibration suppression performance condition was already demonstrated by the authors by using a simple cantilever beam example with the surface bonded piezo-elements, experimentally and numerically [5, 6, 7]. Numerical simulation for the three-story structure with base excitation example will be shown to demonstrate the availability of the proposed design method.

FORMULATION OF DAMPING SYSTEM DESIGN

Governing Equation

Consider the piezoelectrically coupled electro-mechanical system described by

$$M\ddot{z} + C\dot{z} + Kz = f + \Theta q , \quad L\ddot{q} + R\dot{q} + C_p^{-1}q = \Theta^T z + v \tag{1}$$

based on Lagrangian equations of motion and the constitutive equations of the piezo-electric material [5, 6, 8, 9]. The first equation of Eq.(1) is the equations of motion of the n-story structure equipped with the surface bonded piezoelectric ceramic tiles pairs. The second equation of Eq.(1) is the governing equation of the electric circuits of the hybrid damping system by using the PZT tiles pairs. z is the nodal displacement vector of the host structure. M and K are the mass and stiffness matrices of the host structure considering the PZT tiles. C is the damping matrix of the host structure, in which the proportional damping is assumed. Θ is the electro-mechanical coupling matrix. f is the dynamic loading vector applied to the host structure. q is the electric charge vector at the electrode of the PZT tiles. L and R are diagonal matrices whose diagonal elements are the inductance and resistance of the tuned RL shunting circuits for the passive damping. C_p is also diagonal matrix whose diagonal elements are the capacitance of the PZT tiles. v is the external control voltage for the active damping.

In the case of the three-story structure with base excitation shown in Figure 1, each component of Eq.(1) is given by follows. In this case,

$$z = \{ \begin{matrix} z_1 & z_2 & z_3 \end{matrix} \}^T \quad (z_1 = x_1 - x_0, \ z_2 = x_2 - x_0, \ z_3 = x_3 - x_0) \tag{2}$$

is the relative displacement of the i-th story to the base motion x_0 shown in the figure.

$$M = \begin{bmatrix} m_1 & 0 & 0 \\ 0 & m_2 & 0 \\ 0 & 0 & m_3 \end{bmatrix} , \quad C = \begin{bmatrix} c_1 + c_2 & -c_2 & 0 \\ -c_2 & c_2 + c_3 & -c_3 \\ 0 & -c_3 & c_3 \end{bmatrix} ,$$

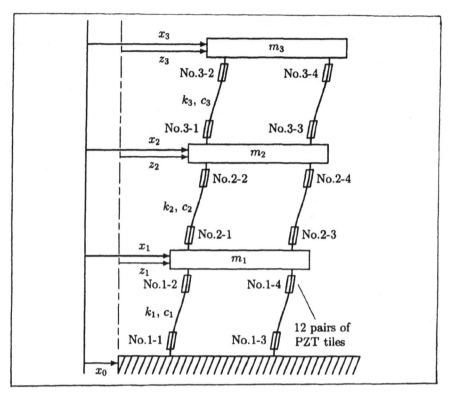

Figure 1. Three-story structure with 12 pairs of PZT tiles

$$K = \begin{bmatrix} k_1 + k_2 & -k_2 & 0 \\ -k_2 & k_2 + k_3 & -k_3 \\ 0 & -k_3 & k_3 \end{bmatrix} \quad, \quad f = -M \left\{ \begin{array}{c} \ddot{x}_1 \\ \ddot{x}_2 \\ \ddot{x}_3 \end{array} \right\} \quad,$$

$$\Theta = \begin{bmatrix} \theta_1 & -\theta_1 & \theta_1 & -\theta_1 & -\theta_2 & \theta_2 & -\theta_2 & \theta_2 & 0 & 0 & 0 & 0 \\ 0 & 0 & 0 & 0 & \theta_2 & -\theta_2 & \theta_2 & -\theta_2 & -\theta_3 & \theta_3 & -\theta_3 & \theta_3 \\ 0 & 0 & 0 & 0 & 0 & 0 & 0 & 0 & \theta_3 & -\theta_3 & \theta_3 & -\theta_3 \end{bmatrix} \quad,$$

$$q = \left\{ \begin{array}{cccccccccccc} q_{11} & q_{12} & q_{13} & q_{14} & q_{21} & q_{22} & q_{23} & q_{24} & q_{31} & q_{32} & q_{33} & q_{34} \end{array} \right\}^{\mathsf{T}} \quad,$$

$$L = \begin{bmatrix} L_1 & 0 & 0 \\ 0 & L_2 & 0 \\ 0 & 0 & L_3 \end{bmatrix} \quad, \quad L_i = \begin{bmatrix} L_{i1} & 0 & 0 \\ 0 & L_{i2} & 0 \\ 0 & 0 & L_{i3} \end{bmatrix} \quad,$$

$$R = \begin{bmatrix} R_1 & 0 & 0 \\ 0 & R_2 & 0 \\ 0 & 0 & R_3 \end{bmatrix} \quad, \quad R_i = \begin{bmatrix} R_{i1} & 0 & 0 \\ 0 & R_{i2} & 0 \\ 0 & 0 & R_{i3} \end{bmatrix} \quad,$$

$$C_p = \begin{bmatrix} C_{p1} & 0 & 0 \\ 0 & C_{p2} & 0 \\ 0 & 0 & C_{p3} \end{bmatrix} \quad, \quad C_{pi} = \begin{bmatrix} C_{pi1} & 0 & 0 \\ 0 & C_{pi2} & 0 \\ 0 & 0 & C_{pi3} \end{bmatrix} \quad,$$

$$v = \left\{ \begin{array}{cccccccccccc} v_{11} & v_{12} & v_{13} & v_{14} & v_{21} & v_{22} & v_{23} & v_{24} & v_{31} & v_{32} & v_{33} & v_{34} \end{array} \right\}^{\mathsf{T}} \tag{3}$$

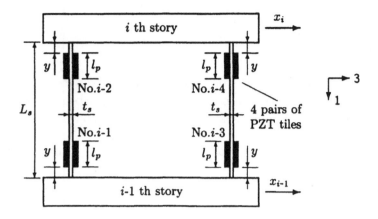

Figure 2. Surface bonded PZT tiles pairs on i-th story

4 pairs of PZT tiles are bonded on the surface of the columns of each story shown in Figure 2. Each column is L_s long, b_s width and t_s thickness. Each PZT tile is l_p long, b_p width and t_p thickness. The strain-displacement relations to No.i-k PZT tiles pair shown in the figure are given by follows.

$$\varepsilon_{pie} = \frac{3t_s}{L_s^3}(l_p + 2y - L_s)(x_i - x_{i-1}) = \psi_i(x_i - x_{i-1}) \tag{4}$$

is for the PZT tiles bonded on the expanded surface of the columns, in which

$$\psi_i = \frac{3t_s}{L_s^3}(l_p + 2y - L_s) . \tag{5}$$

Similarly,

$$\varepsilon_{pic} = -\psi_i(x_i - x_{i-1}) \tag{6}$$

is for the PZT tiles bonded on the opposite contracted surface of the columns. k_i and θ_i of Eq.(3) are given by using ψ_i as follows.

$$k_i = 2\left(\frac{E_s b_s t_s^3}{L_s^3}\right) + 4\left(2l_p t_p b_p E_p \psi_i^2\right) , \quad \theta_i = 2h_{31} t_p \psi_i \tag{7}$$

E_s and E_p are Young's moduli of the column and the PZT tile, respectively. h_{31} is piezoelectric constant. q_{ik}, L_{ik}, R_{ik}, C_{pik} and v_{ik} are the electric charge, inductance and resistance of the tuned RL shunting circuit, capacitance of the piezo-elements and external control voltage related to No.i-k PZT tiles pair.

Above relations are derived by using the following assumptions: (a) the bending deflection, $x_i - x_{i-1}$, of the columns of each story is small (compared with its length: L_s), (b) the PZT tiles displacement and strain are equal to the displacement and strain at the surface of the columns, thus, only uni-axial loading of the piezo-elements in the 1-direction shown in Figure 2 is considered, (c) the poling directions of the PZT tiles are in the 3-direction shown in the figure, and (d) the external control voltage for active damping is uniform.

Hybrid Damping System

The coordinates transformation from a physical space to a modal space is introduced. When No.i-k PZT tiles pair is tuned to the m-th structural vibration mode, the governing equations of the electro-mechanical system are given by follows.

$$\ddot{\eta}_m + 2\zeta_m\omega_m\dot{\eta}_m + \omega_m{}^2\eta_m = \boldsymbol{\phi}_m{}^\mathsf{T}\boldsymbol{f} + \boldsymbol{\phi}_m{}^\mathsf{T}\boldsymbol{\theta}_{ik}q_{ik} ,$$

$$L_{ik}\ddot{q}_{ik} + R_{ik}\dot{q}_{ik} + \frac{1}{C_{pik}}q_{ik} = \boldsymbol{\theta}_{ik}{}^\mathsf{T}\boldsymbol{\phi}_m\eta_m + v_{ik} \tag{8}$$

η_m, ω_m and ζ_m are the m-th modal displacement, natural circular frequency and modal damping ratio, respectively. $\boldsymbol{\phi}_m$ is corresponding to the m-th column of the modal matrix whose columns are the eigenvectors normalized with respect to the mass matrix. $\boldsymbol{\theta}_{ik}$ is a column of $\boldsymbol{\Theta}$ corresponding to No.i-k PZT tiles pair.

The values of the inductance and resistance of the passive tuned RL shunting circuit for the passive damping are usually designed by the analogy with the single degree-of-freedom damped vibration absorber [10, 11], in this study only L_{ik} is designed by using the previous proposed method as follows.

$$L_{ik} = \frac{1}{\omega_m{}^2 C_{pik}} \tag{9}$$

However, R_{ik} is designed by based on the maximum power-transfer theorem.

The external control voltage v_{ik} for driving No.i-k PZT tiles pair as the active damping is generated by using the feedback control,

$$v_{ik} = -G_{ik}\dot{z}_i \tag{10}$$

where G_{ik} is the velocity feedback gain which has to be designed with considering the above designed passive damping system. The integrated design problem is formulated in the following section.

Formulation of Integrated Design Problem

Vibration suppression performance is evaluated by using the gain of the frequency response function of the closed loop system. Active control power requirements of both the hybrid and pure active damping are evaluated by using the power loss at the surface bonded PZT tiles. In the case of the three-story structure, the gain of the frequency response function is given by

$$|F(s)|_{s=j\omega_1} = \left| \frac{\omega_1{}^2 X_3(j\omega_1)}{\omega_1{}^2 X_0(j\omega_1)} \right| \tag{11}$$

in which the frequency response function is measured from the base exciting acceleration $\ddot{x}_0$ to the acceleration response of the 3rd-story $\ddot{x}_3$. $X_i(s)$ is Laplace transform of $x_i(t)$. ω_1

is the 1st resonant frequency of the structure. $|F(s)|_{s=j\omega_1}$ obtained by the hybrid damping is denoted by using $|F_H(s)|_{s=j\omega_1}$, similarly, that obtained by the pure active damping is denoted by using $|F_A(s)|_{s=j\omega_1}$. The power loss at the surface bonded PZT tiles of the i-th story of the structure is given by

$$J_{PLi}(\omega) = \sum_{k=1}^{4}\left\{ G_{ik}^2 \left| \frac{j\omega Z_i(j\omega)}{(j\omega)^2 X_0(j\omega)} \right|^2 \omega C_{pik}\tan\delta \right\} \tag{12}$$

where $\tan\delta$ is the loss tangent of the PZT tiles. $J_{PLi}(\omega)$ obtained by the hybrid damping is denoted by using $J_{PLHi}(\omega)$, similarly, that obtained by the pure active damping is denoted by using $J_{PLAi}(\omega)$.

Integrated design problem of the active / passive hybrid type of the piezoelectric damping system is formulated as follows. "The values of G_{ik} are optimized such that

$$\text{minimize } J_{PLHi} \tag{13}$$

subject to

$$J_{PLHi}(\omega) < \text{minimum } J_{PLAi}(\omega) , \quad |F_H(s)|_{s=j\omega_1} = |F_A(s)|_{s=j\omega_1} , \tag{14}$$

Equations (8), (9), and the lower and upper bounds of the values of G_{ik} ".

DAMPING SYSTEM DESIGN EXAMPLE

Simulation Model

The proposed integrated design method is applied to design the hybrid damping system of the three-story structure shown in Figures 1 and 2 whose properties are listed in TABLE I. β_3^ε is the impermittivity of the PZT tiles. Superscript $(\cdot)^\varepsilon$ indicates that the value is obtained at the constant strain.

TABLE I. SPECIFICATION OF THREE-STORY STRUCTURE AND
SURFACE BONDED PZT TILES

3-story structure		PZT tiles	
$m_1 \sim m_3$	0.4000 (kg)	E_p	1.310×10^{11}(N/m^2)
$k_1 \sim k_3$	1.5479×10^4 (N/m)	h_{31}	-0.7000×10^9(V/m)
$c_1 \sim c_3$	0.5000 (N·s/m)	β_3^ε	4.518×10^7(m/F)
L_s	0.1400 (m)	l_p	0.0300 (m)
b_s	0.0500 (m)	b_p	0.0500 (m)
t_s	0.0010 (m)	t_p	0.0004 (m)

TABLE II. PIEZOELECTRIC DAMPING SYSTEM FOR THREE-STORY STRUCTURE

Case	Number of surface bonded PZT tiles pair											
	1-1	1-2	1-3	1-4	2-1	2-2	2-3	2-4	3-1	3-2	3-3	3-4
1	A[*1]	A	A	A	A	A	A	A	P[*2,*3]	P[*3]	P[*3]	P[*3]
2	A	P[*4]	A	P[*4]	A	P[*3]	A	P[*3]	A	P[*5]	A	P[*5]
3	O[*6]	O	O	O	O	O	O	O	O	O	O	O
4	P[*4]	P[*4]	P[*4]	P[*4]	P[*3]	P[*3]	P[*3]	P[*3]	P[*5]	P[*5]	P[*5]	P[*5]
5	A	A	A	A	A	A	A	A	A	A	A	A

[*1] PZT tuiles pair is used to the **A**ctive Damping system.
[*2] PZT tuiles pair is used to the **P**assive Damping system.
[*3] tuned to 3rd vibration mode.
[*4] tuned to 1st vibration mode.
[*5] tuned to 2nd vibration mode.
[*6] PZT tuiles pair is **O**pen circuit (without electlic damping augmantation).

Result and Discussion

TABLE II shows the design cases. In this example, the passive and active damping systems are separated from each other, therefore, each PZT tiles pair is used to the passive damping system, alternatively used to the active damping system. The hybrid damping system is designed by Cases 1 and 2. In Case 1, each story is equipped with the active damping system, alternatively with the passive damping system. On the contrary, in Case 2, each story is equipped with both the active and passive damping system. Non-electric damping, the pure passive and active damping cases are designed by Cases 3, 4 and 5, respectively. The results of Case 3~5 are comparing with that of Cases 1 and 2 from the view point of both the vibration suppression performance and the active control power requirement.

Figures 3 and 4 summarize the integrated design of the hybrid damping system. Figure 3 shows the gain of the frequency response function measured from the base exciting acceleration to the acceleration response of the 3rd-story. The values of $|F_H(s)|_{s=j\omega_1}$ for Cases 1 and 2 are equal to that for Case 5 under the constraint given in the second equation of Eq.(14). From Figure 4, less active control power requirement is successfully achieved by Case 1.

CONCLUSIONS

In this paper, an integrated design method of the active / passive hybrid type of piezoelectric damping system for the n-story structure was proposed. Governing equations for the piezoelectrically coupled electromechanical system was derived and the hybrid damping system design problem was formulated in which the objective function was the active control power requirement of the damping system. Availability of the proposed design method was numerically demonstrated. The hybrid piezoelectric damping system for the three-story structure with base excitation was successfully designed. The numerical simulation result indicated that the hybrid damping system was effective comparing with

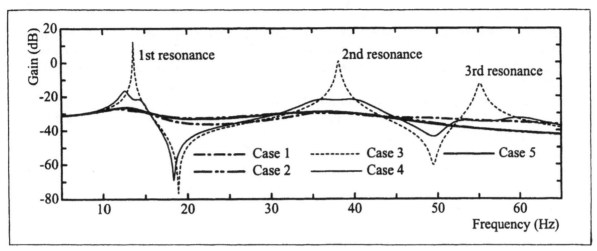

Figure 3. Vibration suppression performance

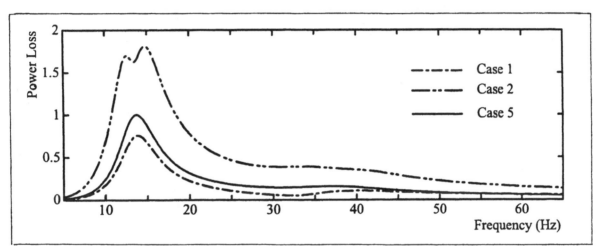

Figure 4. Active control power requirement

the pure active damping system from the viewpoint of the active control power requirement under the equal vibration suppression performance condition.

REFERENCES

1. Fur,L.S., H.T.Y.Yang, and S.Ankireddi. 1996. "Vibration Control of Tall Buildings under Seismic and Wind Loads," ASCE, *Journal of Structural Engineering*, 122(8): 948-957.
2. Ikeda,Y. 1997. "Effect of Weighting a Strike of an Active Mass Damper in the Linear Quadratic Regulator Problem," *Earthquake Engineering and Structural Dynamics*, 26(11): 1125-1136.
3. Sarbjeet,S. and T.K.Datta. 2000. "Nonlinear Sliding Mode Control of Seismic Response of Building Frames," ASCE, *Journal of Engineering Mechanics*, 126(4): 340-347.
4. Adachi,K., Y.Yamashita, and T.Iwatsubo. 2000. "Hybrid Vibration Suppression System Design based on Closed Loop System Redesign for Multi-Story Structure," *Proceedings of the Fifth International Conference on Motion and Vibration Control*, University of Technology, Sydney, Sydney,

Australia, Vol.1 of 2: 145-150, December 4-8, 2000.

5. Adachi,K., Y.Awakura, and T.Iwatsubo. 2000. "Experimental investigation of hybrid damping for flexible structures by using surface bonded piezo-elements," presented at the Smart Structures and Materials 2000: Damping and Isolation, T.Tupper Hyde, Editor, Proceedings of SPIE Vol.3989: 312-321, March 2000.

6. Adachi,K. 2000. "Structural Vibration Suppression by Using Surface Bonded Piezo-Elements," *Proceedings of the Third Korea-Japan Symposium of Frontiers in Vibration Science and Technology*, Yonsei University, Seoul, Korea, 44-51, June 15-16, 2000.

7. Adachi,K., Y.Awakura, and T.Iwatsubo. 2001. "Variable Hybrid Piezoelectric Damping based on Control Power Requirement," presented at the Smart Structures and Materials 2001: Damping and Isolation, Daniel J.Inman, Editor, Proceedings of SPIE Vol.4331, March 2001.

8. Crandall,S.H., D.C.Karnopp, E.F.Kurtz Jr., and D.C.Pridmore-Brown. 1968. *Dynamics of Mechanical and Electromechanical Systems*, McGraw-Hill Book Co., pp.300-308.

9. IEEE Standard on Piezoelectricity, ANSI/IEEE Std., 176-1987.

10. Hagood,N.W. and A.von Flotow. 1991. "Damping of Structural Vibrations with Piezoelectric Materials and Passive Electrical Networks," *Journal of Sound and Vibration*, 146(2): 243-268.

11. Hollkamp,J.J. 1994. "Multimodal Passive Vibration Suppression with Piezoelectric Materials and Resonant Shunts," *Journal of Intelligent Material Systems and Structures*, 5: 49-57.

ELECTRORHEOLOGICAL DAMPER ANALYSIS USING AN EYRING CONSTITUTIVE RELATIONSHIP

Lionel Bitman, Young-Tai Choi, and Norman M. Wereley

ABSTRACT

We validate and assess the applicability of an Eyring constitutive model. The Bingham plastic model has a zero shear rate discontinuity, which leads to inaccuracies in modeling and simulation, and is characterized by two rheological constants for a constant field: yield stress and postyield viscosity. The Eyring model has a smooth transition through the zero shear rate condition, and also has two rheological constants for a constant field. An ER (electrorheological) damper having a damping level comparable to a shock absorber for a small-sized passenger car damper is manufactured, and its performance is predicted using the Eyring model. To accurately identify the rheological parameters of the Eyring model, a parameter identification method was used. The damping force versus piston displacement and velocity behaviors of the ER damper are experimentally measured with respect to applied electric field and excitation frequency. To validate the Eyring model, the experimental results are compared with predictions in the damping force versus piston displacement and velocity behaviors.

INTRODUCTION

ER (electrorheological) or MR (magnetorheological) fluids are known to produce yield shear stresses, when external field is applied to the fluid. These yield stresses can be adjusted by changing the intensity of applied field and its response is rapid. Advantages such as continuous control of yield stress and fast response have led to substantial applied research of ER and MR fluids to various systems such as dampers, valves, clutches and brakes [1-4]. In developing ER or MR applications, it is important to accurately model device characteristics, so as to save developing time and increase cost-effectiveness. So far, Bingham plastic model has been widely used in the evaluation and prediction of ER or MR devices. Even though Bingham plastic model is mathematically simple, it has some deficiencies particularly in the representation of preyield behavior. However, the Bingham plastic model exhibits discontinuous characteristics in shear stress versus shear rate behavior at the zero shear rate condition. Thus, the Bingham plastic model cannot accurately account for the practical behavior of the progressive and continuous yield shear stress that is usually observed as the shear rate increases in the preyield region. Consequently, this paper explores an alternative constitutive model to improve modeling of ER or MR applications in a practical sense.

We validate and assess the applicability of an Eyring constitutive model. The Bingham plastic model has a zero shear rate discontinuity, which leads to inaccuracies in modeling and simulation, and is characterized by two rheological constants for a constant field: yield

Lionel Bitman, Graduate Student, University of Maryland, College Park, MD 20742, USA
Young-Tai Choi, Research Assistant, University of Maryland, College Park, MD 20742, USA
Norman M. Wereley, Associate Professor, University of Maryland, College Park, MD 20742, USA

stress and postyield viscosity. The Eyring model has a smooth transition through the zero shear rate condition, and also has two rheological constants for a constant field. An ER damper having a damping level comparable to a shock absorber for a small-sized passenger car damper is manufactured, and its performance is predicted using the Eyring model. To accurately identify the rheological parameters of the Eyring model, a parameter identification method was used. The force versus piston displacement and velocity behavior of the ER damper are experimentally measured with respect to applied electric field and excitation frequency. To validate the Eyring model, the experimental results are compared with predictions in the damping force versus piston displacement and velocity behaviors.

EYRING MODEL

When external field is applied to ER or MR fluids, typical behavior is shown in Figure 1. The Bingham plastic model exhibits a discontinuity in preyield (low shear rate) region. However, the Eyring model has a smooth transition behavior through the preyield region that is much closer to the practical situation [3,4]. The key difference between the alternative models is the low shear rate behavior – the high shear rate behavior is similar. Maximum shear stress at maximum shear rate is the same in both models.

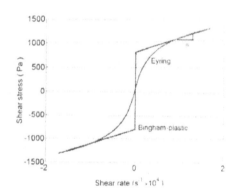

Figure 1. Comparison of Eyring to Bingham plastic model in terms of shear stress versus shear rate.

The constitutive equation of Eyring model for the relationship between the shear stress τ and the shear rate (du/dy) can be represented as follows:

$$\tau = \frac{1}{K} \operatorname{asinh}\left(\frac{1}{\xi}\frac{du}{dy}\right) \tag{1}$$

where u is the fluid velocity, y is the lateral coordinate, and K and ξ are the two rheological parameters of the Eyring model. The parameter K accounts for the magnitude of the shear stress and ξ for the steepness of the shear rate gradient through the preyield region. It is noted that the parameter ξ partially affects the magnitude of the shear stress. In addition, as shown in Eq. (1), the mathematical form of the Eyring model is simpler than the Bingham plastic model, and it does not contain the notions of yield shear stress τ_y and postyield viscosity μ.

PARAMETER IDENTIFICATION

To establish the Eyring model structure, the parameters K and ξ are identified by using experimental data. In this study, for more accurate parameter identification, an optimization method is used on the basis of the measured damping force characteristics of an ER damper. It is noted that MR dampers exhibit qualitatively similar damping force characteristics upon application of magnetic field, so that the Eyring model would be equally applicable. For ER dampers, electric voltages are used to create external fields to the fluid domain. In this study, experimental data from an ER damper is used to identify the parameters K and ξ.

Manufacture of Flow-Mode ER Dampers

A flow-mode ER damper with the damping force level comparable to that needed for a small-sized passenger car damper was designed and manufactured as shown in Figure 2. The ER damper used in this study is a monotube damper with single rod. The damper consists of hydraulic and pneumatic reservoirs separated by a floating piston. Inside the hydraulic reservoir, the piston rod is attached to a piston head holding two concentric tubular electrode lengths. During piston rod motion, ER fluids pass through the electrode gap between two concentric tubes in the piston head and can be energized by applied electric fields. The plastic spacer electrically insulates the two concentric tube electrodes to create electrical voltage potential in the gap. The important design parameters of the ER damper are as follows: electrode length is 43 mm and electrode gap is 0.65 mm. The ER fluid, made in the laboratory, consisted of corn starch and peanut oil. The composition ratio of the ER fluid is chosen by 40% corn starch and 60% peanut oil by weight.

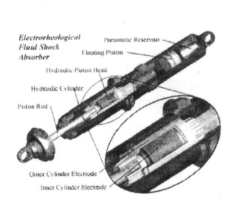
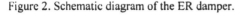

Figure 2. Schematic diagram of the ER damper.

Figure 3. Experimental setup for the damping force test of the ER damper.

Damping Force Test of Flow-Mode ER Dampers

Experimental setup for the damping force test of the ER damper is shown in Figure 3. The ER damper is placed between the load cell and hydraulic excitation actuator. The load cell is fixed and one end of the ER damper is connected to the hydraulic actuator. When the hydraulic actuator moves up and down by a command signal, the ER damper produces the damping force and it is measured by the load cell [3]. In this study, the excitation used is a sinusoidal displacement with constant amplitude of 0.5 inches. Four different excitation frequencies were tested: 0.25, 0.50, 0.75 and 1.0 Hz. The various electric field ranged from 0 to 7.0 kV/mm.

Dynamic Equation of Flow-Mode ER Dampers

In order to derive dynamic equation of the ER damper based on Eyring model, it is assumed that the electrode gap between two cylinders is approximated by parallel plates. It implies that the fluid velocity profile in the gap is symmetric [4]. In addition, the fluid flow in the gap is assumed to be quasi-steady, so that dynamic effects, such as inertia, are not considered in this study. It is noted that above assumptions are not valid at high excitation frequency.

The governing equation in electrode gap of the ER damper is given by

$$\frac{\partial \tau}{\partial y} = \frac{\Delta P}{L} \tag{2}$$

where ΔP is the pressure drop between both ends of the electrode and L is the electrode length. If substituting Eq.(1) into Eq.(2) and integrating twice with respect to y, then, the fluid velocity profile in the gap can be obtained on the basis of Eyring model as follows:

$$u(y) = \left(\frac{L\xi}{\Delta PK} \right) \left\{ \cosh\left(\frac{\Delta P}{L} Ky \right) - \cosh\left(\frac{\Delta P}{2L} Kd \right) \right\} \tag{3}$$

where d is the electrode gap. Typical fluid velocity profile of Eyring model in the electrode gap is shown in Figure 4. As observed in this figure, Eyring model exhibits a smooth and continuous velocity profile, unlike Bingham plastic model where there is a large discontinuity called the "plug" region over which the velocity gradient is zero.

Eyring model   Bingham plastic model

Figure 4. Typical fluid velocity profiles in the electrode gap.

In order to obtain the relationship between the damping force and piston velocity on the basis of Eyring model, integrating Eq. (3) with respect to the gap yields an equation for the flow rate in the gap, Q_g.

$$Q_g = b \int_0^d \left(\frac{L\xi}{\Delta PK} \right) \left\{ \cosh\left(\frac{\Delta P}{L} Ky \right) - \cosh\left(\frac{\Delta P}{2L} Kd \right) \right\} dy \tag{4}$$

where b is electrode width. Using conservation of flow rate, $Q_g = A_p V_p$, and $\Delta P = -F / A_p$, relates the damping force F to the piston velocity V_p.

$$V_p = \frac{b\xi Ld}{FK} \left\{ \cosh\left(\frac{FKd}{2LA_p} \right) - \left(\sinh\left(\frac{FKd}{2LA_p} \right) \right) \cdot \left(\frac{2LA_p}{FKd} \right) \right\} \tag{5}$$

Here, A_p is the effective cross-sectional piston area. The expression of the piston velocity V_p is obtained as a function of the damping force, F in Eq. (5). Consequently, in theoretical analysis, the damping force, F, is the independent variable. However, in the experiments, the independent variable is the piston velocity V_p. Given the damping force, F, geometric properties of the ER damper, and the parameters K and ξ, the piston velocity is easily calculated from Eq. (5). However, the parameters K and ξ are not directly found from the experimental data. Consequently, parameter identification of K and ξ is needed.

Optimization Method

In this study, the parameters K and ξ are obtained by the minimization of error functions for more accurate parameter identification, and an optimization method is used to construct the parameter functions. On the other hand, the shear stresses of ER or MR fluids are known to be dependent on applied field and shear rate. For the ER damper, the damping force depends on applied field and excitation frequency. Therefore, the parameters K and ξ should be obtained with respect to applied field and excitation frequency. In this study, each value of the parameters with respect to applied field and excitation frequency is determined by minimizing the following error function, δV.

$$\delta V(K,\xi) = \frac{1}{N}\sqrt{\sum_{i=1}^{N}\left[V_p(t_i)-V_{pm}(t_i)\right]^2} \tag{6}$$

where V_{pm} is the measured piston velocity and N is the number of data measured. The parameters K and ξ obtained with respect to applied field and excitation frequency are shown in Figure 5.

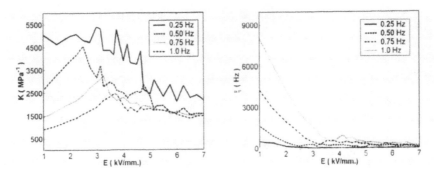

Figure 5. The parameters K and ξ.

Based on above work, the parameters K and ξ are functions of electric field and frequency and behave according to the functions below:

$$K(E,f) = \frac{a_K}{\sqrt{1+\left(b_K\ (E-c_K)\right)^2}} \tag{7}$$

$$\xi(E,f) = \frac{a_\xi}{\sqrt{1+\left(b_\xi\ (E-1)\right)^2}} \tag{8}$$

where E is the applied electric field, f is the excitation frequency, and a_K, b_K, c_K, a_ξ, and b_ξ are coefficients of functions for the parameters K and ξ functions. It is noted that these coefficients are dependent on the excitation frequency only and can be expressed as second degree polynomials of the excitation frequency, f. The coefficients $a_K, b_K, c_K, a_\xi, b_\xi$ can be identified by minimizing the following cost function, J.

$$J(a_K, b_K, c_K, a_\xi, b_\xi) = \frac{\sum_f \sum_E (G_1 + G_2 + 2G_3 + 3G_4)/7}{N_f * N_E} \qquad (9)$$

with

$$G_1 = \frac{1}{N_i}\sqrt{\sum_i \left(\frac{V_p(t_i) - V_{pm}(t_i)}{\overline{V}_{pm}}\right)^2}, \qquad G_2 = \frac{1}{N_i}\sqrt{\sum_i \left(\frac{X_p(t_i) - X_{pm}(t_i)}{\overline{X}_{pm}}\right)^2}$$

$$G_3 = \left|1 - \frac{\overline{V}_p}{\overline{V}_{pm}}\right|, \qquad G_4 = \left|1 - \frac{\overline{X}_p}{\overline{X}_{pm}}\right|$$

where, X_{pm} are the measured piston displacement. $\overline{V}_{pm}$ and $\overline{X}_{pm}$ are the maximums of the measured piston velocity and displacement, and $\overline{V}_p$ and $\overline{X}_p$ are the maximums of the piston velocity and displacement, respectively. In addition, N_f is the number of excitation frequencies tested, N_E is the number of electric fields tested, and N_i is the number of sampling points over one period of time. In this study, $N_f = 4$, $N_E = 28$, and N_i is nominally 4000.

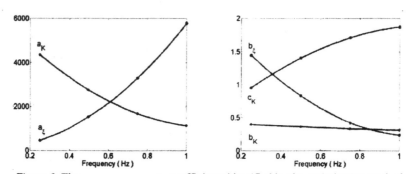

Figure 6. The $a_K, b_K, c_K, a_\xi, b_\xi$ coefficients identified by the optimization method.

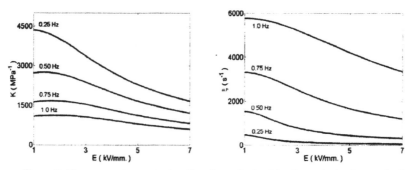

Figure 7. The parameters K and ξ functions obtained by the optimization method.

The minimization of the cost function given by Eq. (9) will give the polynomial expressions for $a_K, b_K, c_K, a_\xi, b_\xi$ coefficients as functions of excitation frequency. These optimal $a_K, b_K, c_K, a_\xi, b_\xi$ coefficients obtained in this study are shown in Figure 6. As observed, the coefficients are strong functions of excitation frequency. Given an excitation frequency, we get the value of these coefficients. Then, the parameters K and ξ functions are obtained by substituting the coefficients into Eqs. (7) and (8) and shown in Figure 7.

EXPERIMENTAL VALIDATION

In order to prove the validity of the Eyring model, the experimental data are compared with theoretical predictions. Figure 8 presents experimental and theoretical results for the damping force characteristics of the ER damper. In this case, the excitation frequency is 0.75 Hz and the applied electric fields shown are 1.0, 3.5, 5.0 and 6.5 kV/mm. In addition, the solid line represents the experimental data and the dashed line represents theoretical data. As clearly observed, the theoretical data obtained using Eyring model are in a good accordance with the experimental data.

Figure 9 presents the rheological characteristics of the ER fluid used in the ER damper. These results are obtained by substituting the identified rheological parameters K and ξ into the Erying model given by Eq. (1). As clearly observed in the figure, the shear stress shows a smooth transition behavior through the preyield region. Especially, at the same electric field, as excitation frequency increases, the smoothness in preyield region increases.

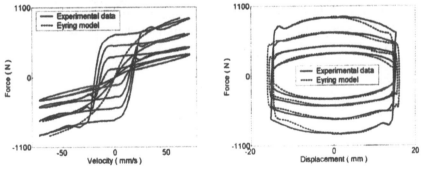

Figure 8. Experimental and theoretical results for the damping force characteristics of the ER damper.

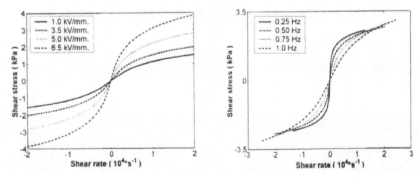

Figure 9. The rheological characteristics of the ER fluid used in the ER damper.

CONCLUSION

The validation and applicability of the Eyring model, as an alternative constitutive model to Bingham plastic model, was evaluated using experimental and theoretical results comparing the performance of an ER damper. In doing so, the ER damper having a damping level comparable to a shock absorber for a small-sized passenger car damper was manufactured, and its performance was predicted using the Eyring model. To accurately identify the rheological parameters of the Eyring model, the parameter identification method was used. The damping force testing of the ER damper was conducted with respect to applied electric field and excitation frequency and its results were compared with predictions in damping force versus piston displacement and velocity behaviors for the validation of the model. It has been demonstrated that the Eyring model provided a good correlation between the experimental and theoretical results.

REFERENCES

1. Williams, S., S.G. Rigby, J.L. Sproston, J.L., and R. Stanway, 1993. "Electrorheological Fluids Applied To As Automotive Engine Mount," J Non-Newtonian Fluid Mechanics, 47:221-238.
2. Han, S.S., S.B., Choi, and C.C. Cheong, 2000. "Position Control X-Y Table Mechanism Using Electro-Rheological Clutches," Mechanism and Machinery Theory, 35:1563-1577.
3. Kamath, G., M. Hurt, and N.M. Wereley, 1996. "Analysis and Testing of Bingham Plastic Behavior In Semi-Active Electrorheological Fluid Dampers," Smart Materials and Structures, 5:576-590.
4. Wereley, N.M. and L. Pang, 1998. "Nondimensional Analysis of Semi-Active Electrorheological and Magnetorheological Dampers Using Approximate Parallel Plate Models," Smart Materials and Structures, 7:732-743.

VIBRATION CONTROL OF TRAIN SUSPENSION SYSTEMS VIA MR FLUID DAMPERS

Wei-Hsin Liao and Dai-Hua Wang

ABSTRACT

This paper is aimed to show the feasibility for improving the ride quality of railway vehicles with semi-active suspension systems using magnetorheological (MR) dampers. A nine degree-of-freedom railway vehicle model, which includes a car body, two trucks and four wheelsets, is proposed to cope with vertical, pitch and roll motions of the car body and trucks. The governing equations of the railway vehicle suspension integrated with MR dampers are developed. To illustrate the feasibility and effectiveness of controlled MR dampers on railway vehicle suspension systems, the LQG control law using the acceleration feedback is adopted, in which the state variables are estimated from the measurable accelerations with the Kalman estimator. The responses of the car body under random track irregularities and periodical irregularities with MR dampers are evaluated and compared with the cases without control. The simulation results show that the vibration control of the train suspension system with semi-active controlled MR dampers is feasible and effective.

INTRODUCTION

The development of high-speed railway vehicles has been a great interest of many countries because high-speed trains have been proven as an efficient and economical transportation means while minimizing air pollution. However, the high speed of the train would cause significant car body vibrations, which induce the following problems: the ride stability, the ride quality, and the cost of track maintenance. Thus the vibration control of the car body is needed to improve the ride comfort and safety of a train. Various kinds of railway vehicle suspensions linking the bogies and the car bodies have been designed to cushion riders from vibrations. In recent years, semi-active suspension systems that utilize controllable devices based on smart fluids have drawn the attention of many researchers. The essential characteristics of the smart fluids are their abilities to reversibly change from a free-flowing fluid to a semi-solid with controllable yield strength in milliseconds when exposed to an electric or magnetic field [1]. Two fluids that are viable contenders for the development of controllable dampers are electrorheological (ER) and magnetorheological (MR) fluids [2].

Peel et al. [3] have developed a mathematical model of a controllable vibration damper employing ER fluids intended for the application to suspension systems of ground vehicles.

Smart Materials and Structures Laboratory, Department of Automation and Computer-Aided Engineering, The Chinese University of Hong Kong, Shatin, N. T., Hong Kong.

The modeling technique was illustrated in an application for controlling the lateral dynamics of a modern rail vehicle. However, it is well known that ER fluids are excited by high electric fields. To produce such levels of field strength requires high voltage, which deters many potential users who perceive safety problems. On the other hand, MR fluids are excited by a magnetic field, which requires only a low voltage source. In addition, MR fluids generate significantly larger dynamic force levels than ER fluids and operate over wide temperature ranges. More recently, the semi-active dampers using MR fluids are developed and applied to control the vibration of automobiles and heavy trucks by some researchers [4,5]. In this paper, a semi-active suspension with MR dampers for a full-size railway vehicle is proposed and the system performance is evaluated.

SEMI-ACTIVE RAILWAY VEHICLE SUSPENSION SYSTEM

In this study, a full-size train model has been formulated for a railway vehicle running on a straight track. A nine degree-of-freedom (DOF) analytical model is considered. This model, which is shown in Figure 1, consists of a car body, two truck frames, and four wheelsets. The wheelsets and truck frames are connected by a primary suspension system that consists of springs and viscous dampers. The car body and truck frames are connected by springs and MR dampers (in vertical direction) / viscous dampers (in lateral direction), which are referred to as the secondary suspension system. In this study, only four MR dampers are installed by replacing the vertical viscous dampers between the car body and two trucks. The schematic configuration of the semi-active control system for the railway vehicle is shown in Figure 2. In this system, four MR dampers, which are used to control the vertical, pitch and roll vibrations, are vertically set on the left and right sides of each truck. In order to realize the feedback control, the accelerations of the car body and trucks are measured with accelerometers, which are installed in vertical and lateral directions as shown in Figure 2.

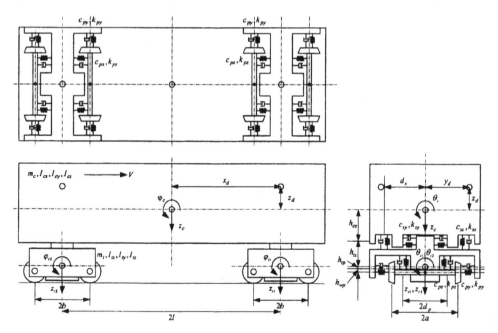

Figure 1. Nine degree-of-freedom train model

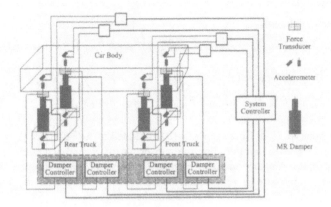

Figure 2. Schematic of semi-active control system for railway vehicle

Analytical Model of Railway Vehicle

The equations of motion for the railway vehicle can be derived using Newton's Laws. The governing equations for the car body (vertical z_c, pitch φ_c, and roll θ_c) can be expressed as

$$m_c \ddot{z}_c = F_{szlr} + F_{szll} + F_{sztr} + F_{sztl} + u_{szlr} + u_{szll} + u_{sztr} + u_{sztl} \tag{1}$$

$$I_{cy}\ddot{\varphi}_c = (F_{sxlr} + F_{sxll})h_{cs} + (F_{sxtr} + F_{sxtl})h_{cs} - (F_{szlr} + F_{szll})l + (F_{sztr} + F_{sztl})l \\ - (u_{szlr} + u_{szll})l + (u_{sztr} + u_{sztl})l \tag{2}$$

$$I_{cx}\ddot{\theta}_c = -(F_{sylr} + F_{syll})h_{cs} - (F_{sytr} + F_{sytl})h_{cs} + (F_{szlr} + F_{sztr})d_s - (F_{szll} + F_{sztl})d_s \\ + (u_{szlr} + u_{sztr})d_s - (u_{szll} + u_{sztl})d_s \tag{3}$$

The governing equations for the front truck (vertical z_{t1}, pitch φ_{t1}, and roll θ_{t1}) can be expressed as follows

$$m_t \ddot{z}_{t1} = -(F_{szlr} + F_{szll}) + (F_{pz1r} + F_{pz2r}) + (F_{pz1l} + F_{pz2l}) - (u_{szlr} + u_{szll}) \tag{4}$$

$$I_{ty}\ddot{\varphi}_{t1} = (F_{sxlr} + F_{sxll})h_{ts} \\ + (F_{px1r} + F_{px1l})h_{tp} + (F_{px2r} + F_{px2l})h_{tp} - (F_{pz1r} + F_{pz1l})b + (F_{pz2r} + F_{pz2l})b \tag{5}$$

$$I_{tx}\ddot{\theta}_{t1} = -(F_{sylr} + F_{syll})h_{ts} - (F_{szlr} - F_{szll})d_s - (F_{py1r} + F_{py2r})h_{tp} - (F_{py1l} + F_{py2l})h_{tp} \\ + (F_{pz1r} + F_{pz2r})d_p - (F_{pz1l} + F_{pz2l})d_p - (u_{szlr} - u_{szll})d_s \tag{6}$$

The governing equations for the rear truck (vertical z_{t2}, pitch φ_{t2}, and roll θ_{t2}) can be expressed as follows

$$m_t \ddot{z}_{t2} = -(F_{sztr} + F_{sztl}) + (F_{pz3r} + F_{pz3l}) + (F_{pz4r} + F_{pz4l}) - (u_{sztr} + u_{sztl}) \tag{7}$$

$$I_{ty}\ddot{\varphi}_{t2} = \left(F_{sxtr} + F_{sxtl}\right)h_{ts}$$
$$+ \left(F_{px3r} + F_{px3l}\right)h_{tp} + \left(F_{px4r} + F_{px4l}\right)h_{tp} - \left(F_{pz3r} + F_{pz3l}\right)b + \left(F_{pz4r} + F_{pz4l}\right)b \qquad (8)$$

$$I_{tx}\ddot{\theta}_{t2} = -\left(F_{sytr} + F_{sytl}\right)h_{ts} - \left(F_{sztr} - F_{sztl}\right)d_s - \left(F_{py3r} + F_{py4r}\right)h_{tp} - \left(F_{py3l} + F_{py4l}\right)h_{tp}$$
$$+ \left(F_{pz3r} + F_{pz4r}\right)d_p - \left(F_{pz3l} + F_{pz4l}\right)d_p - \left(u_{sztr} - u_{sztl}\right)d_s \qquad (9)$$

where the definitions of F in Equations (1)~(9) are listed in APPENDIX II, u represents the damping force of the MR damper. For the subscripts of F and u, s and p represent the secondary and primary suspensions respectively, l represents the leading truck when it is in the third position, otherwise it represents the left side, t represents the trailing truck, r represents the right side. Numbers $1 \sim 4$ represent the wheelsets.

MR Damper Model

The MR damper model proposed by Spencer et al. and depicted in Figure 3 is adopted in this study. The phenomenological model is governed by the following equations [6]

$$F = c_1\dot{y} + k_1\left(x - x_0\right) \qquad (10)$$

$$\dot{z} = -\gamma|\dot{x} - \dot{y}||z|^{n-1}z - \beta(\dot{x} - \dot{y})|z|^n + A(\dot{x} - \dot{y}); \quad \dot{y} = \left[\alpha z + c_0\dot{x} + k_0(x - y)\right]/(c_0 + c_1) \qquad (11~12)$$

$$\alpha = \alpha(u) = \alpha_a + \alpha_b u; \quad c_1 = c_1(u) = c_{1a} + c_{1b}u; \quad c_0 = c_0(u) = c_{0a} + c_{0b}u \qquad (13a~c)$$

$$\dot{u} = -\eta(u - v) \qquad (14)$$

where v is the command voltage sent to the driver for the MR damper. In this model, there are a total of 14 parameters ($c_{0a}, c_{0b}, k_0, c_{1a}, c_{1b}, k_1, x_0, \alpha_a, \alpha_b, \gamma, \mu, A, n, \eta$) to characterize the MR damper. In this paper, the MR dampers that produced by Lord Corporation are used and the model parameters are shown in TABLE I.

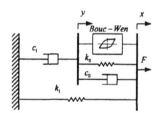

Figure 3. Mechanical model for MR damper

TABLE I. PARAMETERS FOR MR DAMPER MODEL [6]

Parameter	Value	Parameter	Value	Parameter	Value
c_{0a}	21.0 N.s/cm	α_a	140 N/cm	x_0	14.3 cm
c_{0b}	3.50 N.s/cm.V	α_b	695 N/cm.V	η	190 s^{-1}
k_0	46.9 N/cm	γ	363 cm^{-2}	k_1	5.00 N/cm
c_{1a}	283 N.s/cm	β	363 cm^{-2}	n	2
c_{1b}	2.95 N.s/cm.V	A	301		

State Equations

Let $\mathbf{q} = \begin{bmatrix} z_c & \phi_c & \theta_c & z_{t1} & \phi_{t1} & \theta_{t1} & z_{t2} & \phi_{t2} & \theta_{t2} \end{bmatrix}^T$, the governing equations of the railway vehicle with the MR dampers can be written in the matrix form as

$$\mathbf{M}\ddot{\mathbf{q}} + \mathbf{C}\dot{\mathbf{q}} + \mathbf{K}\mathbf{q} = \mathbf{F}_u\mathbf{u} + \mathbf{F}_w\mathbf{w} \tag{15}$$

where $\mathbf{u} = \begin{bmatrix} u_{szlr} & u_{szll} & u_{szrr} & u_{szrl} \end{bmatrix}^T$ is the actuator vector, $\mathbf{w}$ is the disturbance vector determined by track irregularities. Let $\mathbf{w}_1 = \begin{bmatrix} z_{1r} & z_{1l} & z_{2r} & z_{2l} & z_{3r} & z_{3l} & z_{4r} & z_{4l} \end{bmatrix}^T$ and $\mathbf{w}_2 = \begin{bmatrix} \dot{z}_{1r} & \dot{z}_{1l} & \dot{z}_{2r} & \dot{z}_{2l} & \dot{z}_{3r} & \dot{z}_{3l} & \dot{z}_{4r} & \dot{z}_{4l} \end{bmatrix}^T$, then $\mathbf{w} = \begin{bmatrix} \mathbf{w}_1 \\ \mathbf{w}_2 \end{bmatrix}$. By defining the state vector as $\mathbf{x} = \begin{bmatrix} \mathbf{q} \\ \dot{\mathbf{q}} \end{bmatrix}$, we can obtain the equations of motion in state-space form as follows

$$\begin{cases} \dot{\mathbf{x}} = \mathbf{A}\mathbf{x} + \mathbf{B}\mathbf{u}(t) + \mathbf{G}\mathbf{w}(t) \\ \mathbf{y} = \mathbf{C}_o\mathbf{x} + \mathbf{v} \end{cases} \tag{16}$$

where $\mathbf{A} = \begin{bmatrix} \mathbf{0} & \mathbf{I} \\ -\mathbf{M}^{-1}\mathbf{K} & -\mathbf{M}^{-1}\mathbf{C} \end{bmatrix}$; $\mathbf{B} = \begin{bmatrix} \mathbf{0} \\ \mathbf{M}^{-1}\mathbf{F}_u \end{bmatrix}$; $\mathbf{G} = \begin{bmatrix} \mathbf{0} \\ \mathbf{M}^{-1}\mathbf{F}_w \end{bmatrix}$; $\mathbf{C}_o = \begin{bmatrix} -\mathbf{M}^{-1}\mathbf{K} & -\mathbf{M}^{-1}\mathbf{C} \end{bmatrix}$. $\mathbf{A} \in R^{18 \times 18}$; $\mathbf{B} \in R^{18 \times 4}$; $\mathbf{C}_o \in R^{9 \times 18}$; $\mathbf{G} \in R^{18 \times 16}$; $\mathbf{y}$ is the output vector; $\mathbf{v}$ is the sensor noise vector.

SEMI-ACTIVE CONTROLLER DESIGN

In order to evaluate the effectiveness of the suspension system with MR dampers, the Linear Quadratic Gaussian (LQG) control algorithm is employed. The performance index is chosen as

$$J = \lim_{t_f \to \infty} \frac{1}{t_f} E\left\{ \int_0^{t_f} \left[\mathbf{x}^T(t)\mathbf{Q}\mathbf{x}(t) + \mathbf{u}(t)^T\mathbf{R}\mathbf{u}(t) \right] dt \right\} \tag{17}$$

where $\mathbf{Q}$ and $\mathbf{R}$ are symmetric semi-positive-definite and positive-definite matrices. The control law that minimizes Equation (17) is given by

$$\mathbf{u}^*(t) = -\mathbf{K}\hat{\mathbf{x}}(t) \tag{18}$$

where $\mathbf{K} = \mathbf{R}^{-1}\mathbf{B}^T\mathbf{S}$, $\mathbf{S}$ is determined by $\mathbf{S}\mathbf{B}\mathbf{R}^{-1}\mathbf{B}^T\mathbf{S} - \mathbf{S}\mathbf{A} - \mathbf{A}^T\mathbf{S} = \mathbf{Q}$. $\hat{\mathbf{x}}(t)$ is obtained from the following Kalman estimator [7]

$$\dot{\hat{\mathbf{x}}}(t) = \mathbf{A}\hat{\mathbf{x}}(t) + \mathbf{B}\mathbf{u}(t) + \mathbf{K}_f\left[\mathbf{y}(t) - \mathbf{C}_o\hat{\mathbf{x}}(t) \right] \tag{19}$$

where $\mathbf{K}_f = \mathbf{S}_f\mathbf{C}_o^T\mathbf{R}_f^{-1}$, $\mathbf{S}_f$ is determined by $\mathbf{S}_f\mathbf{C}_o^T\mathbf{R}_f^{-1}\mathbf{C}_o\mathbf{S}_f - \mathbf{A}\mathbf{S}_f - \mathbf{S}_f\mathbf{A}^T = \mathbf{G}\mathbf{Q}_f\mathbf{G}^T$. The noise covariance matrices $\mathbf{Q}_f = E(\mathbf{w}\mathbf{w}^T)$ and $\mathbf{R}_f = E(\mathbf{v}\mathbf{v}^T)$ are symmetric, semi-positive-definite and positive-definite respectively.

The desired damping force of the MR damper can be obtained from the LQG controller (Equation 18). However, these forces cannot be commanded directly to the MR dampers that are semi-active devices, thus the following method is proposed

$$v = \frac{1}{2N} \sum_{1 \le i \le N} \left\{ \text{sgn}\left[c \frac{f_d - (1 \pm ki)f_a}{f_a} \right] + 2N \right\} \tag{20}$$

where $\text{sgn}(\cdot)$ is the signum function; f_d and f_a are desired force and measured actual force of the MR damper, respectively; $1 \le i \le N$; N is a positive integer; c and k are constants. In this study, $N = 6$, $c = 10^{-10}$, $k = 0.0005$ are used.

SIMULATION RESULTS AND DISCUSSIONS

In the simulation, the elements of the weighting matrix $\mathbf{Q}(i,j) = 0$ for $i = 1 \sim 18$, $j = 1 \sim 18$ except $\mathbf{Q}(1,1) = 8 \times 10^8$, $\mathbf{Q}(2,2) = 8 \times 10^8$, $\mathbf{Q}(3,3) = 8 \times 10^8$, $\mathbf{Q}(13,13) = 100$, $\mathbf{Q}(16,16) = 100$ and the weighting matrix $\mathbf{R} = 0.01\mathbf{I}_{4 \times 4}$, $\mathbf{Q}_f = 20\mathbf{I}_{16 \times 16}$, $\mathbf{R}_f = \mathbf{I}_{9 \times 9}$ ($\mathbf{I}$ is the identity matrix and its subscript represents the corresponding dimension). The parameters of a railway vehicle are given in TABLE II and the parameters of the MR damper model are shown in TABLE I. It is assumed that the wheels follow the rail perfectly. Two track irregularities are considered here: one is the random track irregularity and the other is the periodical track irregularity.

The power spectrum densities (PSD) of vertical, pitch and roll accelerations of the car body under random track irregularities are illustrated in Figure 4. Figure 5 shows the acceleration responses $\ddot{z}_{fr}$ and $\ddot{z}_{rl}$ at two passenger points of the car body, which are chosen to evaluate the ride quality. The first position is at $(x_d, y_d, z_d) = (9 \text{ m}, 0.75 \text{ m}, -0.2 \text{ m})$ in the right side of the front car body, and the second position is at $(-x_d, -y_d, z_d) = (-9 \text{ m}, -0.75 \text{ m}, -0.2 \text{ m})$ in the left side of the rear car body. Observing Figures 4 and 5, the car body accelerations are significantly attenuated through the semi-active control compared to those with constant 12 voltage (passive on) to the MR dampers.

TABLE II. PARAMETERS FOR RAILWAY VEHICLE MODEL

Parameter	Value	Parameter	Value	Parameter	Value
m_c	39.6 t	m_t	3.25 t	V	200 km/h
I_{cx}	88.5 t.m^2	I_{cy}	2460 t.m^2	I_{cz}	2505 t.m^2
I_{tx}	3.06 t.m^2	I_{ty}	3.02 t.m^2	I_{tz}	4.27 t.m^2
k_{px}	4.0 MN/m	k_{py}	3.25 MN/m	k_{pz}	0.7 MN/m
c_{px}	0	c_{py}	0	c_{pz}	15.00 kNs/m
k_{sx}	0.15 MN/m	k_{sy}	0.15 MN/m	k_{sz}	0.29 MN/m
c_{sx}	0	c_{sy}	50.00 kNs/m	c_{sz}	0
h_{ts}	0.217 m	h_{cs}	1.207 m	h_{tp}	-0.452 m
h_{wp}	0.180 m	l	9 m	a	0.7465 m
d_s	1 m	b	1.25 m	d_p	1 m

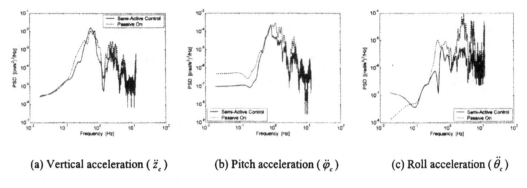

(a) Vertical acceleration ($\ddot{z}_c$)   (b) Pitch acceleration ($\ddot{\varphi}_c$)   (c) Roll acceleration ($\ddot{\theta}_c$)

Figure 4. Acceleration responses of car body under random track irregularities

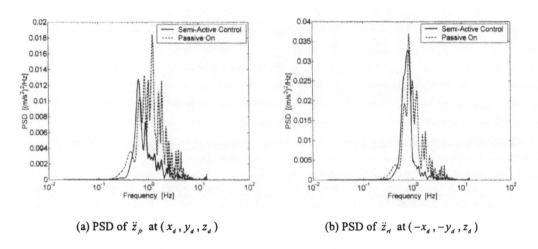

(a) PSD of $\ddot{z}_{fr}$ at (x_d, y_d, z_d)   (b) PSD of $\ddot{z}_{rl}$ at ($-x_d, -y_d, z_d$)

Figure 5. Acceleration responses at passenger points under random track irregularities

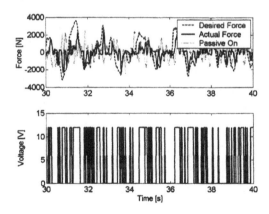

Figure 6. Damping forces and command voltage for u_{szlr} under random track irregularities

In Figure 6, the damping forces and the corresponding command voltage for the MR damper u_{szlr} under random track irregularities are plotted. It can be seen that the actual force of the MR damper can follow the desired damping force on the whole. When the MR damper is in passive on mode (constant 12 V), the damping force of the MR damper has more variation from the desired damping force. Therefore, the semi-active control is superior to the passive on mode of the MR damper, even the controlled damping force cannot perfectly follow the desired damping force.

If the excitation of the front wheelset of the leading truck is represented by z_{1l} and z_{1r}, the periodical track irregularities can be expressed as

$$\begin{bmatrix} z_{1r}(t) \\ z_{1l}(t) \end{bmatrix} = \begin{bmatrix} \dfrac{4A}{\pi}\left[\dfrac{1}{3}\cos\Omega x - \dfrac{1}{15}\cos 2\Omega x + \dfrac{1}{35}\cos 3\Omega x\right] \\ \dfrac{4A}{\pi}\left[\dfrac{1}{3}\cos\Omega x - \dfrac{1}{15}\cos 2\Omega x + \dfrac{1}{35}\cos 3\Omega x\right] \end{bmatrix} \tag{21}$$

where A is the scalar factor of the periodical irregularities of the track, $\Omega = \dfrac{2\pi}{L}$ rad/m is the spatial frequency, L is the spatial length, and $x = Vt$. In this paper $A = 25.4$ and $L = 25$ m. In addition, the other wheelsets $z_{2r} \sim z_{4r}$, $z_{2l} \sim z_{4l}$ are related by

$$\begin{bmatrix} z_{2r}(t) \\ z_{2l}(t) \end{bmatrix} = \begin{bmatrix} z_{1r}(t-\tau_1) \\ z_{1l}(t-\tau_1) \end{bmatrix}; \begin{bmatrix} z_{3r}(t) \\ z_{3l}(t) \end{bmatrix} = \begin{bmatrix} z_{1r}(t-\tau_2) \\ z_{1l}(t-\tau_2) \end{bmatrix}; \begin{bmatrix} z_{4r}(t) \\ z_{4l}(t) \end{bmatrix} = \begin{bmatrix} z_{1r}(t-\tau_3) \\ z_{1l}(t-\tau_3) \end{bmatrix} \tag{22}$$

where $\tau_1 = \dfrac{V}{2b}$, $\tau_2 = \dfrac{V}{2l}$, and $\tau_3 = \dfrac{V}{2(b+l)}$. From Equations (21)~(22), the disturbance vector **w** in Equation (15) can be determined.

Figure 7 shows the time responses of the car body acceleration under the given periodical track irregularities. The controlled accelerations of the car body in the vertical and pitch motions are greatly reduced compared to the passive on cases. However, the roll accelerations of the car body are found to be very small compared to the pitch ones. It is because the given periodical track irregularities do not include cross level track irregularities. While asynchronous tract irregularities are considered, the control effects on the roll motion of the car body would be more significant as shown in Figure 4(c) under the random excitation.

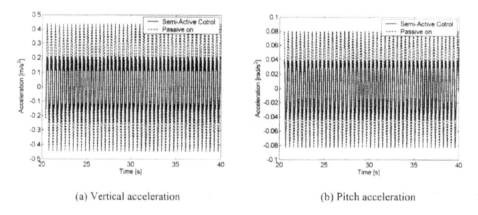

(a) Vertical acceleration   (b) Pitch acceleration

Figure 7. Time responses of car body acceleration under periodical track irregularities

CONCLUSION

In this paper, a semi-active train suspension system with MR dampers has been investigated by considering a full-size railway vehicle, which includes three vibration motions (vertical, pitch and roll) of the car body and trucks. The governing equations of a nine degree-of-freedom railway vehicle model integrated with MR dampers are developed. The performances of the suspension system with MR dampers in passive on and control modes are evaluated. The simulation results show that the proposed means can significantly reduce the accelerations of the car body. However, this paper only presents some preliminary research results that verify the feasibility of applying MR dampers to train suspension systems, more thorough research work is under investigation.

ACKNOWLEDGEMENT

The work described in this paper was fully supported by a grant from Research Grants Council of Hong Kong Special Administrative Region, China (Project No. CUHK4216/01E).

REFERENCES

1. Sims, N. D., R. Stanway, and A. R. Johnson. 1999. "Vibration Control Using Smart Fluids: A State-of-the-Art Review," *The Shock and Vibration Digest*, 31(3): 195-203.
2. Wereley, N. M. and L. Pang. 1998. "Nondimensional Analysis of Semi-Active Electrorheological and Magnetorheological Dampers Using Approximate Parallel Plate Models," *Smart Materials and Structures*, Vol. 7, pp. 732-743.
3. Peel, D. J., R. Stanway, and W. A.. Bullough. 1996. "Dynamic Modeling of an ER Vibration Damper for Vehicle Suspension Applications," *Smart Materials and Structures*, Vol. 5, pp. 591–606.
4. Choi, S. B., B. K. Lee, M. H. Nam, and C. C. Cheong. 2000. "Vibration Control of a MR Seat Damper for Commercial Vehicles," *Proceedings of SPIE Conference on Smart Structures and Materials: Smart Structures and Integrated Systems*, Vol. 3985, pp. 491-496.
5. Simon, D. and M. Ahmadian. 2001. "Vehicle Evaluation of the Performance of Magneto-Rheological Dampers for Heavy Truck Suspensions," *Journal of Vibration and Acoustics*, 123(7): 365-375.
6. Spencer Jr, B. F., S. J. Dyke, M. K. Sain, and J. D. Carlson. 1997. "Phenomenological Model for Magneto-Rheological Dampers," *Journal of Engineering Mechanics*, 123 (3): 230-238.
7. Siouris, G. M. 1996. *An Engineering Approach to Optimal Control and Estimation Theory*. John Wiley & Sons, Inc., New York, USA.

APPENDIX I: NOMENCLATURE

m_c	Mass of car body	I_{cx}	Roll moment of inertia of car body
I_{cy}	Pitch moment of inertia of car body	I_{cz}	Yaw moment of inertia of car body
m_t	Mass of truck	I_{tx}	Roll moment of inertia of truck
I_{ty}	Pitch moment of inertia of truck	I_{tz}	Yaw moment of inertia of truck
k_{px}	Primary longitudinal stiffness	k_{py}	Primary lateral stiffness
k_{pz}	Primary vertical stiffness	c_{px}	Primary longitudinal damping
c_{py}	Primary lateral damping	c_{pz}	Primary vertical damping
k_{sx}	Secondary longitudinal stiffness	k_{sy}	Secondary lateral stiffness
k_{sz}	Secondary vertical stiffness	c_{sx}	Secondary longitudinal damping
c_{sy}	Secondary lateral damping	c_{sz}	Secondary vertical damping
l	Half of truck center pin spacing	b	Half of wheelbase

a	Half of wheelset contact distance	d_p	Half of primary suspension spacing (lateral)

d_s Half of secondary suspension spacing (lateral)

h_{ts} Vertical distance from truck frame center of gravity to secondary suspension

h_{cs} Vertical distance from car body center of gravity to secondary suspension

h_{tp} Vertical distance from truck frame center of gravity to primary suspension

h_{wp} Vertical distance from wheelset center of gravity to primary suspension

V Velocity of railway vehicle

APPENDIX II: DEFINITIONS OF F IN EQUATIONS (1)~(9)

$$F_{sxlr} = -k_{sx}(h_{cs}\varphi_c + h_{ts}\varphi_{t1}) - c_{sx}(h_{cs}\dot{\varphi}_c + h_{ts}\dot{\varphi}_{t1}); \quad F_{sxll} = -k_{sx}(h_{cs}\varphi_c + h_{ts}\varphi_{t1}) - c_{sx}(h_{cs}\dot{\varphi}_c + h_{ts}\dot{\varphi}_{t1})$$

$$F_{sxtr} = -k_{sx}(h_{cs}\varphi_c + h_{ts}\varphi_{t2}) - c_{sx}(h_{cs}\dot{\varphi}_c + h_{ts}\dot{\varphi}_{t2}); \quad F_{sxtl} = -k_{sx}(h_{cs}\varphi_c + h_{ts}\varphi_{t2}) - c_{sx}(h_{cs}\dot{\varphi}_c + h_{ts}\dot{\varphi}_{t2})$$

$$F_{sylr} = k_{sy}(h_{cs}\theta_c + h_{ts}\theta_{t1}) + c_{sy}(\dot{h}_{cs}\dot{\theta}_c + h_{ts}\dot{\theta}_{t1}); \quad F_{syll} = k_{sy}(h_{cs}\theta_c + h_{ts}\theta_{t1}) + c_{sy}(\dot{h}_{cs}\dot{\theta}_c + h_{ts}\dot{\theta}_{t1})$$

$$F_{sytr} = k_{sy}(h_{cs}\theta_c + h_{ts}\theta_{t2}) + c_{sy}(\dot{h}_{cs}\dot{\theta}_c + h_{ts}\dot{\theta}_{t2}); \quad F_{sytr} = k_{sy}(h_{cs}\theta_c + h_{ts}\theta_{t2}) + c_{sy}(\dot{h}_{cs}\dot{\theta}_c + h_{ts}\dot{\theta}_{t2})$$

$$F_{szlr} = -k_{sz}[z_c - l\varphi_c - z_{t1} + d_s(\theta_c - \theta_{t1})] - c_{sz}[\dot{z}_c - l\dot{\varphi}_c - \dot{z}_{t1} + d_s(\dot{\theta}_c - \dot{\theta}_{t1})]$$

$$F_{szll} = -k_{sz}[z_c - l\varphi_c - z_{t1} - d_s(\theta_c - \theta_{t1})] - c_{sz}[\dot{z}_c - l\dot{\varphi}_c - \dot{z}_{t1} - d_s(\dot{\theta}_c - \dot{\theta}_{t1})]$$

$$F_{sztr} = -k_{sz}[z_c + l\varphi_c - z_{t2} + d_s(\theta_c - \theta_{t2})] - c_{sz}[\dot{z}_c + l\dot{\varphi}_c - \dot{z}_{t2} + d_s(\dot{\theta}_c - \dot{\theta}_{t2})]$$

$$F_{sztl} = -k_{sz}[z_c + l\varphi_c - z_{t2} - d_s(\theta_c - \theta_{t2})] - c_{sz}[\dot{z}_c + l\dot{\varphi}_c - \dot{z}_{t2} - d_s(\dot{\theta}_c - \dot{\theta}_{t2})]$$

$$F_{px1r} = -k_{px}h_{tp}\varphi_{t1} - c_{px}h_{tp}\dot{\varphi}_{t1}; \quad F_{px1l} = -k_{px}h_{tp}\varphi_{t1} - c_{px}h_{tp}\dot{\varphi}_{t1}; \quad F_{px2r} = -k_{px}h_{tp}\varphi_{t1} - c_{px}h_{tp}\dot{\varphi}_{t1}$$

$$F_{px2l} = -k_{px}h_{tp}\varphi_{t1} - c_{px}h_{tp}\dot{\varphi}_{t1}; \quad F_{px3r} = -k_{px}h_{tp}\varphi_{t2} - c_{px}h_{tp}\dot{\varphi}_{t2}; \quad F_{px3l} = -k_{px}h_{tp}\varphi_{t2} - c_{px}h_{tp}\dot{\varphi}_{t2}$$

$$F_{px4r} = -k_{px}h_{tp}\varphi_{t2} - c_{px}h_{tp}\dot{\varphi}_{t2}; \quad F_{px4l} = -k_{px}h_{tp}\varphi_{t2} - c_{px}h_{tp}\dot{\varphi}_{t2}$$

$$F_{py1r} = k_{py}h_{tp}\theta_{t1} + c_{py}h_{tp}\dot{\theta}_{t1}; \quad F_{py1l} = k_{py}h_{tp}\theta_{t1} + c_{py}h_{tp}\dot{\theta}_{t1}; \quad F_{py2r} = k_{py}h_{tp}\theta_{t1} + c_{py}h_{tp}\dot{\theta}_{t1}; \quad F_{py2l} = k_{py}h_{tp}\theta_{t1} + c_{py}h_{tp}\dot{\theta}_{t1}$$

$$F_{py3r} = k_{py}h_{tp}\theta_{t2} + c_{py}h_{tp}\dot{\theta}_{t2}; \quad F_{py3l} = k_{py}h_{tp}\theta_{t2} + c_{py}h_{tp}\dot{\theta}_{t2}; \quad F_{py4r} = k_{py}h_{tp}\theta_{t2} + c_{py}h_{tp}\dot{\theta}_{t2}; \quad F_{py4l} = k_{py}h_{tp}\theta_{t2} + c_{py}h_{tp}\dot{\theta}_{t2}$$

$$F_{pz1r} = -k_{pz}(z_{t1} - b\varphi_{t1} + d_p\theta_{t1} - d_p/a\, z_{1r}) - c_{pz}(\dot{z}_{t1} - b\dot{\varphi}_{t1} + d_p\dot{\theta}_{t1} - d_p/a\, \dot{z}_{1r})$$

$$F_{pz1l} = -k_{pz}(z_{t1} - b\varphi_{t1} - d_p\theta_{t1} - d_p/a\, z_{1l}) - c_{pz}(\dot{z}_{t1} - b\dot{\varphi}_{t1} - d_p\dot{\theta}_{t1} - d_p/a\, \dot{z}_{1r})$$

$$F_{pz2r} = -k_{pz}(z_{t1} + b\varphi_{t1} + d_p\theta_{t1} - d_p/a\, z_{2r}) - c_{pz}(\dot{z}_{t1} + b\dot{\varphi}_{t1} + d_p\dot{\theta}_{t1} - d_p/a\, \dot{z}_{2r})$$

$$F_{pz2l} = -k_{pz}(z_{t1} + b\varphi_{t1} - d_p\theta_{t1} - d_p/a\, z_{2l}) - c_{pz}(\dot{z}_{t1} + b\dot{\varphi}_{t1} - d_p\dot{\theta}_{t1} - d_p/a\, \dot{z}_{2l})$$

$$F_{pz3r} = -k_{pz}(z_{t2} - b\varphi_{t2} + d_p\theta_{t2} - d_p/a\, z_{3r}) - c_{pz}(\dot{z}_{t2} - b\dot{\varphi}_{t2} + d_p\dot{\theta}_{t2} - d_p/a\, \dot{z}_{3r})$$

$$F_{pz3l} = -k_{pz}(z_{t2} - b\varphi_{t2} - d_p\theta_{t2} - d_p/a\, z_{3l}) - c_{pz}(\dot{z}_{t2} - b\dot{\varphi}_{t2} - d_p\dot{\theta}_{t2} - d_p/a\, \dot{z}_{3l})$$

$$F_{pz4r} = -k_{pz}(z_{t2} + b\varphi_{t2} + d_p\theta_{t2} - d_p/a\, z_{4r}) - c_{pz}(\dot{z}_{t2} + b\dot{\varphi}_{t2} + d_p\dot{\theta}_{t2} - d_p/a\, \dot{z}_{4r})$$

$$F_{pz4l} = -k_{pz}(z_{t2} + b\varphi_{t2} - d_p\theta_{t2} - d_p/a\, z_{4l}) - c_{pz}(\dot{z}_{t2} + b\dot{\varphi}_{t2} - d_p\dot{\theta}_{t2} - d_p/a\, \dot{z}_{4l})$$

TESTING AND MODELING OF MAGNETORHEOLOGICAL VIBRATION ISOLATORS

Young-Sik Jeon, Young-Tai Choi, and Norman M. Wereley

ABSTRACT

This paper presents experimental and theoretical analysis of a vibration isolation system using MR fluid-based semi-active isolators. In doing so, a vibration isolator using MR fluids is designed and manufactured in this study. A new nonlinear hysteresis model with simplicity in form is proposed to describe the hysteresis force characteristics of the MR isolator. The damping forces of the MR isolator with different excitation frequency and current input are measured and compared with that resulting from the hysteresis model for the verification of the theoretical analysis. After then, a vibration isolation system with the MR isolator is constructed and its dynamic equation of motion is derived. From the equation, a simple skyhook controller is formulated to attenuate the vibration of the system. Controlled performances of the vibration isolation system are experimentally and theoretically evaluated in frequency and time domains.

INTRODUCTION

Vibration isolators are used to isolate systems from various sources, so that systems should not be exposed to significant dynamic stress and fatigue damage. There are three types of potential vibration isolators: passive, semi-active and active. The passive isolators featuring elastic materials and hydraulic oils provide design simplicity and cost-effectiveness. However, performance limitations are inevitable. On the other hand, the active isolators featuring electromagnetic, servovalve and piezoceramics may provide high control performance in wide frequency range. However, the active isolators require high power sources and complex configurations. Recently, in order to solve the disadvantage of the active isolators, the semi-active isolators have been introduced. Among them, we focus on magnetorheological (MR) fluid-based semi-active isolators. MR isolators are known to have advantages such as continuously damping force and fast response. However, MR isolators exhibit nonlinear force hysteresis characteristics dependent on excitation frequency and current input.

So far, Bingham model is widely used to evaluate the behavior of MR fluid-based applications, because of its simplicity in form. However, the Bingham plastic model cannot explain the hysteresis characteristics of MR isolators. Therefore, a nonlinear hysteresis model is needed to accurately capture the force behavior of MR isolators. For theoretical modeling of nonlinear hysteresis characteristics of MR dampers, a few literatures have been already reported [1-4]. However, even though it is able to model the hysteresis characteristics of MR dampers, it is difficult to directly apply nonlinear hysteresis models to control applications

Young-Sik Jeon, Visiting Professor, University of Maryland, College Park. MD 20742, USA
Young-Tai Choi, Research Assistant, University of Maryland, College Park, MD 20742, USA
Norman M. Wereley, Associate Professor, University of Maryland, College Park, MD 20742, USA

since its form is relatively complex and not directly correlated with control current input. Consequently, the main contribution of this paper is to propose a nonlinear hysteresis MR isolator model having simplicity and accuracy, and study its impact on the performance predictions of MR isolators in open and closed loop time and frequency domains.

In this study, a vibration isolation system using MR fluid-based semi-active isolators is investigated through experimental and theoretical works. In doing so, a MR fluid-based vibration isolator to cope with the vibration isolation problem of elevator systems is designed and manufactured in this study. A new nonlinear hysteresis model with simplicity in form is proposed to describe the hysteresis force characteristics of the MR isolator. The damping forces of the MR isolator with different excitation frequency and current input are measured and compared with that resulting from the hysteresis model for the verification of the theoretical analysis. After then, a vibration isolation system with the MR isolator is constructed and its dynamic equation of motion is derived. From the equation, a simple skyhook controller is formulated to attenuate the vibration of the system. Controlled performances of the vibration isolation system under constant current input and simple skyhook controller are experimentally and theoretically evaluated in frequency and time domains on the basis of the nonlinear hysteresis model.

MR VIBRATION ISOLATORS

MR isolators can produce the damping force in response to magnetic field proportional to applied current input and the magnitude of the damping force can be continuously controlled by just changing the current input. Therefore, MR isolators have additional advantage like simple mechanical configuration other than continuous force control ability and fast response. A MR isolator designed and manufactured in this study is shown in Figure 1. This MR isolator was designed to have the damping level appropriate for coping with the vibration isolation problem of the elevator system. For light weight of the isolator, most parts of the isolator were manufactured from the aluminum material other than the magnetic activation areas. In addition, the size of the MR isolator is compact as 79 mm height and 69 mm outer diameter.

Figure 1. Photograph of MR isolator.   Figure 2. Experimental setup for damping force test of MR isolator.

Damping Force Test

For the damping force test of the MR isolator, experimental setup is constructed as shown in Figure 2. The MR isolator is placed between the load cell and excitation actuator. The load cell is fixed and the bottom part of the MR isolator is moved through the excitation actuator. In the absence of current input, the MR isolator just produces passive damping forces due to fluid viscosity and seal friction. However, upon applying current input to the MR isolator, the MR isolator produces additional damping force due to the yield stress of the MR fluid. The additional damping force can be continuously controlled according to applied current input. In this study, the excitation input is used to be the sinusoidal function with constant displacement amplitude and single excitation frequency. In addition, in case of the vibration isolation of the elevator system, the displacement amplitude of the excitation is as small as 1-2 mm. In this study, the displacement amplitude is 1.3 mm.

Theoretical Modeling

We will consider two theoretical models: (1) Bingham plastic model which ignores preyield compression loop and (2) new nonlinear hysteresis model. On the basis of the Bingham plastic model, the damping force of the MR isolator can be expressed as follows:

$$f_d = c\dot{x}_e + kx_e + (f_y + f_f)\,\text{sgn}(\dot{x}_e) \tag{1}$$

Here, f_d is the damping force of the MR isolator, c is the viscous damping, k is the stiffness due to gas, f_f is the seal friction force, and f_y is the yield force due to MR fluids of which magnitude is exponentially proportional to applied current input. In addition, x_e is the excitation displacement and sgn($\cdot$) is the signum function. The typical behavior of the Bingham plastic model in force versus piston velocity plot is shown in Figure 3a.

On the other hand, the damping force of the MR isolator can be also expressed by means of new nonlinear hysteresis model proposed in this study as follows:

$$f_d = c\dot{x}_e + kx_e + (f_y + f_f)\,\tanh\{(\dot{x}_e + \lambda_1 x_e)\lambda_2\} \tag{2}$$

Here, tanh($\cdot$) is the hypertangent function, λ_1 and λ_2 are characteristic parameters of the

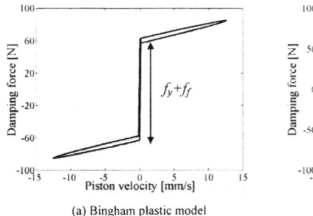
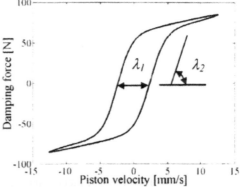

(a) Bingham plastic model　　　　　　(b) nonlinear hysteresis model

Figure 3. Typical behavior of theoretical models in force versus piston velocity plot.

nonlinear hysteresis model to capture the hysteresis loop. The parameter λ_1 accounts for the width of the hysteresis loop in force vs. piston velocity and λ_2 accounts for the slope of the hysteresis loop (see Figure 3b). It is noted that the larger the parameter λ_1, the wider the width of the hysteresis loop. In addition, the larger the parameter λ_2, the steeper the slope of the hysteresis loop. To establish the hysteresis model structure, the parameters λ_1 and λ_2 are identified by curve fitting method. The identified parameters are plotted over excitation frequency and constant current input in Figure 4. As shown in this figure, the parameters λ_1 and λ_2 are strong functions of excitation frequency, but weak functions of current input.

Experimental Verifications

Figure 5 presents the damping force versus piston velocity of the MR isolator. In this figure, the solid line stands for experimental data and dashed line is theoretical data. And the data were collected at the conditions such as excitation frequency of 5.5 Hz and constant current input of 0.4 A. As clearly observed, the Bingham plastic model cannot capture hysteresis behavior of the MR isolator. However, the nonlinear hysteresis model proposed in this study captures the behavior relatively well. It is noted that even though the Bingham plastic model cannot capture the hysteresis loop, but predict the magnitude of the damping force. Because of this fact, the Bingham plastic model could be widely used in basic performance predictions of ER/MR applications.

Figure 6 presents the damping force versus piston velocity at the different conditions that excitation frequency is 8.5 Hz and constant current inputs are 0.2 A and 1.0 A. As shown in this figure, the nonlinear hysteresis model also well captures the behavior of the MR isolator at different conditions. Similar results are obtained at the damping force versus piston displacement shown in Figure 7. Therefore, these results validate that the nonlinear hysteresis model proposed in this study is able to more accurately capture the behavior of the MR isolator than Bingham plastic model.

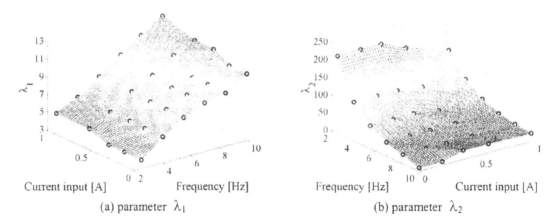

(a) parameter λ_1 (b) parameter λ_2

Figure 4. The identified parameters of the nonlinear hysteresis model.

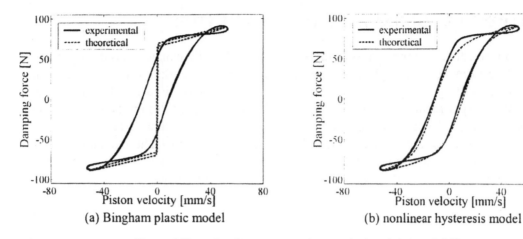

Figure 5.Damping force versus piston velocity (0.4 A at 5.5 Hz).

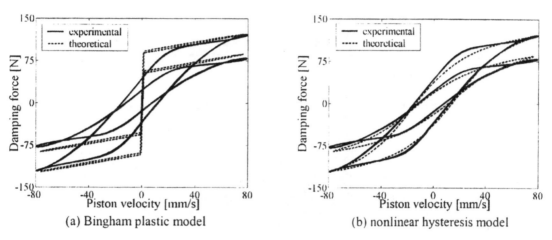

Figure 6.Damping force versus piston velocity at the different conditions (0.2 A and 1.0 A at 8.5 Hz).

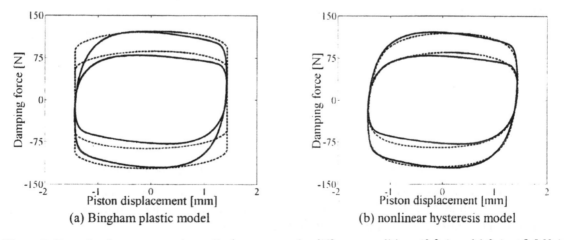

Figure 7. Damping force versus piston displacement at the different conditions (0.2 A and 1.0 A at 8.5 Hz).

VIBRATION ISOLATION SYSTEM WITE MR ISOLATOR

We now construct a vibration isolation system with the MR isolator in order to evaluate its controlled performances through experimental and theoretical works. In theoretical work, two models such as Bingham plastic and nonlinear hysteresis models are still considered. The photograph of experimental setup of SDOF MR vibration isolation system is shown in Figure 8. The MR isolator is placed between system mass and excitation actuator. To prevent the static deflection of the mass, coil spring is installed on the MR isolator. The excitation actuator is driven by hydraulic power system and its excitation displacement is measured by a LVDT. In addition, the acceleration of the mass is measured by an accelerometer.

Mass

MR isolator

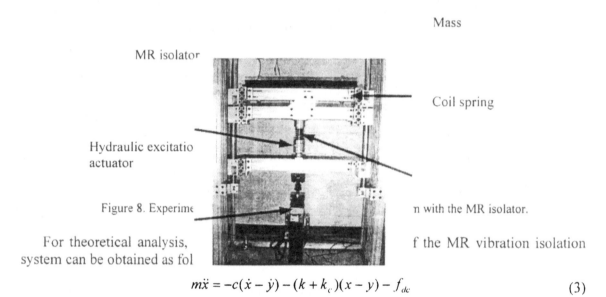

Coil spring

Hydraulic excitatio actuator

Figure 8. Experime   n with the MR isolator.

For theoretical analysis,   f the MR vibration isolation
system can be obtained as fol

$$m\ddot{x} = -c(\dot{x} - \dot{y}) - (k + k_c)(x - y) - f_{dc} \qquad (3)$$

Here, m is the system mass, k_c is the stiffness of the coil spring, x is the displacement of the mass and y is the excitation displacement. In addition, f_{dc} is the controlled damping force and can be expressed by

$$f_{dc} = \begin{cases} (f_{yc} + f_f)\,\mathrm{sgn}(\dot{x} - \dot{y}) & \text{at Bingham plastic model} \\ (f_{yc} + f_f)\tanh\{[(\dot{x} - \dot{y}) + \lambda_1(x - y)]\lambda_2\} & \text{at nonlinear hysteresis model} \end{cases} \qquad (4)$$

Here, $f_{yc} = \alpha i_c^{\beta}$ is controlled yield damping force, α and β are characteristic values of MR fluids, and i_c is applied control current input. In this study, it was chosen by α=80.9 and β=0.38. A simple skyhook control algorithm is adopted in order to improve the vibration isolation performance of the MR vibration isolation system. From the simple skyhook control algorithm, the control current input is determined as follows:

$$i_c = \begin{cases} 0.54\ A & \text{if } \dot{x} \cdot (\dot{x}\text{-}\dot{y}) > 0 \\ 0.0\ A & \text{if } \dot{x} \cdot (\dot{x}\text{-}\dot{y}) \leq 0 \end{cases} \tag{5}$$

It is noted that since the MR vibration isolation system is semi-active type control system, the semi-active actuating condition to ensure the increment of the dissipation energy for the system was superposed in the above [4,5].

CONTROL PERFORMANCES

Figure 9 presents experimental results of the MR vibration isolation system under the constant current input and the simple skyhook controller. The transmissibility computing from root-mean-square values of the accelerations of the mass and excitation is used to predict the performance of the system. As shown in Figure 9a, constant current input can control effectively the vibration of the system in just resonance region. Especially, in the higher frequency region above resonance frequency, constant current input deteriorates the vibration isolation performance of the system. It is due to the fact that the excessive damping in higher frequency range makes a bad effect on the performance of the vibration isolation system. However, from the Figure 9b, the simple skyhook control algorithm can improve the vibration isolation performance of the MR vibration isolation system at overall frequency regions.

Figure 10 presents theoretical results of the MR vibration isolation system on the basis of the Bingham plastic model and nonlinear hysteresis model. As observed in Figure 10a, the Bingham plastic model does not predict well the vibration isolation performance of the MR vibration isolation system at resonance frequency region. However, the nonlinear hysteresis model can accurately predict the vibration isolation performance of the system as shown in Figure 10b.

Figure 11 presents comparison of experimental and theoretical results of the MR vibration isolation system in time domain. Experimental and theoretical results under constant current input are shown in Figure 11a and, for skyhook control input, it is shown in Figure 11b. In this case, control current input is applied to the MR isolator on the time of 1.5 sec and excitation frequency is 3.8 Hz. As clearly observed in this figure, the nonlinear hysteresis model can more accurately capture the overall behavior of the MR vibration isolation system than

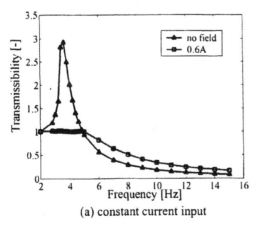

(a) constant current input

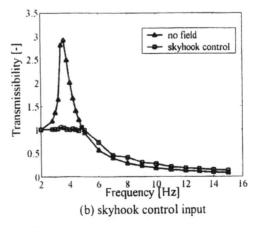

(b) skyhook control input

Figure 9. Experimental results of the MR vibration isolation system.

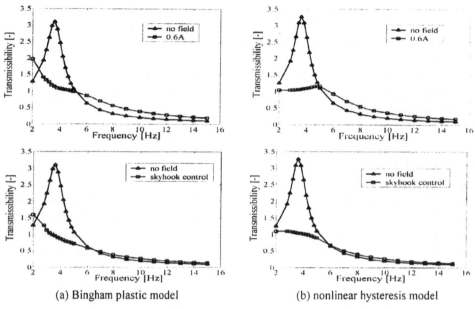

(a) Bingham plastic model   (b) nonlinear hysteresis model

Figure 10. Theoretical results of the MR vibration isolation system.

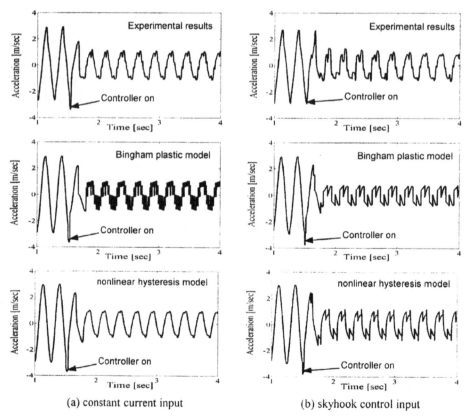

(a) constant current input   (b) skyhook control input

Figure 11. Theoretical results of the MR vibration isolation system.

Bingham plastic model under both constant and skyhook control current inputs. Especially, it is noted that, in case of the skyhook control input, the Bingham plastic model overestimates the controlled performance of the system.

CONCLUSIONS

Experimental and theoretical analysis of a vibration isolation system using MR fluid-based semi-active isolators was presented under constant and simple skyhook control algorithm. A MR vibration isolator applicable to elevator vibration isolation systems was designed and manufactured in this study. To describe the practical hysteresis force characteristics of the MR isolator, a new nonlinear hysteresis model with simplicity in form is proposed. The damping forces of the MR isolator with different excitation frequency and current input are measured and compared with that resulting from the hysteresis model. It was shown that the nonlinear hysteresis model could capture the force hysteresis characteristics of the MR isolator in the presence of current inputs. After then, a vibration isolation system with the MR isolator was constructed and modeled on the basis of two models such as the Bingham plastic model and nonlinear hysteresis model. Controlled performances of the MR vibration isolation system were experimentally and theoretically analyzed in both frequency and time domain. It has been demonstrated that the vibration isolation performance can be much improved by employing the simple skyhook control algorithm and that the nonlinear hysteresis model can predict more accurately the vibration isolation performance than the Bingham plastic model in open and closed loop time and frequency domains.

REFERENCES

1. Kamath, G.M. and N.M. Wereley. 1997. "A Nonlinear Viscoelastic-Plastic Model for Electrorheological Fluids," Smart Materials and Structures, 6(3):351-359.
2. Kamath, G.M., N.M. Wereley and M.R. Jolly. 1999. "Characterization of Magnetorheological Helicopter Lag Dampers," J. the American Helicopter Society, 44(3):234-248.
3. Sims, N.D., D.J. Peel, R. Stanway, A.R. Johnson, and W.A. Bullough. 2000. "The Electrorheological Long-Stroke Damper:A New Modelling Technique With Experimental Validation," J Sound and Vibration, 229(2):207-227.
4. Spencer, B.F., S.J. Dyke, M.K. Sain, and J.D. Carlson. 1997. "Phenomenological Model of a Magnetorheological Damper," ASCE J. Engineering Mechanics, 123(3):230-238.
5. Choi, S.B., Y.T. Choi, and Y.S. Jeon. 1999. "Performance Evaluation of a Mixed Mode ER Engine Mount Via Hardware-in-the-Loop-Simulation," J Intelligent, Material Systems and Structures, 10(8):671-677.

Identification and Control

MODEL TO DETERMINE THE PERFORMANCE REQUIREMENTS OF ADAPTIVE MATERIALS USED TO MORPH A WING WITH AN AERODYNAMIC LOAD

Greg Pettit, Harry Robertshaw, and Daniel J. Inman

ABSTRACT

A computational model is presented which predicts the force, stroke, and energy needed to overcome aerodynamic loads encountered by morphing wings during aircraft maneuvers. This low-cost model generates wing section shapes needed to follow a desired flight path; computes the resulting aerodynamic forces using a unique combination of conformal mapping and the vortex panel method; computes the longitudinal motion of the simulated aircraft; and closes the loop with a zero steady-state-error control law. The aerodynamic force prediction method has been verified against two more expensive codes. This overall model will be used to predict the performance of morphing wings and the requirements for the active material actuators in the wings.

INTRODUCTION

This work addresses the energy needs for active materials to be used to morph, to change the shape of, wings envisioned for future aircraft. Eliminating discrete control surfaces on airfoils in favor of subtly reshaping the airfoil to perform maneuvers is an attractive idea. Assuming that these shape changes can be produced, this work presents a tool that will allow us to predict the force, stroke, and energy required for a distributed system of active material actuators.

While there are two important energy considerations, strain energy needed to morph the structure and energy needed to overcome the aerodynamic forces, this work is concerned with the latter. We cannot predict the configuration of the active materials that will make up the morphing wing; therefore, we are not able to predict how much energy will be required to strain the wing structure. We can envision some designs that will have very little strain energy in the structure after morphing. Those designs could have variable length links or members that cannot be back driven by wing forces. We can also envision some designs that will have significant amounts of strain energy in the structure. These designs could have actuators straining the structure of a typical wing box. Therefore, we are addressing the aerodynamic loads only.

Center for Intelligent Material System and Structures, Virginia Polytechnic Institute and State University, Blacksburg, VA 24061-0261

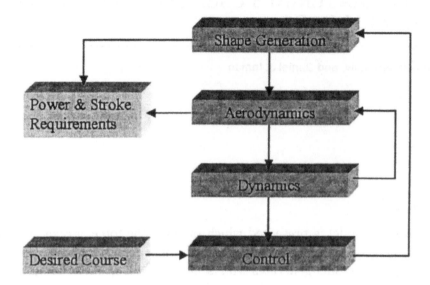

Figure 1: Block diagram of model and layout of paper.

MODEL

The model is broken into four major parts: shape generation, aerodynamics, dynamics, and control, Figure 1. When the four parts are integrated together the model becomes a versatile tool that allows for the input of a desired course with the output being time and power required by the actuators to fly the course. The model is unique for three reasons. First, the shape generation allows for a vast array of shapes suitable for flight along with an easy method of changing this shape. Second, the aerodynamics uses the combination of two methods that will allow for the air loads to quickly and accurately be determined for a wing with a finite span. Third, the air loads can be found on the upper and lower surfaces not just on a mean camber line, which aids in the calculation of the energy required to change the shape.

Shape Generation

In order to meet the needs of a planform that can quickly and easily change shapes a method of conformal mapping was chosen. The method allows for many parameters of the planform to be changed on the go such as: thickness, camber, reflex, chord, taper, sweep, dihedral, and twist.

The method begins by mapping a unit circle into the shape of an airfoil section. This mapping is achieved through four steps beginning with the complex unit circle, s, and ending with the airfoil section, ζ. Figure 2a shows the results from the four transformations. While the basic mapping known as Joukowski's transformation has been around for a long time, it has been expanded to incorporate many NACA airfoils [1].

The wing is then generated by using the mapping technique described above in succession along each span wise location, η, of the wing. In order to produce the desired chord, taper, sweep, dihedral, and twist the mapping results can be scaled, shifted, and rotated at each span wise location. Figure 2b shows the result of one such created wing.

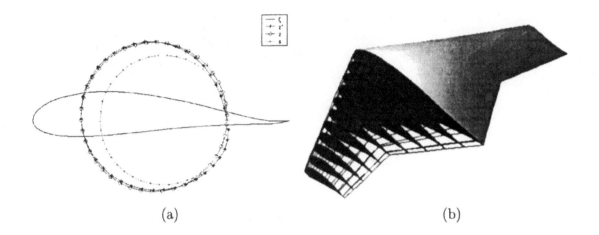

(a)   (b)

Figure 2: Wing created from a conformal mapping technique: (a) wing section created from conformal mapping; and (b) wing created by using the mapping process at each spanwise station.

Aerodynamics

There are several methods to model the flow around a wing. Each method usually does one thing well but will lack in other areas. In addition, these different methods vary greatly in computation time. Many useful methods such as the vortex lattice method are based on thin airfoil theory. This thin airfoil theory implies the results are generally based on a mean camber line and not on the outer surface of the wing. Since the end goal of this model is to predict the amount of work necessary to change the shape of the wing, we must know the forces on the outer surface of the wing at different localized areas. For these reasons we have chosen to use a combination of three methods: a vortex lattice method, potential flow around a circle, and conformal mapping [2].

To begin, the vortex lattice method is used to capture the three dimensional flow effects around the wing [3]. The three dimensional flow effects are caused by the high pressure on the bottom surface of the wing trying to equalize with the low pressure on the upper surface. This equalization causes a vortex which induces a downward velocity on the wing as shown in Figure 3, which in turn causes a loss of lift. The vortex lattice method then uses a set of linear equations to describe this velocity induced by each vortex at certain locations on the wing called control points. Then applying the boundary conditions, flow normal to the wing is equal to zero, the unknown strength of each vortex, Γ, can be determined.

The lift is then directly proportional to the strengths of these vortices as shown in equation 1:

$$l_n = \rho_\infty U_\infty \Gamma_n \tag{1}$$

However, this lift is on the mean camber line and we are concerned with the forces on the outer surface of the wing. Therefore, we continue the aerodynamics by again using a method based on potential flow.

Using potential flow techniques we can describe the around a circle by combining

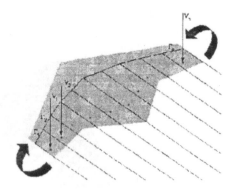

Figure 3: Figure representing the vortex lattice method.

simple flow elements as shown in equation 2 [4]:

$$\frac{d\mathrm{w}}{ds} = V\left(e^{-i\alpha} - \frac{a^2 e^{i\alpha}}{s^2}\right) + \frac{i\Gamma}{2\pi s} \tag{2}$$

where $\frac{d\mathrm{w}}{ds}$ is a complex number representing the velocity of the fluid. These elements making up the equation: free stream velocity, source, sink, and a vortex are represented in Figure 4a. The vortex strength, Γ, is determined from the vortex lattice method while the angle of attack, α, is calculated by applying the Kutta condition which implies there is a stagnation point at the trailing edge of the airfoil. Each wing section was originally created from a unit circle denoted by s and then transformed into the airfoil section by three differentiable functions. Therefore, we can apply the chain rule shown in equation 3 to determine the flow around the wing section as shown in Figure 4b:

$$\frac{d\mathrm{w}}{d\zeta} = \frac{d\mathrm{w}}{ds}\frac{ds}{dz}\frac{dz}{dz'}\frac{dz'}{d\zeta} \tag{3}$$

This process is then repeated for each spanwise station, as explained in the shape generation section.

Dynamics

Once the aerodynamic forces acting on the body have been determined, the rigid body equations of motion can be applied. The dynamics become quite complex when all six degrees of freedom are used. Therefore, for this model we are concerned only with the longitudinal dynamics, which have been decoupled from the lateral dynamics. These equations are well developed in many sources and are represented below [6]:

$$\dot{u}(t) = -q(t)w(t) + \frac{F_x(t, y_c, y_t, \theta_t, T)}{m} \tag{4}$$

$$\dot{w}(t) = q(t)u(t) + \frac{F_z(t, y_c, y_t, \theta_t, T)}{m} \tag{5}$$

$$\dot{q}(t) = \frac{M(t, y_c, y_t, \theta_t, T)}{J_y} \tag{6}$$

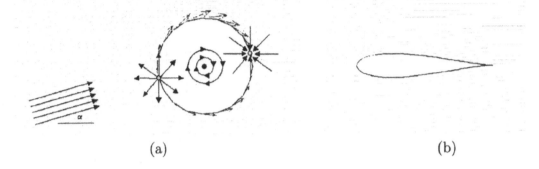

(a)   (b)

Figure 4: Potential flow combined with conformal mapping: (a) potential flow around a circle that is mapped into the wing section; and (b) flow mapped to the wing.

$$\dot{\theta}(t) = q(t) \tag{7}$$

$$\dot{x}_e(t) = u(t)\cos\theta(t) + w(t)\sin\theta(t) \tag{8}$$

$$\dot{z}_e(t) = -u(t)\sin\theta(t) + w(t)\cos\theta(t) \tag{9}$$

$$\dot{y}_c(t) = \frac{U_1(t) - y_c(t)}{\tau_1} \tag{10}$$

$$\dot{y}_t(t) = \frac{U_2(t) - y_t(t)}{\tau_2} \tag{11}$$

$$\dot{\theta}_c(t) = \frac{U_3(t) - \theta_c(t)}{\tau_3} \tag{12}$$

$$\dot{T}(t) = \frac{U_4(t) - T(t)}{\tau_4} \tag{13}$$

The states relative to the rotating coordinate system fixed to the wing are: u and w, velocities in the x and z direction; q, angular velocity; and θ, pitch, while x_e and z_e represent position relative to the inertial coordinate system. The forces and moments that are calculated by the aerodynamics module are functions of four inputs y_c, camber; y_t, reflex; θ_t, twist; and T, thrust. These inputs are the actual inputs used to generate the shape discussed above and are modeled as having first order dynamics with a time constant of τ_n.

Control

The objective of the control system is to track the desired input of x_e, z_e, and θ. In order to achieve this objective a full state feedback controller was used as shown in Figure 5 [5]. Since the reference inputs would most likely be greater than first-order and a zero error controller is desired, the system type was raised by adding integral error feedback [6]. Figure 5 also shows the plant as a linear model; however, when the system is simulated the non-linear model is used with the linear controller determined as follows.

To begin the design of the compensator, the system was first linearized by calculating

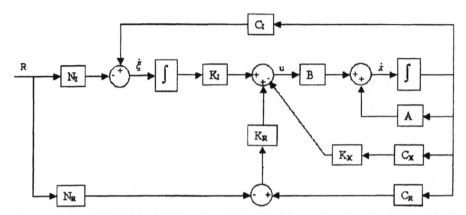

Figure 5: Block diagram of the full state and integral error feedback controller.

the Jacobian about the desired trimmed condition as shown in equations 14 and 15:

$$A = \frac{\partial f}{\partial x}\Big|_{x^o, u^o} \tag{14}$$

$$B = \frac{\partial f}{\partial u}\Big|_{x^o, u^o} \tag{15}$$

With the plant now modeled as a linear time invariant system as shown in Figure 5, we can represent the system in algebraic form so that the new augmented state matrix can be determined. Again referring to Figure 5, the open loop state matrices can be written as:

$$\dot{\xi}(t) = C_I x(t) - N_I R(t) \tag{16}$$
$$\dot{x}(t) = A x(t) + B U(t) \tag{17}$$

In matrix form these equations become:

$$\left\{ \begin{array}{c} \dot{\xi}(t) \\ \dot{x}(t) \end{array} \right\} = \begin{bmatrix} 0 & C_I \\ 0 & A \end{bmatrix} \left\{ \begin{array}{c} \xi(t) \\ x(t) \end{array} \right\} + \begin{bmatrix} 0 \\ B \end{bmatrix} u(t) - \begin{bmatrix} N_I \\ 0 \end{bmatrix} r(t) \tag{18}$$

Using the augmented matrices derived above in equation 18, an lqr algorithm was used to determine the control gains K. While this method does not guarantee the best tracking versus actuator power, it does serve as a good starting point.

Power

With the four modules, shape generation, aerodynamics, dynamics, and control running in closed loop, information from both the shape generation module and the aerodynamics module can be used to determine the necessary power requirements. Three separate power values are of concern. First referring to Figure 6, the power necessary to twist each section, P_η, is calculated from the moment about the twist axis times the angular velocity of the particular section as shown in equation 19:

$$P_\eta = \sum_n (r_n \times F_n)\dot{\theta} \tag{19}$$

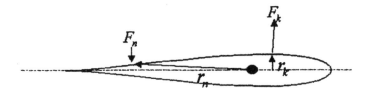

Figure 6: Model used to calculate the power requirements at each section.

Second, power to morph any location, k, on the wing is calculated by the dot product between the force and velocity of the given section as shown in equation 20:

$$P_k = F_k \bullet \dot{r}_k \tag{20}$$

The velocity of point k is determined relative to a plane that runs from the leading edge to the trailing edge of each wing section as shown in Figure 6. The third power is the total power assuming reversibility. This power shown in equation 21 is used as a metric that will be used to help compare different control schemes:

$$P_{total} = \sum_k P_k + \sum_\eta P_\eta \tag{21}$$

RESULTS

In order to simulate the system, the desired course, planform configuration, and the control weights, q and r matrices, are specified in a configuration file. Using MATLAB the configuration file is simulated while outputting the results in several forms. The objective of the simulation for this paper was to transition from straight and level flight to a 3000 feet per minute climb with minimal effort. This objective can be achieved by using the existing energy in the airstream to help morph the wing. The planform used is a flying wing configuration as shown in Figure 7 with a span of 30 feet and a weight of 10,000 lbs. There are a total of 1600 actuators split evenly between the upper and lower surfaces and 40 actuators used for twisting.

Several runs were made varying the lqr weights and timing of the pitching with a wide range of outcomes. Two such cases are shown in Figures 8a-d. These particular cases are shown because they are effective at using energy from the airstream which is provided by the thrust. Since we are concerned with minimizing the amount of energy needed, so that adaptive materials can be used, the thrust is allowed to vary freely and is not of major concern.

Figure 8a shows the total power, reversible work, and work as a function of time. As seen from the figure the net reversible work at the end of the maneuver is positive, showing what is possible if we were able to store energy from the actuators. Another output shown in Figure 8b is the force displacement plot of the most costly actuator. In other words, the actuator with the most negative peak power is used for this plot. By adding more cost to the control effort greatly changes these power requirements as shown in Figures 8c and d.

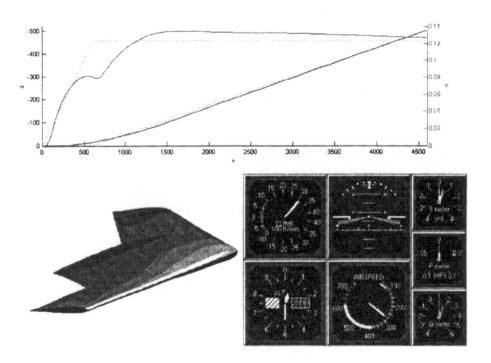

Figure 7: Screen shot from one of the movies output by the simulation allowing for the visualization of the tracking, power, and shape changes throughout the maneuver.

Continued investigations will be done by varying shape changes, number and location of actuators, and the control weights in order to minimize power consumption. By the use of an optimization routine such as a genetic algorithm applied to the existing model, should allow for the configuration of such a wing that would allow for adaptive materials to be used.

CONCLUSIONS

A computational model has been presented which predicts the force, stroke, and energy needed to overcome aerodynamic loads encountered by morphing wings during aircraft maneuvers. The aerodynamic load algorithms have been verified against more time-expensive codes. The overall model allows for desired flight-path inputs and variable control algorithms. The four modules that make up the model integrate well to make a tool that enables the study of needed actuator requirements for morphing wings.

ACKNOWLEDGMENTS

We acknowledge AFOSR for the support of the work under Grant number F49620-99-1-0294. We especially appreciate the advise of our contract monitor Brian Sanders.

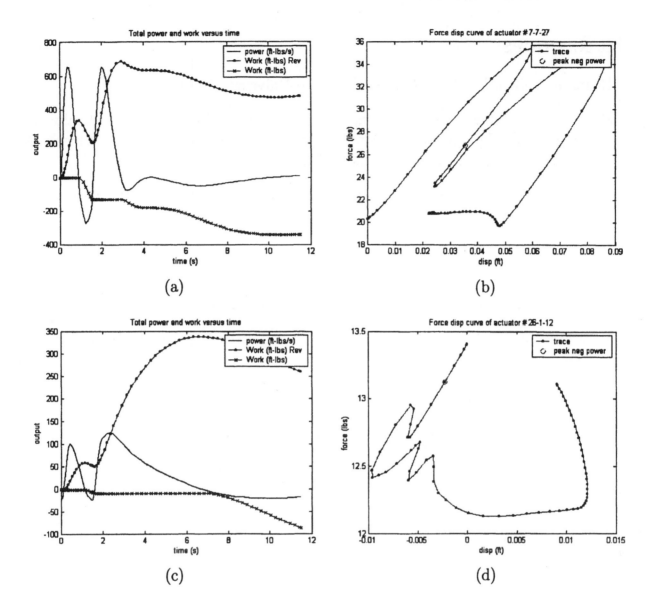

Figure 8: Results from timed based simulation: (a) power and work requirements of simulation run one; (b) force displacement curve of simulation run one; (c) power and work requirements of simulation run two; and (d) force displacement curve of simulation run two.

References

[1] Jones, Robert T., *Wing Theory*, Princeton University Press, 1990.

[2] Pettit, G., Robertshaw, H., Gern, F. and Inman, D. J., "A Model to Evaluate the Aerodynamic Energy Requirements of Active Materials in Morphing Wings", 2001 ASME Design Engineering Technical Conferences.

[3] Bertin, John J. and Smith, Michael L., *Aerodynamics for Engineers*, Prentice-Hall, Inc, 1998.

[4] Abbott, H. Ira and Doenhoff, von E. Albert, *Theory of Wing Sections*, Dover Publications, Inc, 1959.

[5] Ogata, Katsuhiko, *Modern Control Engineering*, Printice-Hall, Inc, 1990.

[6] Bryson, Authur E. Jr., *Control of Spacecraft and Aircraft*, Princeton University Press, 1994.

SHAPE CONTROL WITH KARHUNEN-LOÈVE-DECOMPOSITION: EXPERIMENTAL RESULTS

Weihua Zhang and Bernd Michaelis

ABSTRACT

The Finite Element Method is a powerful method that can be used to describe the dynamic behavior of mechanical systems. The most crucial problem to use this method is the large degree of freedom in FE-models. It is very difficult or even impossible to design controllers. Consequently, it is necessary to reduce the model. However, the model reduction deals only with the state variables, therefore the model reduction is not sufficient for the shape control, because a great number of reference and measured data are needed for the shape description and control.

In [10], we have developed a new method using the Karhunen-Loève-Decomposition to realize the shape control. In this paper, we will give the experiment results.

1   INTRODUCTION

A powerful method, which can be used to describe the dynamic behavior of the mechanical systems, is the Finite Element Method (FEM). The description of mechanical systems with the FEM can be shown as:

$$M\ddot{X} + D\dot{X} + KX = Fu$$

$$Y = CX$$

(1)

where M, D and K are mass, damping and stiffness matrices (n×n) respectively. C is the measurement matrix (m×n), F the control matrix (n×p), X the (n×1) FEM displacement vector, Y the (m×1) measurement vector and u is the (p×1) control force vector. n is the number of degree of freedom (DOF), m is the number of sensors and p is the number of actuators.

The most crucial problem to use this method is the large degree of freedom (DOF) in FE-models. It is very difficult or even impossible to design controllers. Consequently, it is necessary to reduce the model. There are many well known methods to reduce a large system [2][3][6]. However, the model reduction deals only with the state variables, therefore the model reduction is not sufficient for the shape control, because a great number of reference and measured data are needed for the shape description and control.

Dipl. –Inf. Weihua Zhang and Prof. Dr. –Ing. habil. Bernd Michaelis, IESK Otto-von-Guericke-University Magdeburg, P.O. Box 4120, D-39016 Magdeburg

By using the data compression, the great number of reference and measured data are considerably reduced. One way to perform the data reduction is the Modal-Decomposition (MD)[7][8][9]. However, a relatively great number of modal components are still needed in order to achieve high accuracy. Another way to perform the data reduction is the Karhunen-Loève-Decomposition (KLD)[10]. Using the KLD the system DOF, system input and system output can be simultaneously reduced. This reduction is only dependent on the interest of the user and the reduction ratio may be very high. The reduced system is also decoupled, this simplifies the controller design.

In this paper, the principle of the KLD and how to use it in the shape control are briefly introduced. Then simulation and experiment results are given.

2   KARHUNEN-LOÈVE-DECOMPOSITION

The Karhunen-Loève transformation was independently developed by Karhunen [4] and Loève [5] during the 1940s for the optimal series expansions of stochastic processes. In the shape control, the control process is not stochastic except the measurement noise. In the following we firstly extend the use of the Karhunen-Loève transformation from the stochastic area to the deterministic area.

2.1   The Principle of the Karhunen-Loève-Decomposition

Consider any continuous function $f(r)$ with $\int |f(r)| dr < \infty$ and $r \in D \in \mathfrak{R}^n$, we can always find many different sets of orthogonal functions $w_\mu(r)$, $\mu = 1,..., \infty$ such that

$$f(r) = \sum_{\mu=1}^{\infty} a_\mu w_\mu(r) \quad \text{and} \quad a_\mu = \int_D f(r) w_\mu(r) dr \tag{2}$$

For example if we let the $w_\mu(r)$ equal to sinus function, the Eq. (2) is the Fourier sinus transformation.

The task of KLD is now to obtain the orthogonal functions $w_\mu(r)$, $\mu = 1,...,v$, so that the reconstruction error

$$\varepsilon = \int_D \left(f(r) - \hat{f}(r) \right)^2 dr \tag{3}$$

and v are as small as possible, where $\hat{f}(r)$ is the approximation of function $f(r)$ using first v terms in Eq. (2). Because of the orthogonality of the function $w_\mu(r)$ we can express Eq. (3) as

$$\varepsilon = \sum_{\mu=1+v}^{\infty} (a_\mu)^2 = \sum_{\mu=1+v}^{\infty} \int_D \int_D w_\mu(r) f(r) f(s) w_\mu(s) dr ds \tag{4}$$

where $s \in D$. To minimize the Eq. (4) under the condition of the orthogonality of $w_\mu(r)$, the Lagrange multiplicator method can be used. The solution is

$$\int_D f(r) f(s) w_\mu(s) ds = \lambda_\mu w_\mu(r), \quad \mu = 1,..., \infty \tag{5a}$$

and the reconstruction error

$$\varepsilon = \sum_{\mu=1+v}^{\infty} \lambda_\mu \tag{5b}$$

To solve the eigenfunction and eigenvalue in Eq. (5a) we can sample the function $f(r)$ at $r_1, r_2, ..., r_n$, then Eq. (5a) can be rewritten as

$$f(r_1)\sum_{j=1}^{n} b_j f(r_j) w_\mu(r_j) = \lambda_\mu w_\mu(r_1)$$

$$\vdots \qquad (6)$$

$$f(r_n)\sum_{j=1}^{n} b_j f(r_j) w_\mu(r_j) = \lambda_\mu w_\mu(r_n)$$

where the b_j, $j = 1,...,n$, is the integral constant. Let $X=[f(r_1),...,f(r_n)]^T$, $w_\mu=[w_\mu(r_1),...,w_\mu(r_n)]^T$ and $B = \text{diag}(b_j)$, the Eq. (6) becomes

$$XX^T B w_\mu = \lambda_\mu w_\mu \qquad (7)$$

The Eq. (7) is a normal eigenvalue and eigenvector problem. The solution of this problem is well known. The solution of Eq. (7) is an approximation of the solution of the Eq. (5a). If we let all integral constants in Eq. (7) equal to one, the Eq. (7) is the same as the Eq. (6a) in [10].

2.2 Karhunen-Loève-Decomposition for flexible Structures

A response of a flexible structure is dependent on the space and time and it can be divided in two parts, one depending on the space and one on the time. According to Eq. (2) The response can be decomposed as

$$f(r,t) = f_1(r)f_2(t) = \sum_{\mu=1}^{\infty} a_\mu(t)w_\mu(r) \text{ and } a_\mu(t) = \int_D f(r,t)w_\mu(r)dr \qquad (8)$$

Now we want to find a set of orthogonal functions $w_\mu(r)$, $\mu = 1,...,v$, so that the reconstruction error

$$\varepsilon = \int_T \int_D \left(f(r,t) - \hat{f}(r,t) \right)^2 drdt$$

and v are as small as possible, where $\hat{f}(r,t)$ is the approximation of function $f(r,t)$ using first v terms in Eq. (8). Analogous to Sec. 2.1 the optimal solution is given by

$$\int_D \int_T f(r,t)f(s,t)dt w_\mu(s)ds = \lambda_\mu w_\mu(r), \quad \mu = 1,...,\infty \qquad (9)$$

At this stage we use the FEM nodal points as the sample points in space ($f(r,t)=X(t)$), t_1, t_2, ..., t_s as the sample points in time and let all integral constants equal to one, then the numerical solution of Eq. (9) is

$$XX^T w_\mu = \lambda_\mu w_\mu \text{ with } X = [X(t_1),...,X(t_s)] \text{ and } \mu = 1,...,n \qquad (10)$$

where n is the number of nodal points of the FEM or DOF of the FEM. Substituting the w_μ into Eq. (8) and using the discrete expression in space we have

$$X(t) = \sum_{\mu=1}^{v} a_\mu(t)w_\mu \text{ with } a_\mu(t) = w_\mu^T X(t) \qquad (11a)$$

and the reconstruction error

$$\varepsilon = \sum_{\mu=1+v}^{n} \lambda_\mu \qquad (11b)$$

It is noted that the sampled data may be the reference data and/or the deformation due to the disturbance. These data can be also the stationary value and/or a period of dynamical values.

3   SHAPE CONTROL USING KARHUNEN-LOÈVE-DECOMPOSITION

In Sec. 2 we introduce the KLD to represent some given responses of flexible structures. Using this representation we can first reduce the DOF for the shape control. Substituting Eq. (11a) into Eq. (1) and premultiplying W_v^T, we obtain

$$M_{kl}\ddot{a} + D_{kl}\dot{a} + K_{kl}a = F_{kl}u$$
$$Y = C_{kl}a \tag{12}$$

with $M_{kl} = W_v^T M W_v$, $D_{kl} = W_v^T D W_v$, $K_{kl} = W_v^T K W_v$, $F_{kl} = W_v^T F$, and $C_{kl} = C W_v$. In order to close the control loop we only need to find a R_{kl} that maps the measurement space to the Karhunen-Loève space and a T_{kl} that maps back from Karhunen-Loève space to the real control space. The R_{kl} and T_{kl} can be expressed as [10]

$$R_{kl}C_{kl} = I \text{ or } R_{kl} = C_{kl}^+$$
$$F_{kl}T_{kl} = K_{kl} \text{ or } T_{kl} = F_{kl}^+ K_{kl} \tag{13}$$

where the superscript "+" is in sense of the generalized inverse. At this stage we obtain a control system with only v DOF, v "sensors" and v "actuators".

The stationary accuracy and the stability are two important criteria for the shape control. In addition, how many orthogonal functions w_μ should be used is also an important problem. In the following sections they will be discussed.

3.1   Stationary Accuracy

Assuming that we have a stable controller with the form $G_R(s)/s$, we can get the relationship between X in Eq. (1) and the reference in the Laplace transfer function form as follows:

$$X(s) = W_v\overline{G}(s)F_{kl}T_{kl}\left[s + G_R(s)R_{kl}C_{kl}\overline{G}(s)F_{kl}T_{kl}\right]^{-1}G_R(s)a_{ref}(s)$$
$$\overline{G}(s) = \left[M_{kl}s^2 + D_{kl}s + K_{kl}\right]^{-1}$$

Using the final value theorem of the Lapace transform and Eq. (13) and (11a), we have

$$X(t=\infty) = W_v\overline{G}(s=0)K_{kl}\left[G_R(s=0)\overline{G}(s=0)K_{kl}\right]^{-1}G_R(s=0)a_{ref}(t=\infty)$$
$$= W_v a_{ref}(t=\infty) = W_v W_v^T X_{ref}(t=\infty)$$

Then the stationary error of the closed loop system using the KLD is equal to the reconstruction error of the KLD

$$\varepsilon = (X_{ref}(t=\infty) - X(t=\infty))^T(X_{ref}(t=\infty) - X(t=\infty)) = \sum_{\mu=l+v}^{n}\lambda_\mu \tag{14}$$

3.2   Selection of the Number of the Orthogonal Functions

How many orthogonal functions w_μ should be used is an important problem for using KLD in practice. For a small system (n is small) we can get all eigenvalues and eigenvectors in Eq. (10). Then we sort the eigenvalue and correspondent eigenvector in descent order and get the v, the number of the used orthogonal functions, for the given error ε by using Eq. (14).

For a large system, we can't get all eigenvalues and eigenvectors because of limits of the computer capacity. If we have enough actuators and the actuators are strong enough, so that all given references are realizable, we only need as many orthogonal functions as the references[10].

Because of technique and financial reason we can't always get the actuators as many and strong as what we expect in practice. In this case the given references may not be exactly realizable. Therefore the orthogonal functions calculated by the given references may be also unrealizable.

It is assumed that the deformation due to the control forces lie in the linear area and only the stationary situation will be considered for simplicity. It is assumed again that the control forces can be divided in $U(t) = U_0 u(t)$, where U_0 is the location dependent amplitude and $u(t)$ the time dependent part. This means that the control force has the same form in time and different amplitude for different location.

According to Eq. (1) and for $u(t)$ equal to step function, $X = K^{-1}FU_0$. The rank$(X) \leq$ min(n,p,s), where p is the number of actuator and s the number of the different location dependent amplitude of the control forces. In general, $n \gg p$ and $s \geq p$, then rank$(X) \leq p$. Therefore, there are maximal p eigenvalues that are not equal to zero in Eq. (10)[10]. This means that we only need to calculate p eigenvalues and correspondent eigenvectors, then determine the v using Eq. (14). Table I shows the eigenvalues of Eq. (10) with the sampled data using the different U_0 set. These data are calculated by using the left side model in our experiment (see Sec. 4). There are 5 independent actuators in the model. The U_0 set 1 is the combination of -20 and +20 (V) ($s = 32$). In the U_0 set 2, the actuator is individually activated by -50V (min) and +50V (max) value ($s = 10$). In the U_0 set 3 all actuators are activated by minimal, maximal and random value ($s = 12$). In the sampled Data (X) of the set 1 to 3, we only use the stationary value. In the sampled Data of the set 4, we use the same U_0 set as the set 3, but they include the dynamic parts.

TABLE I: EIGENVALUES OF Eq. (10) WITH SAMPELD DATA USING DIFFERENT U_0 SET

Eigenvalues	Set 1 (cb_20)	Set 2 (eyem502p50)	Set 3 (mirmx)	Set 4 (mirmx w. dyn)
1	159.04	62.124	208.52	2106
2	2.7263	1.0650	0.3232	3.4802
3	0.4594	0.1794	0.0434	0.4349
4	0.06531	0.0255	0.0070	0.0740
5	0.02416	0.0094	0.0033	0.0402
6	1.8968×10^{-14}	9.3853×10^{-15}	3.087×10^{-14}	0.0040
7	1.7571×10^{-14}	7.4947×10^{-15}	2.9795×10^{-14}	3.381×10^{-6}
8	-1.6649×10^{-14}	6.6982×10^{-15}	2.6249×10^{-14}	1.1705×10^{-7}
9	-1.7303×10^{-14}	-7.5516×10^{-15}	2.5260×10^{-14}	1.3800×10^{-8}
10	-1.9373×10^{-14}	-92991×10^{-15}	2.4487×10^{-14}	1.3352×10^{-9}

It is noted that, although the dynamic changes the number of non-zero eigenvalues, this change is not significant for practice.

3.3   Stability

Now we extend the expression of X in Eq. (11a) from v to n and write in two parts as follows

$$X(t) = [W_v, W_{n-v}] \, a(t) = W a(t) \qquad (15)$$

where $a(t) = [a_1(t), a_2(t),..., a_n(t)]^T$. Substituting Eq. (15) into Eq. (1), premultiplying W^T and using KLD we can get the closed loop transfer function as

$$\begin{bmatrix} G_{cc}(s) & G_{cd}(s) \\ G_{cd}^T(s) & G_{dd}(s) \end{bmatrix} a(s) = \begin{bmatrix} W_v^T FT_{kl} \\ W_{n-v}^T FT_{kl} \end{bmatrix} a_{ref}(s) \qquad (16)$$

where $G_{cc}(s) = M_{kl}s^2 + D_{kl}s + K_{kl}s + F_{kl}T_{kl}G_R(s)R_{kl}C_{kl}$

$\quad\quad G_{cd}(s) = W_v^T MW_{n-v}s^2 + W_v^T DW_{n-v}s + W_v^T KW_{n-v}s + F_{kl}T_{kl}G_R(s)R_{kl}CW_{n-v}$

$\quad\quad G_{dd}(s) = W_{n-v}^T MW_{n-v}s^2 + W_{n-v}^T DW_{n-v}s + W_{n-v}^T KW_{n-v}s + W_{n-v}^T FT_{kl}G_R(s)R_{kl}CW_{n-v}$

$G_R(s)$ is the transfer function of controller. The system (16) is stable if and only if all roots of

$$det\left(\begin{bmatrix} G_{cc}(s) & G_{cd}(s) \\ G_{cd}^T(s) & G_{dd}(s) \end{bmatrix}\right) = 0 = det(G_{cc}(s))det(G_{dd}(s) - G_{cd}^T(s)G_{cc}(s)^{-1}G_{cd}(s))$$

lie in the left side of the image axis. Our controller design process can guarantee that all roots of $det(G_{cc}(s)) = 0$ lie in the left side of the image axis, but there is no guarantee about all roots of $det(G_{dd}(s) - G_{cd}^T(s)G_{cc}(s)^{-1}G_{cd}(s)) = 0$ lie in the left side of the image axis, therefore there is no guarantee that system (16) will be stable. It should be noted that this is not only valid for KLD, but also for all model reduction methods. It should be also noted that no stable guarantee does not imply that we can't get stable system by controller design using a reduced subsystem.

In practice we first design the controller $\overline{G}_R(s)$ using the reduced model $G_{cc}(s)$, then adjust the gain k in $u(s) = T_{kl}k\overline{G}_R(s)(a_{ref}(s) - R_{kl}CX(s))$ $\forall s$, so that the u is within the limitation of the actuators. By this way, we can get a stable controller $G_R(s) = k\overline{G}_R(s)$ for the whole closed loop system.

3.4   Diagonalization of the reduced System in K-L Space

In addition, the reduced system (12) can be also diagonalized by performing the modal transformation[10]. After this process we get a decoupled system

$$m_i\ddot{q}_{kl_i} + d_i\dot{q}_{kl_i} + k_iq_{kl_i} = k_iu_{mkl_i}, \quad i = 1,...,v \quad\quad (17a)$$

and the modified R and T matrices become

$$u = Tu_{mkl} \text{ and } T = T_{kl}\widetilde{P}$$

$$Q_{kl} = RY \text{ and } R = \widetilde{P}^{-1}R_{kl} \quad\quad\quad\quad (17b)$$

where the u_{mkl} is the modal control forces, Q_{kl} modal coordinates with $a = \widetilde{P}Q_{kl}$ in Karhunen-Loève space.

4   EXPERIMENT AND SIMULATION RESULTS

The test and simulation object is a (one dimensional) parabolic beam ($800 \times 100 \times 1.5$ mm) composed of brass. The beam is fixed on $x = 0$ (center). In its basic form, the parabolic beam has a focal length (line) about 2261 mm. The sketch (a), the photo of top (b) and front (c) view of the test setup are shown in Figure 1. Two cameras are used to measure the shape of the experimental object. Because of the technical reason, only nine independent piezo-ceramic rod device with type P-810.10 are used as actuators and correspondent 9 strain gauges as sensors. It is noted that this configuration (Fig.1 a)) can't measure and overcome the torsion that exists in the test object (Fig. 1 b)).

Because of the existence of the hysteresis of the piezo-ceramic rod device, the method given in [1] is used in our experiment to compensate this phenomena. Figure 2 shows the result with and without compensation.

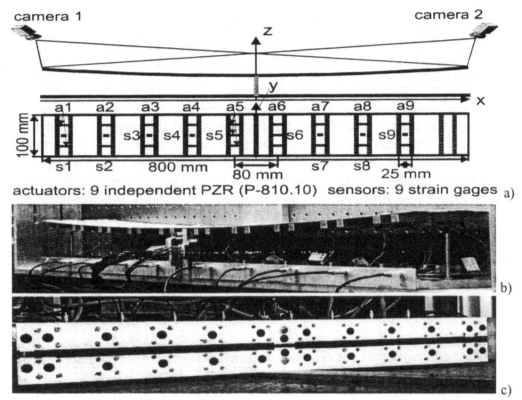

Figure 1: a) sketch of the test setup, b) top view of the test setup, c) front view of the test setup

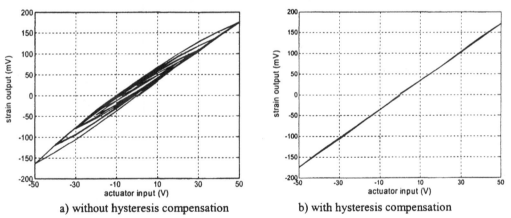

a) without hysteresis compensation   b) with hysteresis compensation

Figure 2: Result of the hysteresis compensation

Using the measured 3D geometrical data of the test object, two FEM models, one for the left side of the test object with 5 system inputs and 5 system outputs, the other for the right side with 4 system inputs and 4 system outputs, both with 2989 DOFs, are established. In order to verify the static behaviors, the independent actuator is individually activated by -50V (min) and +50V (max) value and the deformations calculated by model are compared with the measured results. Table II gives the results after adaptation of the models. The error is calculated by $\varepsilon = \sqrt{\sum \left(\frac{model - measurement}{measurement} \right)^2 / n} * 100\%$. After the verification of the static behaviors the calculated strain signals are also calibrated with the measured ones

TABELLE II: THE RESULTS AFTER MODEL VERIFICATION.

ε(%) activated actuator / actuator voltage	a5a6	a4a7	a3a8	a2a9	a1
+ 50 V (100V)	3.97	2.77	3.09	4.56	2.37
- 50 V (0V)	2.47	3.13	2.87	4.13	2.76

The goal of the control is to change the focal length of the parabolic beam from about 2261 mm to 2240 mm and/or to 2300 mm. Because the configuration of the actuators can't overcome the torsion, the deformation calculated by verified model using the U_0 set 2 (see Sec. 3.2) is used to obtain the KLD orthogonal functions. Two of them, the first and second, are used to reduce the original system. The references in K-L space are calculated by simulation and given in Table III. Using Eq. (13) and (17b) we can get the system (17a) as follows:

Left side:

$$\begin{bmatrix} \ddot{q}_{kl1} \\ \ddot{q}_{kl2} \end{bmatrix} + \begin{bmatrix} 94757 & \\ & 43100 \end{bmatrix}\begin{bmatrix} \dot{q}_{kl1} \\ \dot{q}_{kl2} \end{bmatrix} + \begin{bmatrix} 2.37\times10^5 & \\ & 1.08\times10^7 \end{bmatrix}\begin{bmatrix} q_{kl1} \\ q_{kl2} \end{bmatrix} = \begin{bmatrix} 2.37\times10^5 & \\ & 1.08\times10^7 \end{bmatrix}\begin{bmatrix} u_{mkl1} \\ u_{mkl2} \end{bmatrix}$$

Right side:

$$\begin{bmatrix} \ddot{q}_{kl1} \\ \ddot{q}_{kl2} \end{bmatrix} + \begin{bmatrix} 10325 & \\ & 41976 \end{bmatrix}\begin{bmatrix} \dot{q}_{kl1} \\ \dot{q}_{kl2} \end{bmatrix} + \begin{bmatrix} 2.58\times10^5 & \\ & 1.05\times10^7 \end{bmatrix}\begin{bmatrix} q_{kl1} \\ q_{kl2} \end{bmatrix} = \begin{bmatrix} 2.58\times10^5 & \\ & 1.05\times10^7 \end{bmatrix}\begin{bmatrix} u_{mkl1} \\ u_{mkl2} \end{bmatrix}$$

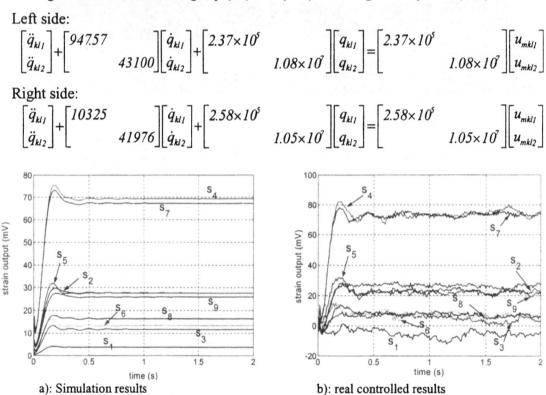

a): Simulation results   b): real controlled results

Figure 3: Dynamic response of simulation and real control for reference 1 (f = 2240 mm)

Up to now, the system DOFs, inputs and outputs are reduced to 2 and we have a decoupled 2×2 system. All design methods for single loop controller can be used to design controller according this system. Here, PID controller is used. Figure 3 shows the simulated and controlled dynamic response of the system for reference 1 (f = 2240 mm). Table IV gives the system accuracy.

TABLE III: REFERENCE VALUES IN K-L SPACE

	Focal length	Left side (x < 0)		Right Side (x > 0)	
Label	(mm)	a_1	a_2	a_1	a_2
2240_1	2240.1	2.3559	0.1731	-2.5065	0.1711
2300_1	2300.0	-4.1253	-0.3032	4.3890	-0.2996

TABLE IV: THE SYSTEM ACCURACY WITH PID CONTROL

Label	Focal length (mm)	
	Calculated by simulation	Measured after control
2240_1	2240.1	2239.0
2300_1	2300.0	2303.4
90% of 2240_1		2240.2
90% of 2300_1		2300.1

It is shown that there is deference between model and test object in static and dynamic. This error will influence the accuracy of the KLD and the accuracy of the shape control in the end. But it can be overcome by the proper adjusting the reference values.

5 CONCLUSION

The method developed in [10] and this paper can be used to realize the shape control. Using the KLD the system DOFs, system inputs and system outputs can be simultaneously reduced. This reduction is dependent on the variance of the needed shape and the reduction ratio may be very high. The reduced system is also decoupled, this simplifies the controller design. The model error can be overcome by the proper change of the reference values. The experiment demonstrates the ability of this method.

ACKNOWLEDGMENTS

The work presented in this paper is supported by the project ADAMES of the Deutsche Forschungsgemeinschaft (DFG) and the Federal State of Sachsen-Anhalt.

REFERENCES

1.   Döschner, C. und Bernert, P.: Modelle zur Hysteresekompensation an Piezotranslatoren. 1. Workshop des Innovationskollegs Adaptive mechanische Systeme, Uni. Magdeburg 11/1996, Preprint Nr.1, 1997, pp. 97-106.

2.   Gawronski,W. and Juang,J.N.: Model Reduction for Flexible Structures, in Control and Dynamic System, Vol.36, Ed. Leondes, C.T.,pp143-222.

3.   Hu, A and Skelton, R: Model Reduction with Weighted Modal Cost Analysis, AIAA GNC Conference, Portland, Oregon, August 1990,pp295-303.

4.   Karhunen, K.: Zur Spektraltheorie stochastischer Prozesse, Ann. Acad. Sci. Fenn., 34, 1946.

5.   Loève, M.: Fonctions alèatoires du second ordre, In P. Lèvy, editor, Suppl. To Processus Stochastique et mouvement Brownien, Gauthier-Villars, Paris, 1948.

6.   O'Callahan, Comparison of Reduced Model Concepts, Eighth International Modal Analysis Conference, Orlando, Florida, January 1990, pp422-430.

7.   Zhang, W.; Lilienblum, T.; Michaelis, B.: Sensorfusion und Datenverdichtung für adaptive mechanische Systeme, 3th Magdebuger Maschinenbau-Tage, Sept. 11-13, 1997, Vol. 2, pp. 213-222.

8.   Zhang, W.; Lilienblum, T.; Michaelis, B.: Data Compression for Control of Adaptive Mechanical Systems, Gabbert, U. (Hrsg.), VDI Fortschritt-Berichte, Reihe 11: Schwingungstechnik Nr. 268, Düsseldorf: VDI-Verlag 1998, pp. 323-332.

9.   Zhang, W. and Michaelis, B.: Data Acquisition and Compression in Adaptive Mechanical Systems, IMTC/98 Instrument and Measurement Technology Conference, Minnesota, USA, May, 1998, pp. 1110-1115.

10.   Zhang, W. and Michaelis, B.: Shape Control with Karhunen-Loève-Decomposition. 11th Int. Conference on Adaptive Structures and Technologies, Nagoya, Japan, Oct. 23-26, 2000, pp. 119-126.

DAMAGE IDENTIFICATION IN AGING AIRCRAFT STRUCTURES WITH PIEZOELECTRIC WAFER ACTIVE SENSORS

Victor Giurgiutiu, Andrei Zagrai, and JingJing Bao

ABSTRACT [1]

Piezoelectric wafer active sensors can be applied on aging aircraft structures to monitor the onset and progress of structural damage such as fatigue cracks and corrosion. Two main detection strategies are considered: (a) the electro-mechanical (E/M) impedance method for near-field damage detection; and (b) the wave propagation method for far-field damage detection. These methods are developed and verified on simple-geometry specimens, and then tested on realistic aging-aircraft panels with seeded cracks and corrosion. The instrumentation of these specimens with piezoelectric-wafer active sensors and ancillary apparatus is presented. The experimental methods, signal processing, and damage detection algorithms, tuned to the specific method used for structural interrogation, are discussed. In the E/M impedance method approach, the high-frequency spectrum is processed using overall-statistics damage metrics. The $(1-R^2)^3$ damage metric, where R is the correlation coefficient, was found to yield the best results. In the wave propagation approach, the pulse-echo and acousto-ultrasonic methods were considered. Reflections from seeded cracks were successfully recorded. The simultaneously use of the E/M impedance method in the near field and of the wave propagation method in the far field opens the way for a comprehensive multifunctional damage detection system for aging aircraft structural health monitoring.

PIEZOELECTRIC ACTIVE SENSORS

Piezo-electric active sensors are small, non-intrusive, and inexpensive piezoelectric wafers that are intimately affixed to the structure and can actively interrogate the structure (Giurgiutiu and Zagrai, 2001a). Piezoelectric active sensors are non-resonant devices with wide band capabilities. They can be wired into sensor arrays and connected to data concentrators and wireless communicators. Piezoelectric active sensors have captured the interest of academic and industrial community (Bartkowicz et al., 1996; Boller et al., 1999; Chang 1988, 2001). The general constitutive equations of linear piezoelectric materials given by ANSI/IEEE Standard 176-1987, describe a tensorial relation between mechanical and electrical variables (mechanical strain, S_{ij}, mechanical stress, T_{kl}, electrical field, E_k, and electrical displacement D_j) in the form:

$$S_{ij} = s_{ijkl}^E T_{kl} + d_{kij} E_k$$
$$D_j = d_{jkl} T_{kl} + \varepsilon_{jk}^T E_k \ , \tag{1}$$

where s_{ijkl}^E is the mechanical compliance of the material measured at zero electric field ($E = 0$), ε_{jk}^T is the dielectric permittivity measured at zero mechanical stress ($T = 0$), and d_{kij} is the

University of South Carolina, Columbia, SC 29208, 803-777-8018, FAX: 803-777-0106, *victorg@sc.edu*

piezoelectric coupling between the electrical and mechanical variables. Figure 1a shows an active sensor consisting of a Lead Zirconate Titanate (PZT) piezoceramic wafer affixed to the structural surface. In this configuration, mechanical stress and strain are applied in the 1 and 2 directions, i.e. in the plane of the surface, while the electric field acts in the 3 direction, i.e., normal to the surface. Hence, the significant electro-mechanical couplings for this type of analysis are the 31 and 32 effects. The application of an electric field, E_3, induces surface strains, S_{11} and S_{22}, and vice-versa. As the PZT sensor is activated, interaction forces and moments appear (Figure 1b):

$$M_a = F_{PZT}\frac{h}{2}, \quad F_{PZT} = \hat{F}_{PZT}e^{i\omega t}, \quad N_a = F_{PZT} \tag{2}$$

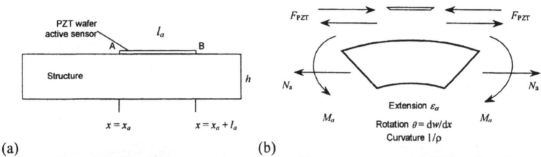

(a)   (b)

Figure 1   Piezoelectric active sensor interaction with host structure: (a) PZT wafer affixed to the host structure; (b) interaction forces and moments; (Giurgiutiu, 1999)

ELECTRO-MECHANICAL IMPEDANCE DAMAGE IDENTIFICATION

The impedance method is a damage detection technique complementary to the wave propagation techniques. Ultrasonic equipment manufacturers offer, as options, mechanical impedance analysis (MIA) probes and equipment (Staveley NDT Technologies, 1998). The mechanical impedance method consists of exciting vibrations of bonded plates using a specialized transducer that simultaneously measures the applied normal force and the induced velocity. Cawley (1984) extended Lange's (1978) work on the mechanical impedance method and studied the identification of local disbonds in bonded plates using a small shaker. Though phase information was not used in Cawley's analysis, present day MIA methodology uses both magnitude and phase information to detect damage.

The electro-mechanical (E/M) impedance method (Giurgiutiu and Rogers, 1997, 1998; Giurgiutiu and Zagrai, 2001a) is an emerging technology that uses in-plane surface excitation to measure the pointwise mechanical impedance of the structure through the real part of the electrical impedance measured at the sensor terminals. The principles of the E/M impedance technique are illustrated in Figure 2. The effect of a piezo-electric active sensor affixed to the structure is to apply a local strain parallel to the surface that creates stationary elastic waves in the structure. The drive-point impedance presented by the to the active sensor structure can be expressed as a frequency dependent quantity, $Z_{str}(\omega) = i\omega m_e(\omega) + c_e(\omega) - ik_e(\omega)/\omega$. Through the mechanical coupling between the PZT active sensor and the host structure, on one hand, and through the electro-mechanical transduction inside the PZT active sensor, on the other hand, the drive-point structural impedance is reflected directly in the electrical impedance, $Z(\omega)$, at the active sensor terminals:

$$Z(\omega) = \frac{1}{i\,\omega \cdot C}\left[1 - \kappa_{31}^2\left(1 - \frac{1}{\varphi\cot\varphi + r(\omega)}\right)\right]^{-1},\tag{3}$$

where C is the zero-load capacitance of the PZT active sensor, κ_{31} is the electro-mechanical cross coupling coefficient of the PZT active sensor ($\kappa_{31} = d_{13}/\sqrt{\overline{s}_{11}\overline{\varepsilon}_{33}}$), $r(\omega)$ is the impedance ratio between the pointwise structural impedance, $Z_{str}(\omega)$, and the impedance of the PZT active sensor, Z_{PZT}, while $\varphi = \frac{1}{2}\gamma l_a$, with γ being the active sensor wave number and l_a its linear dimension. The electro-mechanical impedance method is applied by scanning a predetermined frequency range in the high kHz band and recording the complex impedance spectrum. By comparing the impedance spectra taken at various times during the service life of a structure, meaningful information can be extracted pertinent to structural degradation and the appearance of incipient damage. It must be noted that the frequency range must be high enough for the signal wavelength to be significantly smaller than the defect size.

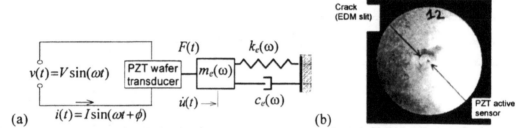

Figure 2   (a) Electro-mechanical coupling between PZT active sensor and structure. (b) circular plate with a 7-mm dia. piezoelectric active sensor at its center

SYSTEMATIC E/M IMPEDANCE EXPERIMENTS ON CIRCULAR-PLATES

A series of experiments on thin-gage aluminum plates were conducted to validate and calibrate the E/M impedance technique (Giurgiutiu and Zagrai, 2001b). Twenty-five plate specimens (100-mm diameter, 1.6-mm thick) were constructed from aircraft-grade aluminum stock. Each plate was instrumented with one 7-mm diameter PZT active sensor placed at its center (Figure 2b). A 10-mm circumferential slit EDM was used to simulate an in-service crack. The crack was placed at increasing distance from the sensor (Figure 3). Thus, 5 groups of five identical plates were obtained. E/M impedance data was taken using an HP 4194A Impedance Analyzer. During the experiments, the specimens were supported on packing foam to simulate free-free conditions. The experiments were conducted over three frequency bands: 10-40 kHz; 10-150 kHz, and 300-450 kHz. The data was process by displaying the real part of the E/M impedance spectrum, and determining a damage metric to quantify the difference between two spectra, "pristine" and "damaged". Figure 4a indicates that the presence of the crack in the close proximity of the sensor drastically modifies the pointwise E/M impedance spectrum. Resonant frequency shifts and the appearance of new resonances are noticed. Several damage metrics were tried: root mean square deviation (RMSD); mean absolute percentage deviation (MAPD); covariance change (CC); correlation coefficient, R, deviation (CCD). The $(1-R^2)^3$ damage metric was found to decrease almost linearly with the distance between the crack and the sensor (Figure 4b). However, in order to obtain consistent results, the proper frequency band (usually in high kHz) and the appropriate damage metric must be used. Further work is needed to systematically investigate the most appropriate damage metric to be used for successful processing of the frequency spectra.

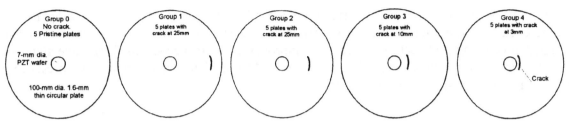

Figure 3   Systematic study of circular plates with simulated cracks (slits) at increasing distance from the E/M impedance sensor

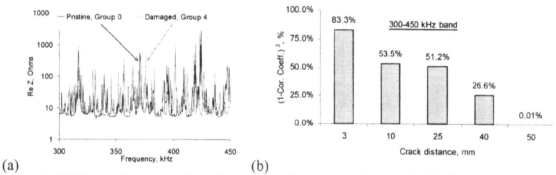

(a)   (b)

Figure 4   E/M impedance results: (a) superposed spectra of groups 1 & 5; (b) variation of $(1-R^2)^3$ damage metric with the distance between the crack and the sensor.

E/M IMPEDANCE EXPERIMENTS ON AGING AIRCRAFT PANELS

Realistic aerospace panels containing simulated crack and corrosion damage representative of aging-aircraft structures were constructed at Sandia National Labs (Figure 5). These panels were instrumented with PZT active sensors and subjected to E/M impedance evaluation. The sensors were applied to the simulated aircraft panels to detect the change of E/M impedance spectrum induced by the proximity of a simulated crack. Figure 6 shows sensors installation: the sensors are placed along a line, perpendicular to a 12.7-mm crack originating at a rivet hole. The sensors are 7-mm square and are spaced at 7-mm pitch. E/M impedance readings were taken of each sensor in the 200 – 2600 kHz range. Figure 7a shows the frequency spectrum of the E/M impedance real part. The spectrum reflects clearly defined resonances that are indicative of the coupled dynamics between the PZT sensors and the frequency-dependent pointwise structural stiffness as seen at each sensor location. The spectrum presented in Figure 7a shows high consistency. The dominant resonance peaks are in the same frequency range, and the variations from sensor to sensor are consistent with the variations previously observed in the circular plate experiments.

Examination of Figure 7a indicates that, out of the four E/M impedance spectra, that of sensor 1 (closest to the crack) has lower frequency peaks, which could be correlated to the presence of the damage. In order to better understand these aspects, further investigations were performed at lower frequencies, in the 50 – 1000 kHz range (Figure 7b). In this range, we can see that the crack presence generated features in the sensor 1 spectrum that did not appear in the other sensors spectra. For example, sensor 1 presents an additional frequency peak at 114 kHz that is not present in the other sensors. It also shows a downward shift of the 400 kHz main peak. These features are indicative of a correlation between the particularities of sensor 1 spectrum and the fact that sensor 1 is placed closest to the crack. However, at this stage of the investigation, these correlations are not self evident, nor are they supported by theoretical analysis and predictive modeling of the structure under consideration. Further signal processing and features extraction improvements are needed to fully understand the correlation between the spectral features of the E/M impedance response and the presence of structural damage in the sensor vicinity.

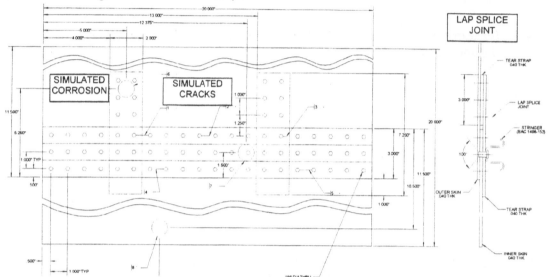

Figure 5   Realistic aging aircraft lap-splice joint panel with simulated cracks (EDM slits) and corrosion (chem-mill areas).

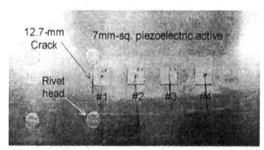

Figure 6   Piezoelectric sensors installed on the aircraft panel with aging damage simulated by a 10-mm crack originating from a rivet.

Based on these results, we formulated a damage detection strategy to be used with the E/M impedance method. The real part of the E/M impedance (Re Z) reflects the pointwise mechanical impedance of the structure, and the E/M impedance spectrum is equivalent to the pointwise frequency response of the structure. As damage (cracks, corrosion) develops in the structure, the pointwise impedance in the damage vicinity changes. Piezoelectric active sensors placed at critical structural locations detect these near-field changes. In addition, due to the sensing localization property of this method, far-field influences will not be registered in the E/M impedance spectrum. The integrity of the sensor itself, which may also modify the E/M impedance spectrum, is independently confirmed using the imaginary part of E/M impedance (Im Z), which is highly sensitive to sensor disbond, but much less sensitive than the real part to structural resonances (Giurgiutiu and Zagrai, 2001a).

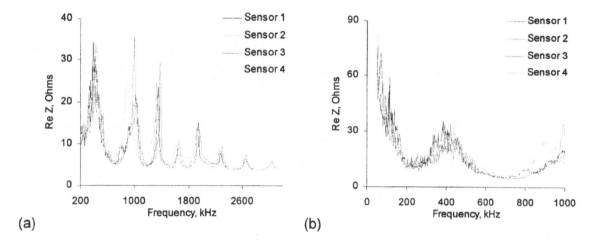

Figure 7 Real part of impedance for sensors bonded on aging aircraft structure: (a) 200-2600 kHz range; (b) zoom into the 50-1000 kHz range.

GUIDED PLATE WAVES (LAMB WAVES) DAMAGE IDENTIFICATION

Lamb waves (a.k.a., guided plate waves) are ultrasonic elastic waves that travel inside and along thin plates. Lamb waves can propagate in two modes, symmetrical and anti-symmetrical. The wave velocity depends on the product of frequency and material thickness. Investigations on Lamb and leaky Lamb waves have been pursued theoretically and experimentally for a variety of applications, ranging from seismology, to ship construction industry, to acoustic microscopy, and to non-destructive testing and acoustic sensors (Krautkramer, 1990; Rose, 1999; Lemistre *et al.* 1999). The Lamb wave speed is obtained by solving the Rayleigh-Lamb equation (Viktorov, 1967). First, define $\xi = \sqrt{c_S^2 / c_P^2}$, $\zeta = \sqrt{c_S^2 / c_L^2}$, and $\bar{d} = k_S d$; where c_L is the Lamb wave speed, and d is the half thickness of the plate. In addition, also define Lamb wave number $k_L = \omega/c_L$, and the variables, $q = \sqrt{k_L^2 - k_P^2}$, $s = \sqrt{k_L^2 - k_S^2}$. For symmetric motion (Figure 8a), the Rayleigh-Lamb frequency equation

$$\frac{\tan(\sqrt{1-\zeta^2}\,\bar{d})}{\tan\sqrt{\xi^2-\zeta^2}} + \frac{4\zeta^2\sqrt{1-\zeta^2}\sqrt{\xi^2-\zeta^2}}{(2\zeta^2-1)^2} = 0 \tag{4}$$

yields the value of ζ. Hence, one can write the two components of the particle motion, as plotted in Figure 8a:

$$U(x,z,t) = \text{Re}[Ak_L\left(\frac{\cosh(qz)}{\sinh(qd)} - \frac{2qs}{k_L^2+s^2}\frac{\cosh(sz)}{\sinh(sd)}\right)e^{i(k_L x-\omega t\frac{\pi}{2})}] \tag{5}$$

$$W(x,z,t) = \text{Re}[Aq\left(\frac{\sinh(qz)}{\sinh(qd)} - \frac{2k_L^2}{k_L^2+s^2}\frac{\sinh(sz)}{\sinh(sd)}\right)e^{i(k_L x-\omega t)}] \tag{6}$$

For anti-symmetric motion (Figure 8b), similar equations can be derived.

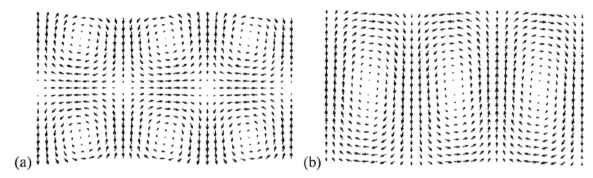

Figure 8 Simulation of Lamb wave particle motion: (a) S_0 symmetric mode; (b) A_0 anti-symmetric mode

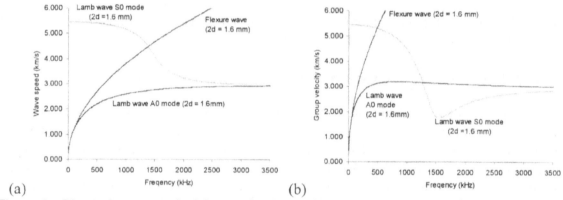

Figure 9   Dispersion curves in 1.6 mm aluminum plates: (a) wave speed; (b) group velocity

Figure 9 presents the dispersive (frequency dependant) symmetric and anti-symmetric (S_0 and A_0) Lamb wave speed and group velocity in 1.6-mm aluminum plates. Also shown in Figure 9 are the conventional (Bernoulli-Euler) flexural waves. One notices that, at low frequencies, the conventional flexural wave and the A_0 Lamb wave speeds tend to coincide. At high frequencies, the Lamb wave speeds reach a horizontal asymptote, while the flexural wave speed would continue to increase.

LAMB WAVE EXPERIMENTS ON RECTANGULAR PLATES

To understand and calibrate the Lamb-waves damage-detection method, active-sensor experiments were conducted on thin metallic plates of regular geometries (Giurgiutiu *et al.*, 2001). A 1.6 mm thick, 2024-aluminum alloy plate (914 mm x 504 mm) was instrumented with an array of eleven 7-mm x 7-mm PZT wafer active sensors positioned on a rectangular grid. The sensors were connected with thin insulated wires to a 16-channels signal bus and two 8-pin connectors (Figure 10). An HP33120A arbitrary signal generator was used to generate a smoothed 300 kHz tone-burst excitation with a 10 Hz repetition rate. The signal was sent to active sensor #11, which generated a package of elastic waves that spread out into the entire plate. A Tektronix TDS210 two-channel digital oscilloscope, synchronized with the signal generator, was used to collect the signals captured at the remaining 10 active sensors. A digitally controlled switching unit and a LabView data acquisition program were used. A Motorola MC68HC11 microcontroller was tested as an embedded stand-alone option.

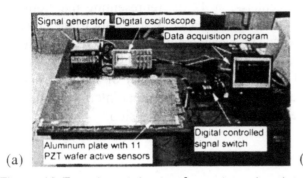
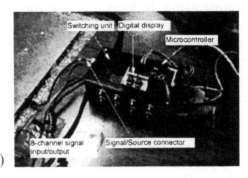

(a)

(b)

Figure 10 Experimental setup for rectangular plate wave propagation experiment: (a) overall view showing the plate, active sensors, and instrumentation; (b) detail of the microcontroller and switch box.

These systematic experiments proved several important points (Giurgiutiu *et al.*, 2001): (a) High-frequency Lamb waves could be excited in an aircraft-grade metallic plate using small inexpensive non-intrusive PZT-wafer active sensors. (b) The elastic waves generated by this method had remarkable clarity and showed a 99.99% distance-time correlation that permitted accurate wave speed determination. (c) The pulse-echo method was successfully verified.

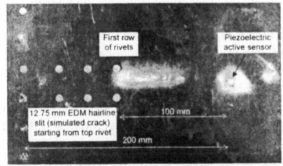
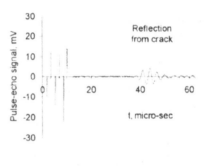

(a)

(b)

Figure 11 Crack detection experiment on aging aircraft panel: (a) damaged panel featuring 3 rows of rivets, a 12.75 mm EDM hairline slit (simulated crack) and an active sensor; (b) echo reflection from the crack detected by the active sensor

LAMB WAVE EXPERIMENTS ON AGING-AIRCRAFT PANEL

Lamb-wave active-sensor damage detection strategy stems from the ultrasonic and acousto-ultrasonic damage-detection methodologies (Blitz *et al.*, 1996; Duke, 1988, respectively). Wave propagation experiments were conducted on the aging aircraft panels of Figure 5 using a number of PZT active sensors affixed at various locations. Several experiments were performed to verify the wave propagation properties, and to identify the reflections due to the construction features of the panels (rivets, splice joints, etc.) Damage detection of cracks and corrosion damage was studied and successfully verified. For illustration, Figure 11a shows a region of the aging aircraft panel displaying a horizontal double row of stiffener rivets, a vertical row of splice rivets in the far left, and a simulated crack (12.75 mm EDM hairline slit) starting from the first rivet in the horizontal top row. A piezoelectric active sensor was placed at 200 mm from the vertical row of rivets, i.e., at 100 mm from the start of the horizontal row of rivets. The pulse-echo method was used. Figure 11b shows the result of subtracting the signal received on a pristine panel from the signal received on the panel with crack. The resulting signal features a strong wave pack centered at 42 μs, representing the "reflection from the crack". The time of flight is consistent with the distance from sensor to the crack and the group velocity for 300-kHz S_0 Lamb waves. The cleanness of the crack detection feature and the quietness of the signal ahead of it are remarkable. This indicates that this method permits good and un-ambiguous detection of structural cracks. An array of active sensors can be used in a round-robin fashion to generate and detect elastic waves that interrogate the aging aircraft structure in order to determine the presence of cracks and corrosion.

SUMMARY AND CONCLUSIONS

Piezoelectric-wafer active sensors are small inexpensive non-intrusive sensors that can be applied on existing aging aircraft structures to monitor the onset and progress of structural damage (cracks and corrosion). Two complementary methods can be simultaneously used with the same active-sensor installation: (a) electro-mechanical (E/M) impedance; and (b) elastic wave propagation. Through systematically conducted experiments, this paper has attained a double objective: (a) to develop and validate the methodology using simple-geometry specimens; and (b) to illustrate its practical application using realistic structural specimens representative of aging aerospace structures with seeded crack and corrosion.

The E/M impedance method was used for near-field damage detection. The E/M impedance experiments showed that the real part of the E/M impedance spectrum is clearly influenced by the presence of damage (simulated crack). This behavior was explained in terms of the direct correlation between the pointwise mechanical impedance of the structure at the sensor location and the real part of the E/M impedance measured at the sensor terminals. Experiments performed on 100-mm diameter thin-gage circular discs showed that the distance between sensor and a simulated crack can be directly correlated with the $(1-R^2)^3$ damage metric. Experiments performed on realistic aging aircraft panel showed that a left shift in the natural frequencies and the appearance of a new frequency peak at around 114 kHz were created by the presence of a 12.7-mm crack in the sensor proximity. However, the complete understanding of the relationship between the sensor location and the changes in the E/M spectrum for complex built-up panels has not yet been fully achieved. Additional efforts in advanced signal processing, identification of spectrum features that are sensitive to the crack presence, and adequate modeling and simulation are required.

Elastic wave propagation was studied for far-field damage detection. Simple-geometry specimens were used to clarify the Lamb wave propagation mechanism, verify the group-velocity dispersion curves, and illustrate the pulse-echo method using the reflections from the specimen boundaries. Realistic aging-aircraft specimens were used to demonstrate how a 12.7- mm crack emanating from a rivet hole can be detected with the pulse-echo method using a 300 kHz wave generated and received by the same piezoelectric-wafer active sensor placed at 100 mm from the crack.

This study has shown that the E/M impedance method and the wave propagation approach are complementary techniques that can be simultaneously used for damage detection in aging aircraft structures instrumented with an array of piezoelectric-wafer active sensors. Since one method works in the near field, while the other acts in the far field, their simultaneous utilization will ensure that the aging aircraft structure is fully covered during the health monitoring process.

ACKNOWLEDGMENTS

Financial support of US DOE through the Sandia National Laboratories, contract doc. # BF 0133. Sandia National Laboratories is a multi-program laboratory operated by Sandia Corporation, a Lockheed Martin Company, for the US DOE, contract DE-AC04-94AL85000.

REFERENCES

Bartkowicz, T. J., Kim, H. M., Zimmerman, D. C., Weaver-Smith, S. (1996) "Autonomous Structural Health Monitoring System: A Demonstration", *Proceedings of the 37th AIAA/ASME/ASCE/AHS/ASC Structures, Structural Dynamics, and Materials Conference*, Salt-Lake City, UT, April 15-17, 1996

Blitz, J.; Simpson, G. (1996) *Ultrasonic Methods of Non-Destructive Testing*, Chapman & Hall, 1996

Boller, C., Biemans, C., Staszewski, W., Worden, K., and Tomlinson, G. (1999) "Structural Damage Monitoring Based on an Actuator-Sensor System", *Proceedings of SPIE Smart Structures and Integrated Systems Conference*, Newport CA. March 1-4, 1999

Cawley, P. (1997) "Quick Inspection of Large Structures Using Low Frequency Ultrasound", *Structural Health Monitoring – Current Status and Perspective*, Fu-Kuo Chang (Ed.), Technomic, Inc., 1997.

Cawley, P. (1984) "The Impedance Method for Non-Destructive Inspection", *NDT International*, Vol. 17, No. 2, pp. 59-65.

Chang, F.-K. (1998) "Manufacturing and Design of Built-in Diagnostics for Composite Structures", *52nd Meeting of the Society for Machinery Failure Prevention Technology*, Virginia Beach, VA, March 30 – April 3, 1998.

Chang, F.-K. (2001) "Structural Health Monitoring: Aerospace Assessment", *Aero Mat 2001, 12th ASM Annual Advanced Aerospace Materials and Processes Conference*, 12-13 June 2001, Long Beach CA.

Duke, J. C. Jr., *Acousto-Ultrasonics – Theory and Applications*, Plenum Press, 1988.

Giurgiutiu, V., and Rogers, C. A. (1997) " Electro-Mechanical (E/M) Impedance Method for Structural Health Monitoring an Non-Destructive Evaluation", *Int. Workshop on Structural Health Monitoring*, Stanford University, CA, Sep. 18-20, 1997, pp. 433-444

Giurgiutiu, V., and Rogers, C. A. (1998) "Recent Advancements in the Electro-Mechanical (E/M) Impedance Method for Structural Health Monitoring and NDE", *Proceedings of the SPIE's 5th International Symposium on Smart Structures and Materials*, 1-5 March 1998, Catamaran Resort Hotel, San Diego, CA, SPIE Vol. 3329, pp. 536-547

Giurgiutiu, V., Zagrai, A. (2001a), "Embedded Self-Sensing Piezoelectric Active Sensors for On-Line Structural Identification", Submitted to: *Transactions of ASME, Journal of Vibration and Acoustics*, January 2001.

Giurgiutiu, V.; Zagrai, A. (2001b) "Electro-Mechanical Impedance Method for Crack Detection in Metallic Plates", *SPIE's 6th Annual International Symposium on NDE for Health Monitoring and Diagnostics*, 4-8 March 2001, Newport Beach, CA, paper #4335-22 (in press)

Giurgiutiu, V.; Bao, J.; Zhao, W. (2001) "Active Sensor Wave Propagation Health Monitoring of Beam and Plate Structures", *SPIE's 8th Annual International Symposium on Smart Structures and Materials*, 4-8 March 2001, Newport Beach, CA, paper #4327-32, pp. 234-245

Krautkramer, J.; Krautkramer, H. (1990) *Ultrasonic Testing of Materials*, Springer-Verlag, 1990

ANSI/IEEE Std. 176 (1987) *IEEE Standard on Piezoelectricity*, The Institute of Electrical and Electronics Engineers, Inc., 1987

Lange (1978), "Characteristics of the Impedance Method of Inspection and of Impedance Inspection Transducers", *Sov. J. NDT*, 1978, pp. 958-966.

Lemistre, M.; Gouyon, R.; Kaczmarek, H.; Balageas, D. (1999) "Damage Localization in Composite Plates Using Wavelet Transform Processing on Lamb Wave Signals", *2nd International Workshop of Structural Health Monitoring*, Stanford University, September 8-10, 1999, pp. 861-870

Rose, J. L. (1999) *Ultrasonic Waves in Solid Media*, Cambridge University Press, 1999

Staveley NDT Technologies (1998) "Sonic Bondmaster™ Product Description", Kennewick, WA 99336, http://www.staveleyndt.com/prodset4.htm

Viktorov, I. A., (1967) *Rayleigh and Lamb Waves*, Plenum Press, New York, 1967

ADAPTIVE STIFFNESS DESIGN FOR MULTI-MATERIAL STRUCTURAL SYSTEM

Masao Tanaka, Masahiro Todoh, and Akihisa Naomi

ABSTRACT

Stiffness is one of basic characteristics of structural system, and is an important feature for structural design problems. Although it has been often recruited as the design objective or constraints, the structural optimality attained has guarantee only for the mechanical conditions considered at the design stage. The adaptive stiffness design for force-dependent stiffness is a concept to tailor the structural stiffness in accordance with the external load applied. The passive adaptive stiffness has been realized by using single conventional material only. In this article, it is evolved to the active one that enables us to switch the force-dependent stiffness actively in part in the context of the multi-material structural system. The topology design problem is formulated for the distribution of conventional and functional materials under the constraints of mean compliance and its switching point as the structure. The effectiveness of the proposed structural design concept is examined through numerical case studies solved by the SIMP method of topology design.

INTRODUCTION

The multi-material design has a long tradition in composite materials e.g. [1], [2] and is recently recognized as a key technology in adaptive mechanical systems of micro electromechanical systems [3], microactuators [4] and so on. The classic composite materials have recruited various materials, but of conventional ones only, for its compositions traditionally. On the other hand, functional materials such as piezoelectric materials has been often invited to adaptive structural systems as guest materials in order to realize certain advances in adaptive/smart features, as reported, for instance, for flextensional actuators [5] and transducers [6]. The structural design has been the approach straightforward to realize the mechanical features expected for the multi-material structural system, in which the topological design plays an essential role to determine the distribution of different materials in the system accompanying geometry design for individual material portions.

Stiffness is the fundamental characteristics of structural systems, and has been used as design

[1]Professor, [2]Research Associate, [3]Graduate Student, Division of Mechanical Science, Department of Systems and Human Science, Graduate School of Engineering Science, Osaka University, 1-3 Machikaneyama, Toyonaka, Osaka 560-8531, Japan

188

objective and/or constraints for optimization in structural design problems. The authors submitted a concept of passive adaptive stiffness, that is, the force-dependent variable stiffness of piecewise linearity by using a conventional host material only in the context of traditional structural design [7] and of material structural design [8]. In this concept, a certain gap has been introduced as an internal boundary in the structure and/or in the internal structure. Such a gap has been expected to work as the switching mechanism, which brings the force/stress-dependent stiffness in the structure of conventional linear elastic materials as the result of gap opening/closure at the internal boundary depending on working force/stress. In this article, the concept of force-dependent stiffness is extended to the multi-material structural system of conventional linear elastic material as the host material and piezoelectric material as the guest material, and the structural design problem is discussed for active adaptive stiffness of piecewise linearity. This extension enables us to adjust the switching of force-dependent stiffness actively by changing the structural topology by means of internal gap control.

FORCE-DEPENDENT ADAPTIVE STIFFNESS

Concept

The force-dependent stiffness means that the stiffness designed for a structure depends on the level of external load. Even when the structure consists of single/multiple conventional materials only, such a feature is realized by arranging the structural geometry and topology appropriately with a certain internal gap introduced in the structure [7, 8]. It, however, is of passive one, since the gap opening and closing is determined a priori dependent on working load only and the structure has no way to alter the force-dependency from that designed. When some functional materials are included as a constitutive material, such a structure has a tool to adjust the force-dependency of structural stiffness independent of the external load working in part. That is, the force-dependent stiffness is now of active. In this study, the passive stress-dependent stiffness of piecewise linearity is extended to the active one in the context of the adjustment of internal gap distance by using the active effects of piezoelectric material imbedded in the conventional material. The design problem for the active stress-dependent stiffness is studied by means of the topology design technique determining the spatial arrangement of conventional and functional materials.

Active Switching Mechanism

Let consider the structure of linear elastic conventional material in which some piezoelectric materials are imbedded as shown in Figure 1 (a). An internal gap ω_c of empty domain is introduced in the material domain Ω, where the internal boundary is denoted by Γ_c. As far as the external load remain under a certain threshold level, the topology of deformed structure maintains the original topology without external load (Figure 1 (b)). When the external load elevated beyond the threshold, the deformation becomes larger and some contact will happen at a certain part of the internal boundary, and the structural topology becomes different one from the original (Figure 1 (c)). This topology change with gap closure brings the change of structural stiffness, that is, the passive force-dependent adaptive stiffness (Figure 2 (a)). Activating the piezoelectric materials imbedded

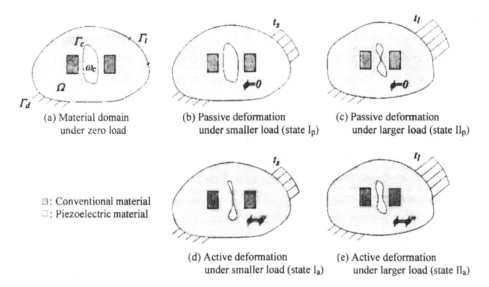

(a) Material domain under zero load

(b) Passive deformation under smaller load (state I$_p$)

(c) Passive deformation under larger load (state II$_p$)

☐ : Conventional material
☐ : Piezoelectric material

(d) Active deformation under smaller load (state I$_a$)

(e) Active deformation under larger load (state II$_a$)

Figure 1. Topological change in structure

in the conventional material has the potential to alter these opening or closure at the internal boundary dependent on the external load. That is, the activation of piezoelectric elements may causes some deformation that will change the gap open/close status at the internal gap, as shown in Figure 1 (d) and (e).   The resultant topology change of the structure enables us to adjust the structural stiffness actively (Figure 2 (b)).   Thus, the force-dependent stiffness feature becomes active one.

Description of Topology and Stiffness Switching

This section provides the fundamental description of the topology and following stiffness change due to external load and activation of piezoelectric materials.   Structural stiffness is represented as the mean compliance L of the domain Ω, that is written as

$$L(t_i, u_i) = \int_{\Gamma_i} t_i u_i d\Gamma, \quad u_i \in U_0 = \{u_i \mid u_i = 0 \text{ on } \Gamma_d\} \tag{1}$$

where Γ_i and Γ_d denote traction and displacement boundaries, and t_i and u_i represents the prescribed surface traction and displacement, respectively.   The symbol U_0 stands for the admissible displacement field.

In order to characterize the change of the structural topology, the proportional loading is assumed for the external load applied, that is, $t_i = \alpha \, t_i^0$, where t_i^0 is the reference.   The threshold force is, thus,

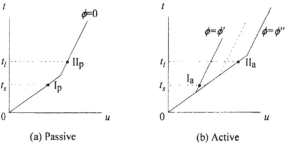

(a) Passive

(b) Active

Figure 2. Force-dependent stiffness

denoted by using the threshold loading parameter α^*. The displacement of material point x_i in the domain Ω is written by $u_i(x_i) = \alpha u_i^0(x_i)$ using the displacement u_i^0 under the reference force t_i^0, as far as the linear elastic deformation. Then, the force beyond the threshold force of $\alpha > \alpha^*$ yields some numerical overlapping $\Omega^+(\alpha)$ in the deformed domain $\Omega'(\alpha)$, and the threshold loading parameter is defined by

$$\Omega'(\alpha) = \{X_i \mid X_i = x_i + \alpha u_i^0(x_j), x_i \in \Omega\} \tag{2}$$

and the loading coefficient α^* at the threshold force is defined as

$$\alpha^* = max\left(arg \min_{\alpha} \int_{\Omega^+(\alpha)} d\Omega \right) \tag{3}$$

$$\Omega^+(\alpha) = \{X_i \in \Omega'(\alpha) \mid X_i = Y_i, Y_i \in \Omega'(\alpha)\} \tag{4}$$

Two parts on the inner boundary Γ_c is then distinguished for the inner boundary of closed status, that is, Γ_{c+} and Γ_{c-} as;

$$\Gamma_{c+} \subset \Gamma_c, \Gamma_{c-} \subset \Gamma_c$$
$$\Gamma_{c-} = \{y_i \mid y_i \in \Gamma_c, y_i \notin \Gamma_{c+}, \exists x_i \in \Gamma_{c+}, y_i + \alpha^* u_i^0(y_j) = x_i + \alpha^* u_i^0(x_j)\} \tag{5}$$
$$\Gamma_{c+} \cap \Gamma_{c-} = \phi$$

and the fixed contact condition is assumed between them as

$$\alpha^*(u_i^0(y_j) - u_i^0(y_j)) = x_i - y_i, x_i \in \Gamma_{c+}, y_i \in \Gamma_{c-}, \alpha \geq \alpha^* \tag{6}$$

beyond the threshold force.

Piezoelectric material introduced into the continuum structure yields certain strain under electric potential, and the strain works to alter topology and loading coefficient α above-mentioned (Figure 2). The field equation of linear elastic material with piezoelectricity under external load t_i and electric potential ϕ is written as

$$a(u_i, v_j) + b(\phi, v_j) = L(t_i, v_j) \quad \forall v_j \in U_0 \tag{7}$$

where

$$a(u_i, v_j) = \int_\Omega C_{ijkl}^E \frac{\partial u_k}{\partial x_l} \frac{\partial v_i}{\partial x_j} d\Omega, \quad b(\phi, v_j) = \int_\Omega e_{kij} \frac{\partial v_i}{\partial x_j} \frac{\partial \phi}{\partial x_k} d\Omega, \quad L(t_i, v_j) = \int_{\Gamma_i} t_i v_j d\Gamma \tag{8}$$

$$e_{ikl} = d_{imn} C_{mnkl}^E$$

and C_{ijkl}^E, e_{kij}, d_{imn} are the elastic, piezoelectric stress, piezoelectric strain constants of the material, respectively. This field equation gives the mean compliance under the given electric potential ϕ applied to the piezoelectric part of the domain Ω, where the second term of left hand side of Eq. (7), of course, does not exist for the conventional material part. The applicable electric potential has their lower limit ϕ_{min} and the upper limit ϕ_{max}, and the threshold loading parameter α^* has the range for active adjustment between

$$\alpha^*(\phi_{max}) = \alpha_{max}^*, \alpha^*(\phi_{min}) = \alpha_{min}^* \tag{9}$$

for the active adjustment of adaptive structural stiffness.

DESIGN PROBLEM FORMULATION

The structural design problem for active force-dependent adaptive stiffness of piecewise linearity is defined for the specified piecewise linearity with the prescribed threshold loading parameter $\alpha^{(n)*}$ and the prescribed mean compliance $L^{(m)}$ for individual linear part of $m = 0$ for $\alpha < \alpha^{(n)*}$ and $m = 1$ for $\alpha > \alpha^{(n)*}$, and n distinguishes external electric potential status, $n = 0$ for $\phi = \phi_{min}$, $n = 1$ for $\phi = 0$, and $n = 2$ for $\phi = \phi_{max}$. This problem is formulated under the constraints of mean compliance before and after the threshold force prescribed for individual external potential applied. The topology design problem is, then, written as the volume minimization of the structure under the constraints of prescribed mean compliance. The volume fraction ρ is the design variable in the context of the SIMP method for topology design problem, and the optimization problem is written as follows:

$$
\begin{aligned}
&\textit{minimize} && \int_\Omega \rho\, d\Omega \\
&\textit{with respect to} && \rho(x_i) \\
&\textit{subject to} && \rho_{min} \leq \rho \leq \rho_{max} \\
& && max\left(arg\min_\alpha \int_{\Omega^*(\alpha;\phi^{(n)})} d\Omega\right) = \alpha^{(n)*} \\
& && a(u_i^{(m)(n)}, v_j^{(m)(n)}) + b(\phi^{(n)}, v_j^{(m)(n)}) = L(t_i, v_j^{(m)}) \ \forall v_j^{(m)} \in U_0 \\
& && L(t_i, u_i^{(m)(1)}) = L^{(m)*} \\
& && \alpha^{(n)*}(u_i^{(0)(n)}(y_j^{(n)}) - u_i^{(0)(n)}(x_j^{(n)})) = (x_i^{(n)} - y_i^{(n)}) \\
& && x_i^{(n)} \in \Gamma_{c-}, y_i^{(n)} \in \Gamma_{c+}, \alpha \geq \alpha^{(n)*}
\end{aligned}
\tag{10}
$$

A penalty is imposed on the intermediate volume fraction apart from its lower limit ρ_{min} and upper limit ρ_{max}, because this is beneficial to make the obtained structural topology more distinguishable. The constraints of the volume minimization problem of Eq.(10) are treated by means of the penalty method, and the augmented functional Π_T is written as follows:

$$
\begin{aligned}
\Pi_T =\ & \int_\Omega \rho\, d\Omega \\
& + \sum_{m=0}^{1}\left[-\sum_{n=0}^{2}\left\{a(u_i^{(m)(n)}, v_j^{(m)(n)}) + b(\phi^{(n)}, v_i^{(m)(n)}) - L(t_i, v_i^{(m)(n)})\right\}\right.\\
& \left. + r^{(m)}\left\{\int_{\Gamma_t} t_i u_i^{(m)(1)} d\Gamma - L^{(m)*}\right\}^2\right] \\
& + \sum_{n=0}^{2}\left[r_\alpha^{(n)}\int_{\Gamma_c}\left\{\alpha^{(n)*}(u_i^{(0)(n)}(y_i^{(n)}) - u_i^{(0)(n)}(x_j^{(n)})) - (x_i^{(n)} - y_i^{(n)})\right\}^2 d\Gamma\right] \\
& + r_\rho\int_\Omega(\rho - \rho_{min})(\rho_{max} - \rho)\, d\Omega
\end{aligned}
\tag{11}
$$

where $r^{(m)}$, $r_\alpha^{(n)}$ and r_ρ are penalty parameters. This functional Π_T is discretized into the finite elements of square voxels. This topology determination problem is solved by using the standard sequential linear programming supported by the moving limits strategy for the stability of the convergence process [9, 10] and the SIMP method without filtering techniques.

NUMERICAL CASE STUDIES

This section illustrates a couple of 2-dimensional case studies for structure with tensile and compressive loads. Figure 3 shows the design domain for the topology design and the slit arrangement in x_1 and x_3 coordinates. The isotropic host material used has Young's modulus of $E=70.6$ GPa and Poisson's ratio of $v=0.33$. The piezoelectric material used has Young's moduli of $E_1=79.0$ GPa and $E_3=65.0$ GPa, Shear modulus of $G=37.2$GPa, Poisson's ratios of $v_{13}=v_{31}=0.32$, piezoelectric strain constants of $d_{311}=-133\times 10^{-9}$ mm/V, $d_{333}=302\times 10^{-9}$ mm/V and $d_{113}=419\times 10^{-9}$ mm/V. The upper and lower limits of external electric potential are equivalent to the uniform electric field of -5.0V/mm and 5.0V/mm along coordinates, respectively. The gap and length of slit are 0.001 mm and 40 mm for tensile structure and 0.001 mm and 50 mm for compressive structure, respectively, and the unit value is assumed for the initial volume fraction, $\rho=1.0$. The domain is discretized by 50×50 square elements.

Adaptive Stiffness under Tension

The first case is devoted for an adaptive stiffness structure under tensile load. The mean compliance are prescribed for the tension under threshold as $L^{(0)}=4.00\times 10^{-3}$ N/mm^2 and for the tension over threshold as $L^{(1)}=2.00\times 10^{-3}$N/mm^2 with the threshold loading coefficients $\alpha^{(0)*}=1.8$ for the minimum external electric potential, $\alpha^{(1)*}=2.0$ for zero electric potential, and $\alpha^{(2)*}=2.2$ for maximum one. Figure 4 (a) shows the initial arrangement of conventional and piezoelectric material of unit volume fraction in solid and crossed elements, and Figure 4 (b) illustrates the optimal topology, that is, the distribution of volume fraction obtained for the prescribed conditions as listed in Table I. Figure 4 (c) and (d) illustrate the deformations of the element of host and guest materials in binary format demonstrating opening and closure of the gap collinear to x_1 axis. Such a structural topology is not unique for the current requirement. In fact, the initial arrangement of materials shown in Figure 5 (a) gives us another material distribution of Figure 5 (b). The active adaptive stiffness realized is almost identical each other (Table I), while the obtained optimal structural topology is completely different.

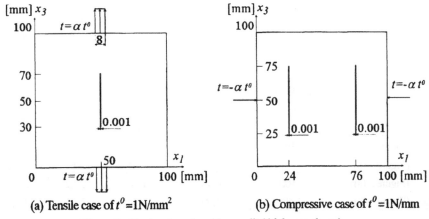

(a) Tensile case of $t^0=1$N/mm^2 (b) Compressive case of $t^0=1$N/mm

Figure 3. Design domain with gap slit (thickness=1mm)

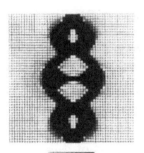

(a) Initial arrangement
(■:conventional, ⊠:piezoelectric)

(b) Optimal topology
(Gray-scaled volume fraction)

0.005   1.000

(c) Deformed shape, $\alpha < \alpha^{(n)*}$
(■:conventional, ⊠:piezoelectric)

(d) Deformed shape, $\alpha \geq \alpha^{(n)*}$
(■:conventional, ⊠:piezoelectric)

Figure 4. Adaptive stiffness under tension

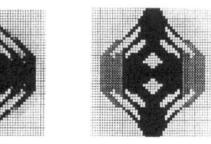

0.005   1.000

(a) Initial arrangement
(■:conventional, ⊠:piezoelectric)

(b) Optimal topology
(Gray-scaled volume fraction

(c) Binary format topology
(■:conventional, ⊠:piezoelectric)

Figure 5. Adaptive stiffness under tension (alternative design)

TABLE I. ADAPTIVE STIFFNESS UNDER TENSION

		Volume (%)	$L^{(0)}(\times 10^3)$	$L^{(1)}(\times 10^3)$	$\alpha^{(0)}$	$\alpha^{(1)}$	$\alpha^{(2)}$
	Target		4.00	2.00	1.80	2.00	2.20
Initial	(Figure 4 (a))	100	0.90	0.89	22.5	32.1	51.0
Optimal	(Figure 4 (b))	36	4.00	2.00	1.81	1.99	2.20
Initial	(Figure 5 (a))	100	0.93	0.92	37.2	30.2	25.5
Optimal	(Figure 5 (b))	49	4.00	2.00	1.81	1.99	2.21

L: [N/mm²]

Adaptive Stiffness under Compression

An adaptive stiffness structure under compression is chosen as the second case study. In this case, two vertical slits are introduced in the design domain as shown in Figure 3 (b). The specified values for mean compliance are $L^{(0)}=2.00\times10^{-3}$ N/mm^2 and $L^{(1)}=1.00\times10^{-3}$ N/mm^2. The threshold loading coefficients assumed for this case are $\alpha^{(0)*}=0.9$ for the minimum external electric potential, $\alpha^{(1)*}=1.0$ for zero electric potential, and $\alpha^{(2)*}=1.1$ for maximum one. From the initial material layout of unit volume fraction shown in Figures 6 (a) and 7 (a), two different topologies are obtained for the specified conditions as is found in Figures 6 (b) and 7 (b) respectively. The distribution of two materials is not unique in this case again as is in the previous case.

 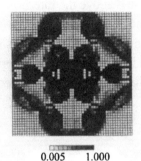

0.005   1.000

(a) Initial arrangement   (b) Optimal topology
(■:conventional, ⊠:piezoelectric)   (Gray-scaled volume fraction)

Figure 6. Adaptive stiffness under compression

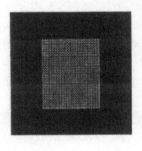 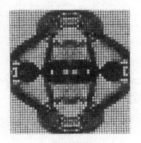

0.005   1.000

(a) Initial arrangement   (b) Optimal topology
(■:conventional, ⊠:piezoelectric)   (Gray-scaled volume fraction)

Figure 7. Adaptive stiffness under compression (alternative design)

TABLE II. ADAPTIVE STIFFNESS UNDER COMPRESSION

		Volume (%)	$L^{(0)}(\times10^3)$	$L^{(1)}(\times10^3)$	$\alpha^{(0)}$	$\alpha^{(1)}$	$\alpha^{(2)}$
Target			2.00	1.00	0.90	1.00	1.10
Initial	(Figure 6 (a))	100	0.17	0.13	7.65	8.80	10.4
Optimal	(Figure 6 (b))	51	2.00	1.00	0.90	1.00	1.11
Initial	(Figure 7 (a))	100	0.17	0.12	7.36	8.82	11.0
Optimal	(Figure 7 (b))	44	2.00	1.00	0.90	1.00	1.12

L: [N/mm^2]

CONCLUSIONS

In this article, the concept of active force-dependent adaptive stiffness is investigated in the context of the structural optimization problem for multi-material system. The topology change of the structure is introduced by utilizing the open/close transition at the internal boundary of the structure. This topology change is dependent on the magnitude of the external load only when the structure consists of conventional materials only. By inviting some functional materials as a constitutive material of the structure, such a force-dependent adaptive stiffness are evolved to an active one, which enables us to alter the stiffness switching load and select the structural stiffness partially independent of the external load. The material distribution problem of conventional and functional materials is formulated as the topology optimization problem under the constraints of mean compliance below and above the threshold level of external load. The optimization problem is solved by SIMP method, the most simple topology optimization method, and the design capability of the active force-dependent adaptive stiffness of multi-material system is demonstrated through numerical cases. Remained for the future work involves the selection of initial arrangement of functional material and the internal gap assignment.

REFERENCES

1. Sigmund, O. and Torquato, S., 1999, "Design of smart composite materials using topology optimization," *Smart Materials and Structures*, Vol. 8, pp.365-379.

2. Fujii, D., Chen, B., C. and Kikuchi, N., 2001, "Composite material design of two-dimensional structures using the homogenization design method," *International Journal for Numerical Methods in Engineering*, Vol.50, pp.2031-2051.

3. Sigmund, O., 1998, "Topology optimization in multiphysics problems," *American Institute of Aeronautics and Astronautics, AIAA-98-4905*, pp.1942-1500.

4. Silva, E., Ono, J., S. and Kikuchi, N., 1997, "Optimal design of piezoelectric microstructure," *Computational mechanics*, Vol.19, pp.397-410.

5. Silva, E., Nishiwaki, S., Ono, J., S. and Kikuchi, N., 1999, "Optimization methods applied to material and flextensional actuator design using the homogenization method," *Computer Methods in Applied Mechanics and Engineering*, Vol.172, pp.241-271.

6. Silva, E. and Kikuchi, N., 1999, "Design of piezoelectric transducers using topology optimization," *Smart Materials and Structures*, pp.350-364.

7. Tanaka, M. and Oki, K., 1999, "Shape design of bilinear stiffness structure," *Jpn. Soc. Mech. Eng. Design & Systems Conference*, No.99-27, pp.346-348 (in Japanese).

8. Tanaka, M., Todoh, M., Oki, K. and Naomi, A., 2000, "Material structural design of stress-dependent adaptive stiffness," in *Eleventh International Conference on Adaptive structures and Technologies*, Edited by Y. Matsuzaki, T. Ikeda, and V. Baburaj, eds. Lancaster, PA: Technomic Publishing Co., Inc., pp.35-42.

9. Sigmund, O., 1997, "On the design of compliant mechanisms using topology optimization," Mechanics of structures and machines, Vol.25, No.4, pp.493-524.

10. Petersson, J. and Sigmund, O., 1998, "Slope topology optimization," *International Journal for Numerical Methods in Engineering*, Vol.41, pp.1417-1434.

Smart Materials

CONSTITUTIVE MODEL OF SHAPE MEMORY ALLOYS BASED ON THE PHASE INTERACTION ENERGY FUNCTION AND ITS APPLICATION TO THERMOMECHANICAL PROCESS

Hisashi Naito, Yuji Matsuzaki, and Tadashige Ikeda

ABSTRACT

This paper presents a summary of a series of our thermomechanical analyses for pseudoelasticity and shape memory effect of shape memory alloys based on a general function of the dissipation potential, called the phase interaction energy function, that we introduced in modeling of the constitutive law. Our analytical model may well predict complex stress-strain-temperature relationship, inner hysteresis loops appearing in a bounding loop and effect of loading cycle. A future direction of R&D of SMA will also be addressed.

INTRODUCTION

Shape memory alloys (SMA) have attracted a great attention from scientific and engineering societies because of their smart characteristics, that is, shape memory effect (SME) and pseudoelasticity (PE). Their thermomechanical and thermodynamic behaviors are, however, very complicated so that they are not fully understood yet. Both SME and PE are, respectively, revealed through irreversible processes of phase transformations (PT) and phase rearrangements (PR) that are associated with hysteresis yielding energy dissipation. During the phase transformations, heat is generated and absorbed in SMA as latent heat phenomenon. Consequently, the heat is transferred to and from media surrounding SMA unless the temperature distribution of the media is kept equal to that of the SMA. In order to use SMA effectively as components of a smart structure, it is necessary to fully understand their static and dynamic characteristics and establish an accurate analytical model. Using such an analytical model, we need to perform system analysis and design of the structure with the SMA components to evaluate the effectiveness of utilization of SMA in the structural system.

Many researchers proposed analytical models on constitutive laws, that is, stress-strain-temperature relationship, by taking a thermomechanical, micromechanical or energy approach on the materials. Among them, introducing the phase interaction energy function (PIEF) which represents a dissipation energy potential during the phase transformation, Kamita and Matsuzaki [1-3] proposed a one-dimensional (actually, lumped-parameter) pseudoelastic model of SMA. Since then, using PIEF, we have exploited the model so that we may treat both pseudoelasticity and shape memory effect, and analyze thermodynamic aspects of behavior of SMA subjected to dynamic loading [8-14, 16, 17]. The thermo-mechanical and -dynamic

Department of Aerospace Engineering, Nagoya University, Chikusa, Nagoya 464-8603, Japan

behaviors of SMA are much more complicated than originally thought, and it appears that little will sufficiently be understood in a near future. We will here summarize our previous studies on SMA and present future research directions, focusing only on stress-induced phenomena.

ANALYTICAL MODELING

Energy Approach

We use a phenomenological approach based on macro-level energy consideration, because little is sufficiently known about material behaviors at microscopic-level during PT and PR to formulate a thermo-micromechanical model. From a viewpoint of the application of SMA to smart structural systems, it is most important to model accurately the constitutive relationship. Without accurate modeling of the constitutive law, we can hardly analyze and design any system installed with SMA components which behaves smartly.

Phase Interaction Energy Function [1-3]

Both PT and PR are thermomechanically irreversible processes, so that the energy is dissipated. In addition to free energies of crystal phases of SMA, we need to take into account dissipation energy. Now we focus only on thermodynamic aspects of a long thin SMA wire subjected to a unidirectional load at a prescribed frequency or strain rate and exposed to a given environmental temperature. We assume that there exists a uniform stress-strain-temperature state along the beam axis, so that we may take a lumped-parameter approach. There are no experimental data from which a two- or three-dimensional constitutive analytical model of SMA can be established.

For PT between austenite and martensite, some investigators such as Falk [4], Mueller and Xu [5], Lexcellent and his coworkers [6, 7], assumed the dissipation potential in a specific form of z(1-z), where z and 1-z stand for the volume fractions of martensite and austenite, respectively. Their analyses predicted that PT was a jump-type unstable phenomenon, however, which did not agree with experimental observations. In our analytical model, therefore, we introduced a general function of the dissipation potential, called the phase interaction energy function (PIEF), ϕ_a. PIEF is numerically determined so that an analytical stress-strain curve derived fits an analytical or experimental reference curve.

We will first discuss PT between austenite (AU) and detwinned martensite (DM), and later PT associated with rhombohedral phases as well as PR. We may write the specific free energy functions, ϕ_i, of AU and DM as

$$\phi_i = \frac{1}{2} E_i \left(\varepsilon_i - \varepsilon_{0i} \right)^2 - \alpha_i E_i \left(\varepsilon_i - \varepsilon_{0i} \right)(T - T_0) + C_i \left(T - T_0 - T \log \frac{T}{T_0} \right) + U_i \left(T_0, \varepsilon_{0i} \right) - TS_i \left(T_0, \varepsilon_{0i} \right)$$

with $\varepsilon_{01} = 0$ for $i = 1$ to 2,   (1)

where suffixes 1 and 2 represent AU and DM, respectively. ε_{02} is a non-zero macro strain associated with the lattice structure of DM. C_i, E_i, α_i and ε_i are, respectively, specific heats, Young's moduli, thermal expansion rates, and strains. U_i, S_i, T and T_0 are specific internal energies, specific entropies, temperature and a reference temperature, respectively. We assume that densities of AU and DM are the same, and

$$E_1 = kE_2, \quad \alpha = \alpha_i, \quad C = C_i \quad \text{for } i = 1 \text{ to } 2.$$ (2)

ϕ_a is formally given [8] by

$$\phi_a = \frac{sign(\dot{z})}{2}\left[\{1+sign(\dot{z})\}\phi_{aM} + \{1-sign(\dot{z})\}\phi_{aA}\right], \tag{3}$$

which is reduced to

$$\phi_a = \begin{cases} \phi_{aM} & \text{for martensitic transformation } (\dot{z} > 0), \\ \phi_{aA} & \text{for reverse transformation } (\dot{z} < 0), \\ 0 & \text{for nontransformation } (\dot{z} = 0). \end{cases} \tag{4}$$

Now, we may define the extended specific free energy of the alloy in a mixed state by

$$\phi = (1-z)\phi_1 + z\phi_2 + \phi_a \tag{5}$$

The extended potential energy of the alloy, π, is given by

$$\pi = \phi - \overline{\sigma}\varepsilon, \tag{6}$$

where $\overline{\sigma}$ is a stress due to an externally applied load. The stress of each phase is given by

$$\sigma_i = \frac{\partial\phi}{\partial\varepsilon_i} = E_i\left\{(\varepsilon_i - \varepsilon_{0i}) - \alpha_i(T - T_0)\right\} \quad \text{for } i = 1 \text{ to } 2. \tag{7}$$

Constitutive Relationship [1-3]

Applying the minimum potential energy theorem to Eq. (6), we obtain the martensitic and the reverse transformation stress as

$$\sigma(z, T, \phi_a) = \frac{-E_1\varepsilon_0 + \sqrt{F(z, T, \phi_a)}}{k-1}, \tag{8}$$

where

$$F(z, T, \phi_a) = E_1^2\varepsilon_0^2 + E_1E_2\alpha^2(k-1)^2(T-T_0)^2 + 2(k-1)E_1G(z,T), \tag{9.1}$$

$$G(z, T) = \frac{\partial\phi_a}{\partial z} + (S_{01} - S_{02})T - (U_{01} - U_{02}). \tag{9.2}$$

To numerically determine $G(z, T)$ of the SMA material with the use of its experimental constitutive data, we will assume $G(z, T)$ [8, 9] as

$$G(z,T) = (e_1T + e_0)\sum_{i=0}^{n}g_iz^i, \tag{10}$$

where e_i for $i=0$ and 1 and g_i for $i=0$ to n are parameters to be estimated from experimental data [10], which are desirably measured in a quasi-static manner at two different temperatures.

Effect of Strain Rate [9, 11]

When the martensitic and the austenitic phase transformation take place, the heat generation and absorption will, respectively, occur as a result of latent heat in the wire together with an energy dissipation due to hysteresis. The dissipated energy changes mainly into heat. Whenever there is a temperature difference between the wire and its surrounding, the heat will transfer between them. It is assumed that during a small phase transformation, dz, in a short time period, dt, the effects of latent heat, dQ_L, heat transfer, dQ_T, and energy dissipation, dE_D, will change the temperature of the wire by dT from T.

The latent heat dQ_L and the heat transfer dQ_T are governed by

$$dQ_L = \rho H_L dz, \tag{11}$$

$$dQ_T = \frac{4h}{d}\{T_S - T(t)\}d, \tag{12}$$

where d, h, H_L, T_S and ρ are, respectively, the diameter of the wire, heat transfer and latent heat coefficients, temperature of the surrounding medium and density of both AU and DM. The energy dE_D dissipated during the transformation process dz, which is accompanied by a strain $d\varepsilon(T,\phi_a)$, is given [9, 12] as

$$dE_D = \{\sigma(\varepsilon,T,\phi_a) - \sigma(\varepsilon,T,0)\} \times \frac{d\varepsilon(T,\phi_a) + d\varepsilon(T,0)}{2}. \tag{13}$$

The dissipated energy represents a difference between the work, $dW(T,\phi_a)$, done for the transformation strain $d\varepsilon$ and a work, $dW(T,0)$, which is done for a hypothetical elastic strain induced on an assumption that no energy is dissipated during the transformation, that is, $\phi_a = 0$. Considering a balance of heat and energy during the phase transformation, we have

$$CdT = dQ_L + dQ_T + dE_D. \tag{14}$$

Effect of Cyclic Loading [13, 14]

SMA show large hysteresis during the austenite-martensite phase transformations in a loading and unloading process above the austenite finish temperature. The stress-strain hysteresis observed in experiments changes considerably in an early stage of cyclic loading. The change in thermomechanical characteristics is considered due to the training effect of stress, strain, or thermal cycling. The training effect was analytically studied by Tobushi and his coworkers [15], who included directly the effect of loading cycle in expressions of residual strain at zero applied stress and the martensite and the reverse transformation start temperature.

In the early phase of cycling, adjustment or annihilation of local irregularity may clearly take place on grain boundaries and interfaces between AU and DM. There occurs also formulation of defects, such as dislocations, cracks, etc. In a domain surrounded by the boundaries and interfaces containing the micromechanical defects, the strained martensite may remain without transforming back to AU during the unloading process. Such a martensitic domain keeps internal stresses and strains as residual. The residual stress is nothing but a pre-stress in the phase of loading. When another loading is applied, the apparent martensitic transformation start stress becomes smaller. As the number of cycles increases, these training effects will diminish gradually. We have, therefore, taken into account the effect of the residual martensite (RM) in the specific potential energy. The specific energy of the alloy material in the mixed state of AU, DM and RM is assumed as

$$\phi = (1 - f(N))(1 - z)\phi_1 + z\phi_2 + f(N)(1 - z)\phi_{2R} + (1 - f(N))\phi_a \tag{15}$$

for the Nth loading, where ϕ_{2R}, $f(N)$ and $1 - f(N)$ are, respectively, the specific energy of RM, and the volume fractions of AU and RM just before the Nth loading starts, and ϕ for the Nth unloading is given by replacing $f(N)$ with $f(N+1)$ in Eq. (15). The transformation stresses are expressed as

$$\sigma(z,T,N) = -\left[I(N,T) + \sqrt{I(N,T)^2 - 4H(N)\{J(N,T) + (1 - f(N))(\partial\phi_a / \partial z)\}} \right] \Big/ 2H(N), \tag{16}$$

where H, I and J are functions of N and T [14]. The form of the function $f(N)$ are determined using stress-strain curves measured in an experiment.

Internal Hysteresis Loop [16]

The loading and unloading between the full AU and DM states, respectively given by $z=0$ and $z=1$, produces a bounding hysteresis loop covering a large strain range. If a loading and unloading is applied with a smaller strain range, then we have a smaller hysteresis loop inside of the bounding loop. A different set of strain range, strain rate, etc., yields a different loop. This feature is more complex than that of the bounding loop. To take into account micro-level PT in crystal grains of SMA, we have introduced a probabilistic approach in the specific potential energy, using the density distribution of martensite volume fraction of crystal grains in the material. As a result, we may predict complicated behavior of an inner loop observed when a cycle of a smaller strain is applied to a given prestrain state at a low strain rate.

Unified Presentation of Shape Memory Effect and Pseudoelasticity [17]

Finally, let us present a unified approach on PT and PR associated with AU, DM, twinned martensite (TM), detwinned rhombohedral (DR) and twinned rhombohedral (TR).   To account for energies dissipated during PT and PR between these phases, we assumed PIEF like Eq. (3) as

$$\phi_a = \phi_a(\phi_{AM}, \ \phi_{MA}, \ \phi_{MM}, \ \phi_{RM}, \ \phi_{MR}, \ \phi_{AR}, \ \phi_{RA}, \ \phi_{RR}), \tag{17}$$

where ϕ_{AM}, ϕ_{MA}, ϕ_{RM}, ϕ_{MR}, ϕ_{AR}, ϕ_{RA}, ϕ_{MM} and ϕ_{RR} denote dissipation potentials for the transformations from AU to DM, from DM to AU, from DR to DM, from DM to DR, from AU to DR and from DR to AU, and the rearrangements from TM to DM and from TR to DR, respectively. Suffixes 1, 2, 3, 4 and 5 represent AU, DM, TM, DR and TR, respectively. The total specific energy is therefore given by

$$\phi = \sum_{i=1}^{5} z_i \phi_i + \phi_a \quad \text{with} \ \sum_{i=1}^{5} z_i = 1. \tag{18}$$

Analytical and numerical procedures are essentially the same as given by Eqs. (1) to (10) for the martensitic transformation.

NUMERICAL EXAMPLES

Now, we examine the effectiveness of the analytical model based on the PIEF and some modifications included in the specific potential energy of SMA as described above. Figure 1 shows a comparison between experimental data and numerical results on the stress and temperature versus strain of a thin SMA wire (Fig. 2 of Ref. 8). The data are plotted from stress-strain curves measured at our laboratory with the frequency of f =0.001Hz and T_s =296K or 313K, and the numerical calculation was performed using the thermomechanical model described by Eqs. (1)-(14). Figures 2 and 3 also show another comparisons between the experiment and the analysis on the stress-strain curves for the first and the 10th cycle at f =0.1Hz and the temperature vs. time for the first ten cycles, respectively. The experimental data are sufficiently well fit by the numerical curves for which n=8 was used in Eq. (10). As shown in Fig. 1, there is some discrepancy between the experimental and the numerical results above the strain of 5%. If we take more temperature terms in the parenthesis of Eq. (10), this discrepancy will decrease. Numerical values used in the calculation are given in Ref. 8.

For the effect of loading cycles, Fig. 4 presents a comparison between measured and analytical results on stress-strain for N=1, 10, 100 and 1000, which are illustrated by curves with small fluctuations and straight lines, respectively. The strain rate was 0.001 sec^{-1} and T_s =293K, and in the numerical calculation the function $f(N)$ was assumed in a simple form of $f(N) = \beta(1 - 1/N^\gamma)$, for which $\beta = 0.974$ and $\gamma = 0.0998$

were selected. The analytical results agree well with the measured data. As for the effect of smaller strain on hysteresis, Figure 5 shows experimental and numerical results of pseudoelastic bounding and two inner loops by triangles and solid curves, respectively. In Fig. 6 open and solid circles represent experimental data which were measure at T=293K and 333K by H. Tobushi and his coworkers on PR and PT associated with rhombohedral phases and the transformation from DM to AM.

A comparison of numerical stress-strain curves calculated using models proposed by Tanaka [18], Liang et al. [19], Brinson [20] and Boyd et al. [21] with experimental curves for shape memory effect and pseudoelasticity is given by Chopra in Fig. 47 of his review paper [22]. It is clear that our model may fit experimental data better than the models proposed in [18-21].

FUTURE RESEARCH DIRECTION

As clearly seen in Fig. 3, the temperature of the SMA wire changes periodically by about 20 degrees in each loading cycle at a loading frequency of 0.1Hz. In addition to stress-strain data, therefore, it is necessary to provide data including temperature and thermal boundary condition. As the strain rate is influential to the change in the alloy's temperature, constitutive data at higher strain rates should be measured. At the same time, it is necessary to establish a thermodynamic analytical model that may treat the constitutive behavior at higher strain rates.

CONCLUSION

We have briefly summarized our thermomechanical model of shape memory alloys based on the phase interaction energy function which we introduced as the general expression of the dissipation potential for phase transformations and rearrangements. The constitutive relationships with large hysteresis for the bounding loop and smaller hysteresis for an inner loop are well modeled. The effect of cyclic loading on the stress-strain is also included in the model. There remains much to be resolved in SMA research.

REFERENCES

1. Kamita, T., and Matsuzaki, Y. 1996. "Pseudoelastic Theory of Shape Memory Alloys," *Proceed. of SPIE's 1996 Symposium on Smart Structures and Materials*, 2717:509-516.

2. Kamita, T., and Matsuzaki, Y. 1997. "Pseudoelastic Hysteresis of Shape Memory Alloys," *Proceed. of First US-Japan Workshop on Smart Materials and Structures*, ed. by K. Inoue, et al., the Minerals, Metals & Materials Society, pp. 117-124.

3. Kamita, T., and Matsuzaki, Y. 1998. "One-Dimensional Pseudoelastic Theory of Shape Memory Alloys," *Smart Materials and Structures*, 7: 489-495.

4. Falk, F. 1983. "One-Dimensional Model of Shape Memory Alloys," *Archieves of Mechanics*, 35:63-84.

5. Mueller, I., and Xu, H. 1991. "On the Pseudo-Elastic Hysteresis," *Acta Metallic Materials*, 39:263-271.

6. Raniecki, B., Lexcellent, C., and Tanaka, K. 1992. "Thermodynamic Model of Pseudoelastic Behaviour of Shape Memory Alloys," *Archieves of Mechanics*, 44:261-284.

7. Leclercq, S. and Lexcellent, C. 1996. "A General Macroscopic Description of TheThermomechanical Behaviour of Shape Memory Alloys," *J. of The Mechanics and Physics of Solids*, 44:953-980.

8. Naito, H., Matsuzaki, Y., Funami, K. and Ikeda T. 2001. "Frequency Characteristics of Pseudoelasticity of Shape Memory Alloys," *Proceed. of 42nd AIAA Structures, Structural Dynamics and Materials Conference*, AIAA-2001-1355.

9. Matsuzaki, Y., Naito, H., Ikeda, T. and Funami, K. 2001. "Thermo-Mechanical Behavior Associated with Pseudoelastic Transformation of Shape Memory Alloys," submitted to *Smart Materials and Strucutre*.

10. Matsuzaki, Y., Kamita, T., and Ishida, A. 1997. "Stress-Strain-Temperature Relationship of Shape Memory Alloys," *Proceed. of SPIE's 1997 Symposium on Smart Structures and Materials*, 3241:230-236.

11. Matsuzaki, Y., Naito, H., and Ikeda, T. 2000. "Pseudoelastic Thermal Behavior of Shape Memory Alloys," to appear in *Proceed. of 10th Int. Conf. Adaptive Structures and Technologies*, Technomic Pub., Lancaster, pp. 208-214.

12. Matsuzaki, Y., Naito, H., and Yamamoto, T. 1998. "Pseudoelastic Behavior of Shape Memory Alloys with Temperature Change," *Proceed. of 40th Structures and Strength Conference,* Japanese Society for Aeronautics and Space Science, pp. 209-212.

13. Naito, H., Matsuzaki, Y., Sato, J., Funami, K. and Ikeda, T. 2001. "Stress-Strain-Relationship of Pseudoelastic Transformation of Shape Memory Alloys during loading Cycles," *Proceed. of 11th Int'l Conf. on Adaptive Structures and Technologies*, Technomic Pub., Lancaster, pp. 344-350, 2001.

14. Naito, H., Sato, J., Funami, K., Matsuzaki, Y. and Ikeda, T. 2001. "Analytical Study on Training Effect of Pseudoelastic Transformation of Shape Memory Alloys in Cyclic Loading," submitted to J. Intel. Mater. Syst. & Struct.

15. Tobushi, H., Iwanaga, H., Tanaka, K., Hori, T., and Sawada, T. 1991. "Stress-Strain-Tempreture Relations of TiNi Shape Memory Alloy Suitable for Thermomechanical Cycling," *Trans. Japan. Soc. Mech. Eng.*, Ser. A, 57:2747-2752.

16. Matsuzaki, Y., Funami, K and Naito, H. 2002. "Inner Loops of Pseudoelastic Hysteresis of Shape Memory Alloys," submitted to SPIE 2002 Symposium on Smart Structures and Materials to be held in March 2002 in San Diego.

17. Naito, H., Matsuzaki, Y. and Ikeda, T. 2001. "A Unified Model of Thermomechanical Behavior of Shape Memory Alloys," *Proceed. of SPIE 2001 Symposium on Smart Structures and Materials*, SPIE 4333-52.

18. Tanaka, K. 1986. "A Thermomechanical Sketch of Shape Memory effect: One-dimensional Tensile Behavior," *Res. Mechanica*, 18:251-263.

19. Liang, C., and Rogers, C.A. 1990. "One-dimensional Thermomechanical Constitutive Relations for Shape Memory Material," *J. Intel. Mater. & Syst.*, 1:207-234.

20. Brinson, L.C. 1993. "One Dimensional Constitutive Behavior of Shape Memory Alloys: Thermo-mechanical Derivation with Non-constant Material Functions," *J. Intel. Mater. & Syst.*, 4:229-242

21. Boyd, J.G. and Lagoudas, D.C. 1996. "A Thermodynamic Constitutive Model for the Shape Memory Materials. Part I. The Monolithic Shape Memory Alloys," *Int'l J. of Plasticity*, 12:805-842.

22. Chopra, I. 2001. "Recent Progress on the Development of a Smart Rotor System," *Proceed. of 11th Int'l Conf. on Adaptive Structures and Technologies*, ed. by Y. Matsuzaki, et al., Technomic Pub., Lancaster, pp. 455-500.

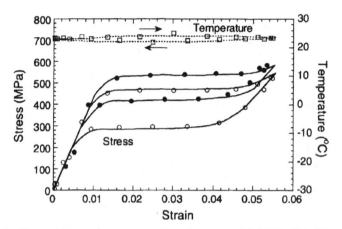

Fig. 1   Stress and temperature vs. strain at frequency of 0.001Hz from [8].
(Experiment: open symbols for T_S=296K and solid for T_S=313K; Analysis: curves)

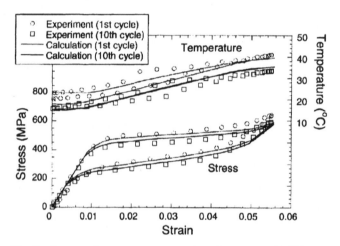

Fig. 2   Stress and temperature vs. strain at frequency of 0.1Hz.

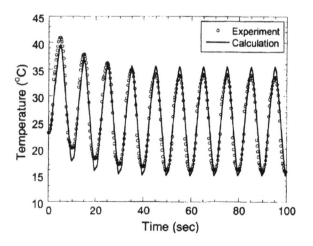

Fig. 3   Time history of temperature at frequency of 0.1Hz.

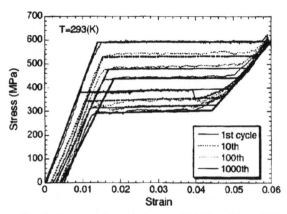

Fig. 4   Stress-strain at strain rate of 0.001 sec^{-1} from [13].
(Experiment: fluctuating curves; Analysis: straight lines)

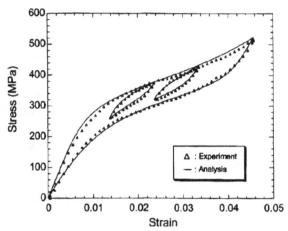

Fig. 5   Inner hysteresis loops [16].

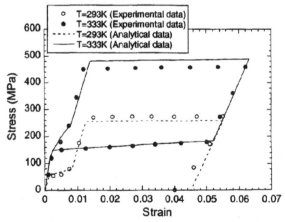

Fig. 6   Stress-strain of phase transformation rearrangement from [17].
(Experiment: symbols; Analysis: lines)

MANUFACTURE OF IONIC POLYMER ACTUATORS USING NON-PRECIOUS METAL ELECTRODES

Matt Bennett and Don Leo

ABSTRACT

Ionic polymer membranes are commonly used in fuel cell power generation, water electrolysis and desalinization, chlorine generation, and other niche applications. A more recent development is the potential use of these materials as electromechanical actuators and sensors. Currently, the performance characteristics and high cost of these actuators has prevented them from experiencing widespread use. This paper discusses novel approaches to overcome these obstacles. At this time ionic polymer actuators are only made using gold or platinum as the electrode in a lengthy and labor-intensive process. The current research focuses on using less costly metals and revising the metal deposition process. Several new methods allowing for faster deposition of metals onto ionic polymer membranes are developed and evaluated including electroless plating and impregnation/reduction. Using these methods, ionic polymer actuators have been made with nickel and copper electrodes and results show that they have the potential to perform as well as gold-electroded actuators. Also evalutated are actuators made with sputter-coated gold electrodes. The importance of the use of noble metals and the formation of dendritic electrodes is evaluated. Although interpenetration of the electrode is desireable for good adhesion and better actuator performance, electromechanical coupling is demonstrated without. Results demonstrate that non-precious metals are very effective as electrodes, but oxidation of the electrode is identified as the key issues with their use. Attempts are made to overcome this issue using thin gold protective coatings. Although the gold coatings improve the performance of the acutators, they are not effective in preventing the oxidation. Performance and longevity metrics for actuators made using these new methods are presented and future work is discussed.

INTRODUCTION

Ionic polymers are materials that exhibit ion selectivity due to their molecular structure. They consist of a hydrocarbon backbone with pendant acidic sidegroups that dissolve when hydrated. This dissolution frees the cation associated with each pendant acidic group to move throughout the polymer matrix while the anion maintains a bond to the hydrocarbon backbone. It is this property of ionic polymers which allows them to transmit cations while blocking anions and makes them useful in water electrolyzers, solid electrolyte fuel cells, and the like.

More recently, ionic polymers have been used as electromechanical actuators and sensors. In the early 1990s, three groups of researchers published papers that demonstrated the use of ion-exchange membranes as electromechanical sensors and actuators. Sadeghipour, Salomon,

---

[1]CIMSS, 307 Durham Hall, Blacksburg, VA 24061-0261, donleo@vt.edu

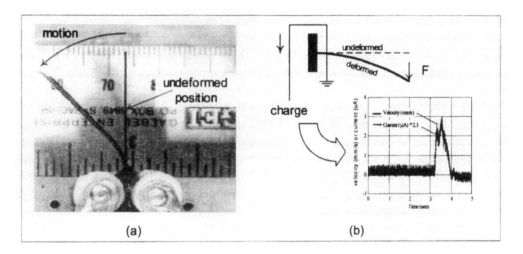

Figure 1: An ionic polymer membrane as (a) an actuator, (b) a sensor.

and Neogi developed an accelerometer using the ion-exchange membrane Nafion (a trademark of DuPont), which is a telfon-based polymer with sulfonic acid side groups [18]. They demonstrated that the voltage produced by a small cantilever constructed from Nafion was proportional to the acceleration. Thus, an electromechanical sensor was produced from an ion-exchange membrane. Almost concurrently with Sadeghipour et al, Oguro et al and Segalman et al published results demonstrating the use of Nafion-based composites as electromechanical actuators [15, 19]. Both groups demonstrated that mechanical deformation is produced in the membrane when an electric field is applied to the material.

The principle of electromechanical transduction is shown in Figure 1. Figure 1a is a picture of an ion-exchange membrane that has been hydrated and coated with a conductive electrode on both sides. Upon application of an electric field, the membrane bends towards the anode due to the motion of the mobile cations and water molecules within the polymer matrix. Also, work by Mallavarapu and Leo has shown that feedback control may be used effectively to control the motion of these actuators in a free bender configuration [11]. The converse property is exhibited in Figure 1b. A cantilever sample of an ion-exchange membrane has deformed by the application of a force to the tip of the sample. It is believed that macro motion of the membrane produces micro motion of the mobile cations, resulting in a charge imbalance across the electrodes. Several researchers have demonstrated that the quasi-static displacement of the polymer is correlated with the voltage that is produced by the membrane [20, 12], and more recently, Newbury and Leo demonstrated that the current induced in the membrane is proportional to the rate of motion caused by the mechanical deformation [13].

Since the early 1990s several other research groups have become involved with ionic polymer membranes as actuators. In 1993, Oguro et al patented the notion of using ionic polymer membranes as actuators [14]. Oguro et al have also demonstrated novel applications of ionic polymer actuators, using them in active catheters [8], underwater microrobots [7], and as elliptical drive elements in rotary and linear motors [23]. Bar-Cohen has focused more on the use of these devices for interplanetary applications, and has demonstrated their ability to operate as low-mass actuators for robotic grippers and dust wipers for camera lenses [2]. Although these represent novel applications for ionic polymer actuators, the state of the current technology has limited the use of these materials to laboratory demonstrations.

To this end, research has also focused on increasing the work output of these actuators.

Shahinpoor has made several developments in this area, showing that more force can be obtained from ionic polymer actuators by increasing the conductivity of the electrode [22] and by decreasing the porosity of the electrode [9]. Work by Shahinpoor and others has also shown that increased performance can be obtained by changing the mobile cation inside the membrane [1, 21].

While these efforts have increased the effectiveness of ionic polymer actuators, they have not addressed the high cost of these devices. At this time ionic polymer actuators are made using gold or platinum as the electrodes. The high cost of these metals (~$285 per ounce for gold and ~$480 per ounce for platinum) have kept the price of these transducers relatively high and prevented them from being able to compete with other actuator technologies. The goal of the current research is to reduce the cost of ionic polymer membrane transducers by employing less expensive metals as the conductive electrodes.

APPROACH

The principle of electromechanical transduction in ionic polymer membrane devices is based on the motion of hydrated mobile cations within the polymer matrix. This motion is induced by electrostatic forces which arise from the application of an electric field across the membrane thickness—See Figure 2.

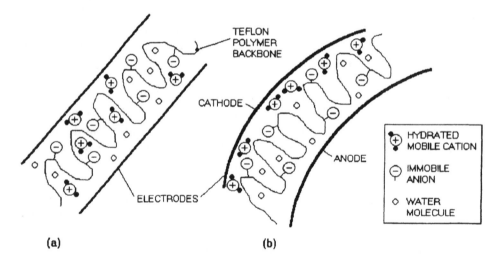

Figure 2: The principle of electromechanical transduction in ionic polymer membranes with conductive electrodes: (a) voltage off, (b) voltage on.

Although no explicit constitutive equations for ionic polymer membranes transducers have been presented to date, there are several key results which most researchers agree on [10]

- The macro motion of the membrane is caused my micro motion of the mobile cations,

- In the case of Nafion and most other ionic polymers, this motion will not occur unless the membrane is hydrated,

- The motion of the cations is caused by imposition of an electric field across the membrane thickness,

- A larger electric field can be created by increasing the conductivity of the electrode and by increasing the interfacial area between the electrode and polymer, and

- Changing the mobile cation inside the membrane will change the characteristics of the bending response of the membrane.

Currently, the main research groups investigating ionic polymers as actuators and sensors are using either gold or platinum as the electrodes. Although many researchers have suggested using these metals as the electrodes [14, 16], no definitive research has been presented to prove that this is necessary. It is the goal of this work to show that any conductive metal may be used as the electrode and to identify several key issues associated with the use of metals other than gold and platinum.

A typical process for the formation of electrodes on Nafion membranes consists of the following basic steps [17, 1]

1. Mechanical roughening of the membrane by abrading with emory paper or sandblasting,

2. Hydrating the membrane in high temperature water,

3. Ultrasonic washing to remove surface particulates,

4. Washing in hydrochloric acid to remove surface oils,

5. Impregnating metal cations into the membrane by soaking in solution of the metal to be plated,

6. Reducing the absorbed metal cations in aqueous bath.

Following this process the mobile cation may be exchanged for any other cation by soaking the plated membrane in a solution rich in that cation. As no mechanism for preventing dehydration of the membrane has been developed yet, tests are typically run either under water or with frequent addition of water by brushing. Dr. Shahinpoor of the University of New Mexico has been making ionic polymer membrane actuators using variations of this process for years and reports the processing time to be around 30 hours [17]. Because Nafion is a teflon-based polymer, adhesion of the membrane is an issue which is difficult to overcome. The process descibed above solves this problem by reducing the metal ions <u>inside</u> the membrane, thus producing an electrode which strongly interlocks with the polymer.

The initial approach of the current research was to determine to what extent this interlocking was necessary. As a first step, gold electrodes were sputter-coated onto dry Nafion membranes. Recently, Zhou et al have used a sputtering process with subsequent electroplating to plate gold onto Nafion membranes [24]. Because the sputter-coating process results in only a surface deposition, the interlocking associated with the impregnation-reduction process is nonexistent. Results with this method showed that the electromechanical response of the actuators was evident, but significantly diminished as compared to gold-plated actuators obtained from Dr. Shahinpoor, which were used as a baseline. One reason for the drop in performance is the high surface resistance of the sputtered electrodes. The membrane must be in a dry state during the sputtering process and then hydrated after. Because of the large amount of water that the Nafion absorbs during hydration, it swells in size and stretches the gold electrode, increasing its resistance. As-plated, the surface resistance was measured to be approximatedly 7.5 Ω/cm; once hydrated, this number may become as high as 100 Ω/cm. Based upon this result, it was apparent that the interlocking of the electrode and the polymer was not necessary for the electromechanical transduction to take place. It was also apparent that the electrode should be plated in an aqueous process after the hydration of the membrane in order to acheive higher conductivity in the electrodes.

A second process for plating electrodes onto Nafion membranes involed the electroless reduction of nickel using a palladium catalyst. This process was derived from a process to plate through-holes on teflon printed circuit boards. Using a proprietary technology of Solution Technology Systems, Inc. (Redlands, CA), a very thin layer of palladium is deposited onto the surface of the Nafion membrane by dipping into several aqueous solutions. Once the palladium layer is deposited, nickel is plated over the palladium in an electroless process. The source of the nickel ions is nickel (II) chloride and the reducing agent is sodium hypophosphite. Because the nickel ions will only reduce at the palladium interface, this process also results in a purely surface deposition of metal. Results with this process showed that the electromechanical response of the actuators was evident, but again the performance was not as good as the baseline material. Also, significant adhesion problems were encounted when using this process. Although electrodes with relatively low surface resistance measurements were obtained (2.0-3.0 Ω/cm), the electrodes would typically flake off upon application of the electric field. The poor performance of the actuators prepared in this way was attributed to this loss of adhesion of the electrode.

Based upon these results, a third process was developed that involved the interpenetration of the metal electrode into the polymer matrix, much like that used by other researchers. Investigation into the available literature revealed that the in the work being done on ionic polymer actuators, only gold and platinum had been used as the conductive electrodes. However, further investigation revealed that researchers in other fields, particularly electrochemistry, had demonstrated success in plating Nafion membranes with many other metals, and that most of this work was being done using impregnation/reduction techniques. For example, Dewulf and Bard and Cook et al have both demonstrated methods to electrode Nafion membranes with copper by dissolving copper ions into the membrane and reducing them in situ [6, 5]. Their work focused on the use of these membranes to perform the electrochemical reduction of CO_2. More recently, Chen and Chou have plated nickel, lead, copper, and silver onto Nafion by a similar method for the reduction of benzaldehyde [4]. Based on these findings, a process was developed to plate Nafion membranes with interpenetrating copper electrodes for the purpose of creating ionic polymer membrane transducers. This process was found to result in actuators with significantly better performance than those obtained through the other two processes. The electrodes made thereby were found to have surface resistance measurements around 2.0-3.0 Ω/cm. Also, using this process a sample can be made in about four hours.

EXPERIMENTAL

Testing of the ionic polymer membranes was performed to evaluate the performance of these devices as compared to the baseline material obtained from UNM. Blocked force and free deflection data was obtained by clamping plated samples of hydrated ionic polymer membranes in a cantilever configuration and applying a voltage across the electrodes. The free displacement or blocked force at the tip was then measured using a high-speed digital camera (MotionScope PCI 2000 S) or 10-gm load cell (Transducer Techniques GSO-10), respectively. Longevity testing of the samples was performed by immersing four samples at a time in a DI water bath and actuating them with a 1.5V, 1.0Hz sine wave continuously for two hours using an HP 8904A function generator through an HP 6825A amplifier. The free displacement at the tip of each actuator

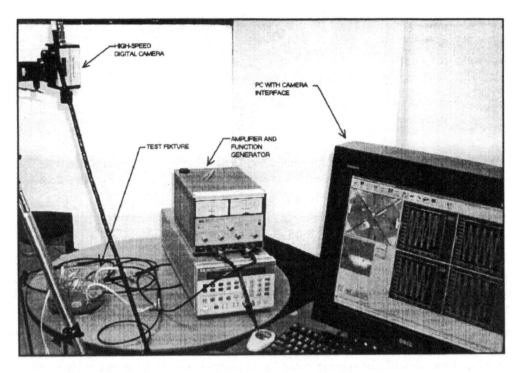

Figure 3: Photo of the setup used to perform the longevity testing.

was measured at regular intervals using the PCI 2000 S high-speed digital camera. See Figure 3 for an illustration of the setup used to perform the longevity testing. Gold sputter-coating was performed using a Bal-Tec SCD005 sputter coater with a current of 30 mA for 3 minute intervals. Cross-sectional micrographs of plated Nafion membranes were obtained by impregnating the samples in epoxy and microtoming to obtain a smooth surface. This was done to get a good cross-section with minimal distortion. Images of the micotomed surface were then captured using a Leo 1550 Field-Emmision Scanning Electron Microscope (FE-SEM) in the back-scatter mode. Surface resistance measurements of the electrodes were made using a Fluke 87 digital multimeter.

RESULTS

Upon development of the plating methods dicussed above, a more thorough study was begun to identify the key issues associated with making ionic polymer actuators in these ways. Surface resistance measurements of the gold sputter-coated samples were measured as a function of the thickness of the gold layer—see Figure 4a. One coating cycle represents sputtering under 30 mA of current for 3 minutes per side. As can be seen in the figure, the resistance of the gold electrode decreases rapidly as the coating becomes thicker, but quickly reaches a minimum value around 7.5 Ω/cm. Tests with gold sputter-coated Nafion actuators showed that electromechanical transduction was possible; blocked tip forces of 5mm wide X 20mm long cantilevered samples were on the order of 0.5 mN for a 4.0V input. Figure 4b shows the free tip displacement of a gold-sputtered sample to a 1.5V square wave input. Note that the tip displacement is on the order of 0.25 mm, whereas the tip displacements copper-plated samples made using the impregnation/reduction process have been observed to exceed 10 mm. Our work with this

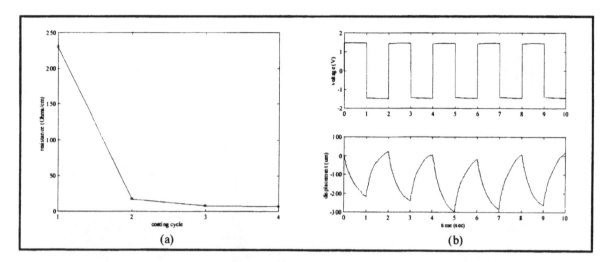

Figure 4: (a) Surface resistance of sputter-coated gold as a function of coating cycle. (b) Free tip deflection of a gold-sputtered Nafion sample to a 1.5V square wave input. The sample is 5mm wide with a 20mm free length.

process was short-lived due to issues with the mechanical stability and high resistance of the electrodes made thereby.

Following the work with the sputter-coated gold samples, the decision was made to move on to other plating processes. A more thorough analysis of the performance and setbacks of actuators made with nickel and copper electrodes was made. The thickness of the copper electrodes was measured using scanning electron microscopy. Figure 5 shows a micrograph of the cross-section of a copper-plated Nafion membrane. From the figure, the penetration of the copper electrode is

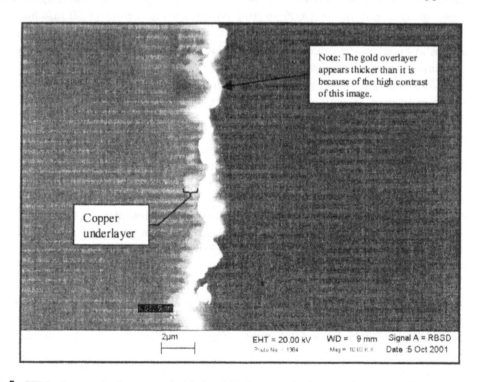

Figure 5: SEM micrograph of a copper/gold-plated Nafion sample showing the copper electrode penetration.

around 0.6 μm. Electrodes made by this method had surface resistances on the order of 2.0-3.0 Ω/cm. Although the copper-plating process works better than any of the other processes which have been discussed, there remains an issue of electrode degredation over time. Instability of the copper was observed early, as the electrodes underwent a color change and an increase in resistance upon application of an actuation voltage to the membrane. In order to understand this change, x-ray photoelectron spectroscopy (XPS) was implemented to identify this new substance. The XPS results indicated that the copper electrodes were oxidizing when a voltage was applied. In a similar manner, XPS was also used to identify oxidation as the cause of the delamination and associated drop in performance of the nickel-plated samples. The loss of adhesion of the nickel electrode can be easily explained, as the growth of an oxide layer will typically generate stresses between the oxide layer and the substrate [3].

It is important to note that although both the copper and nickel electrodes oxidized, the nickel electrodes would flake off of the Nafion membranes upon oxidation, whereas there was no loss of adhesion of the copper electrodes. This is likely due to the strong mechanical linking afforded by the copper impregnation/reduction process. This oxidation is not surprising, as it is well-known that copper and nickel are not highly stable metals and will readily grow oxide films if exposed to air. The oxidation of the copper- and nickel-plated Nafion actuators can be explained by the fact that the ionic polymer-metal composite is effectively an electrochemical cell where the metal electrodes and molecular oxygen are the reactants and the metal oxide is the product. The applied voltage feeds the reaction through a constant current of electrons—the higher the voltage, the faster the oxidation reaction will occur. This problem is not encountered when using gold and platinum as the electrode because these metals are much more stable and hence much harder to oxidize than copper and nickel.

Initial findings with the copper electrodes indicated that they would oxidize overnight if the

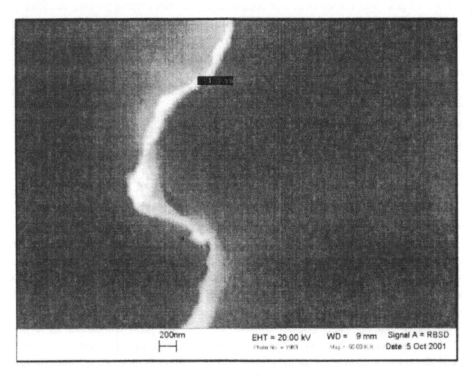

Figure 6: SEM micrograph showing the gold overlayer. Please note that the copper underlayer is not visible due to the low contrast of this image.

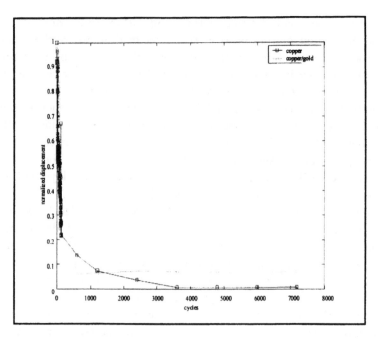

Figure 7: Normalized displacement versus number of actuation cycles for a copper- and copper/gold-plated Nafion actuator. The input was a 1.5V, 1.0Hz sine wave.

sample was stored under DI water. If a voltage was applied to the plated sample, the oxidation would occur much more rapidly. As a first attempt at protecting the copper electrode, a very thin (~55 nm) layer of gold was plated over the copper using an electroless plating method (Alfa Aesar stock no. 42307). The addition of this gold layer improved the surface resistance of the electrode to less than 1.0 Ω/cm. See Figure 6 for a micrograph of the gold layer. Although the gold layer did improve the stability of the copper by preventing spontaneous oxidation under storage conditions, the electrode still oxidized when actuated with an applied voltage. This oxidation leads to a loss of conductivity in the electrode and a corresponding loss of performance of the ionic polymer actuator. Figure 7 shows an illustration of how electrode oxidation can affect the performance of these devices. This plot shows the free tip deflection (as normalized by the initial value) of a copper-plated Nafion sample versus the number of actuation cycles both with and without the gold overlayer. As can be seen, the rate of degradation of both samples is about the same, but the sample with the gold overlayer still moves even after the copper underlayer has oxidized completely, whereas the sample without the gold overlayer ceases to deform entirely by the end of the test. This is because the gold layer does not degrade, and so although most of the effective polymer/metal interface is lost due to the oxidation of the copper, the gold still serves to transmit a charge down the length of the polymer. Without the gold layer this is not the case, and the electrode becomes completely inneffective. This is an interesting point, but belies the fact that the gold layer did not protect the copper electrode from oxidation under actuation conditions. This is likely due to the high oxygen solubility of the Nafion polymer and the porosity of the thin gold layer.

Tip force of the nickel- and copper-plated samples was also measured as a performance metric. A force-deflection curve is a typical measure of the performance of active materials such as piezoelectrics or shape memory alloys. With ionic polymer actuators, the concept of a force-displacement curve is a bit different. When actuated with a step voltage, an ionic polymer actuator will undergo a large initial deformation followed by a slow relaxation. It is believed that this relaxation is caused by the back-diffusion of the hydrated cations as driven by osmotic pressure. This is illustrated in Figure 8. Therefore, the steady-state force or deflection is typically very small. For this reason, force/deflection curves for ionic polymer actuators are typically built using peak values. Figure 9 is a force-deflection plot of a copper/gold-plated Nafion actuator. The performance of this actuator represents values which are comparable to gold-plated samples obtained from Dr. Shahinpoor at the University of New Mexico [13].

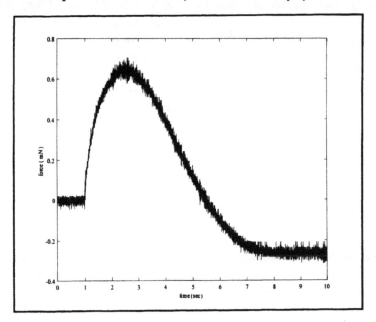

Figure 8: Time response of the blocked tip force of a nickel-plated Nafion actuator to a 3.0V step input

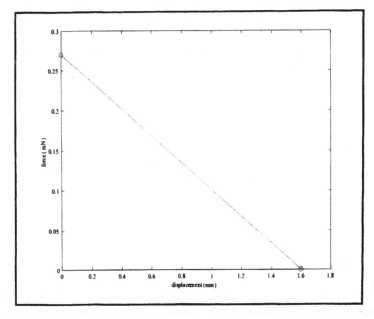

Figure 9: Blocked force and free deflection at the tip of a copper/gold-plated Nafion membrane actuated with a 1.0V step immediately after manufacture. The sample is 5mm wide and has a 17mm free length.

SUMMARY

Although electromechanical coupling was observed in all of the ionic polymer actuators made in this study, the degree to which this coupling was present was found to be dependant on a number of factors. First, the surface resistance of the electrodes should be as low as possible. The gold sputtering process yielded actuators with surface resistances around 7.5 Ω/cm. With the nickel plating process this number was reduced to 2.0-3.0 Ω/cm. Using the copper plating process descibed with the gold overlayer, measurements as low as 1.0 Ω/cm were obtained. Also, strong mechnical linking of the polymer and the metal should be obtained in order to promote adhesion under the large bending strains that occur during actuation. This also leads to a larger polymer/metal interface and a correspondingly better performing actuator. The gold sputtered electrodes were found to rub off of the Nafion membranes if handled too roughly, and the nickel electrodes flaked off upon actuation. Although the copper electrodes oxidized, there was no apparent loss of adhesion to the membrane. Based on these results, the impregnation/reduction process is most favorable for formation of electrodes on Nafion membranes, although the choice of metal used is very important. Non-noble metal electrodes have been found to be relatively unstable and will oxidize under actuation conditions.

CONCLUSIONS

Actuation and sensing technologies represent a relatively new application for ionic polymers membranes. Previous research work has shown that ionic polymer membranes may be used to make large displacement/low force actuators and sensors using gold and platinum electrodes. The current work demonstrates the use of nickel and copper as viable electrodes for these devices. Although the interpenetration of the metal electrode into the polymer is desired, it is not necessary for electromechanical transduction. Preliminary results indicate that ionic polymer actuators made with nickel and copper electrodes have the potential to perform as well as those made with gold and platinum. Oxidation is identified as the key issue associated with the use of non-noble metal electrodes. Actuators made with gold electrodes did not have the oxidation problem, but issues were encountered with the sputtering process involving high surface resistance and poor adhesion of the electrode.

FUTURE WORK

Planned future work by the authors inclues the development of protective barrier coatings for ionic polymer transducers with the goal of preventing the oxidation of non-noble metal electrodes and the dehydration of the polymer. Also possible is the investigation of the use of alloys or conductive non-noble metal oxides as electrodes and the study of new actuation and control schemes for these devices.

ACKNOWLEDGEMENTS

The authors would like to thank Dr. Mohsen Shahinpoor and Dr. Kwang Kim of the University of New Mexico for donating some of the Nafion polymer actuators used in this work. This material is based upon work supported by the National Science Foundation under Grant No. 9975678 and Grant No. CMS 0093889, Program Officer Dr. Alison Flatau.

References

[1] Y. Abe, A. Mochizuki, T. Kawashima, S. Yamashita, K. Asaka, and K. Oguro. Effect on bending behavior of counter cation species in perfluorinated sulfonate membrane-platinum composite. *Polymers for Advanced Technologies*, 9:520–526, 1998.

[2] Y. Bar-Cohen, S. Leary, M. Shahinpoor, J. Harrison, and J. Smith. Flexible low-mass devices and mechanisms actuated by electroactive polymers. In *EAP Actuators and Devices*, volume 3669, pages 51–56. SPIE, 1999.

[3] J. V. Cathcart. *The Structure and Properties of Thin Oxide Films*, chapter 2, pages 27–29. American Society for Metals, 1970.

[4] Y.-L. Chen and T.-C. Chou. Metals and alloys bonded on solid polymer electrolyte for electrochemical reduction of pure benzaldehyde without liquid supporting electrolyte. *Journal of Electroanalytical Chemistry*, 360:247–259, 1993.

[5] R. L. Cook, R. C. MacDuff, and A. F. Sammells. Ambient temperature gas phase CO_2 reduction to hydrocarbons at solid polymer electrolyte cells. *Journal of the Electrochemical Society*, 135:1470–1471, 1988.

[6] D. W. Dewulf and A. J. Bard. The electrochemical reduction of CO_2 to CH_4 and C_2H_4 at Cu/Nafion electrodes (solid polymer electrolyte structures). *Catalysis Letters*, 1:73–80, 1988.

[7] S. Guo, T. Fukuda, N. Kato, and K. Oguro. Development of underwater microrobot using ICPF actuator. In *International Conference on Robotics and Automation*, pages 1829–1834. IEEE, 1998.

[8] S. Guo, T. Fukuda, K. Kosuge, F. Arai, K. Oguro, and M. Negoro. Micro catheter system with active guide wire. In *International Conference on Robotics and Automation*, pages 79–84. IEEE, 1995.

[9] K. J. Kim and M. Shahinpoor. The synthesis of nano-scaled platinum particles (NSPP)-their role in performance improvement of ionic polymer-metal composite (IPMC) artificial muscles. In *EAP Actuators and Devices*, volume 4329, pages 189–198. SPIE, 2001.

[10] D. J. Leo and K. Newbury. Chemoelectric and electromechanical modeling of ionic polymer materials. In *12th International Conference on Adaptive Structures and Technologies*, 2001. to appear.

[11] K. Mallavarapu, K. Newbury, and D. J. Leo. Feedback control of the bending response of ionic polymer-metal composites actuators. In *EAP Actuators and Devices*, volume 4329, pages 301–310. SPIE, 2001.

[12] M. Mojarrad and M. Shahinpoor. Ion-exchange-metal composite sensor films. In *Smart Materials and Structures*, volume 3042, pages 52–60. SPIE, 1997.

[13] K. Newbury and D. J. Leo. Mechanical work and electromechanical coupling in ionic polymer bender actuators. In *IMECE*. ASME, 2001. to appear.

[14] K. Oguro. Actuator element. U.S. Patent 5,268,082, 1993.

[15] K. Oguro, Y. Kawami, and H. Takenaka. Bending of an ion-conducting polymer film-electrode composite by an electric stimulus at low volage. *Journal of Micromachine Society*, 5:27–30, 1992.

[16] K. Onishi, S. Sewa, K. Asaka, N. Fujiwara, and K. Oguro. Bending response of polymer electrolyte actuator. In *EAP Actuators and Devices*, volume 3987, pages 121–128. SPIE, 2000.

[17] T. Rashid and M. Shahinpoor. Force optimization of ionic polymermic platinum composite artificial muscles by means of orthogonal array manufacturing method. In *EAP Actuators and Devices*, volume 3669, pages 289–298. SPIE, 1999.

[18] K. Sadeghipour, R. Salomon, and S. Neogi. Development of a novel electrochemically active membrane and 'smart' material based vibration sensor/damper. *Smart Materials and Structures*, 1:172–179, 1992.

[19] D. Segalman, W. Witkowski, D. Adolf, and M. Shahinpoor. Theory of electrically controlled polymeric muscles as active materials in adaptive structures. *Smart Materials and Structures*, 1:44–54, 1992.

[20] M. Shahinpoor, Y. Bar-Cohen, J. Simpson, and J. Smith. Ionic polymer-metal composites (IPMCs) as biomimetic sensors, actuators and artificial muscles - a review. *Smart Materials and Structures*, 7(6):R15–R30, 1998.

[21] M. Shahinpoor and K. J. Kim. Effect of counter-ions on the performance of IPMCs. In *EAP Actuators and Devices*, volume 3987, pages 110–120. SPIE, 2000.

[22] M. Shahinpoor and K. J. Kim. The effect of surface-electrode resistance on the performance of ionic-polymer metal composite (IPMC) artificial muscles. *Smart Materials and Structures*, 9:543–551, 2000.

[23] S. Tadokoro, T. Murakami, S. Fuji, R. Kanno, M. Hattori, and T. Takamori. An elliptical friction drive element using an ICPF (ion conducting polymer gel film) actuator. In *International Conference on Robotics and Automation*, pages 205–212. IEEE, 1996.

[24] W. Zhou, W. J. Li, N. Xi, and S. Ma. Development of force-feedback controlled nafion micromanipulators. In *EAP Actuators and Devices*, volume 4329, pages 401–410. SPIE, 2001.

CHEMOELECTRIC AND ELECTROMECHANICAL MODELING OF IONIC POLYMER MATERIALS

Kenneth M. Newbury and Donald J. Leo

ABSTRACT

Ion-exchange membranes known as ionic polymers exhibit both chemoelectric and electromechanical transduction properties. Ionic polymers are the base materials for proton-exchange membrane fuel cells, and several years ago it was shown that these materials exhibited electromechanical sensing and actuation properties. In this paper, we develop charge and water transport models for the purpose of modeling both chemoelectric and electromechanical transduction properties in ionic polymers. Expressions for steady-state chemoelectric transduction are developed from previous work in fuel cell modeling. The same continuity and conservation expressions are used to model electromechanical transduction in ionic polymer materials. Analysis of the nonlinear coupled expressions demonstrates that the transient electromechanical response cannot be modeled without assuming water flow within the polymer. It is also shown that the nonlinear expressions reduce to a set of linear expressions only under the assumptions of steady-state operation and zero concentration gradient.

INTRODUCTION

A majority of applications for active material systems require an external power source to operate. Aside from passive implementations of active materials, such as energy-dissipating piezoelectric shunts, most applications require an external energy source to power actuators, sensors, and associated electronics for control and signal processing. Oftentimes the weight, size, and availability of a power source can be a determining factor in whether an active system is feasible for a given application.

Recently we have begun a research program for the development of active material systems that seamlessly integrate chemoelectric power generation with electromechanical transduction. The ability to integrate these different types of energy conversion is dependent on the functionality of the active material. For example, piezoelectric materials are able to operate as both a generator and motor, making it possible to *harvest* energy from an external source. But the composition of the material precludes the conversion of stored chemical energy to electrical energy. The same is true of shape memory materials, electrostrictive materials, and rheological fluids.

One class of material that is able to perform chemoelectric and electromechanical energy conversion is *ionic polymer* materials. Ionic polymer materials consist of a stiff, polymer backbone

[1]CIMSS, 307 Durham Hall, Blacksburg, VA 24061-0261, donleo@vt.edu

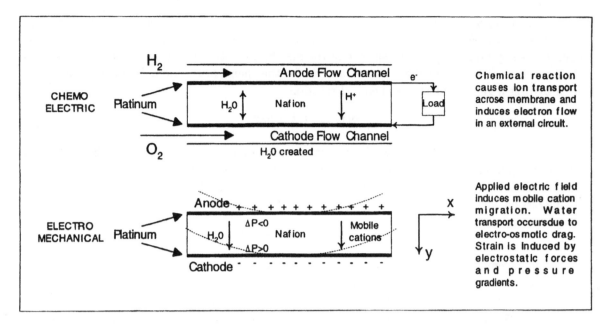

Figure 1: Electrochemical (top) and electromechanical (bottom) coupling associated with ion-exchange membranes.

with negatively-charged acidic sidegroups. Neutralizing the polymer with a positively-charged ion produces an ionic bonded between the sidegroup and the cation. Hydrating the material weakens the bond between the sidegroup and the positive ion due to disassocation of the cation from the polymer backbone.

The key attribute of ionic polymer materials is that the negatively-charged sidegroups remain fixed to the backbone while the cation is free to migrate through the hydrated polymer. Ionic conductivity is the fundamental property in both chemoelectric power generation and electromechanical transduction. When an ionic polymer is properly supplied with hydrogen and air, chemical reactions at the surface of the material produce proton conduction through the thickness of the material. Electron conduction produces DC power generation from the stored chemical energy in the reactants. In a reciprocal manner, applying a field to the polymer initiates ionic migration due to attraction between the mobile cations and the surface charge. Although the exact mechanism for electromechanical coupling remains uncertain, it is believed that ionic migration induces stress in the material due to pressure gradients and electrostatic forces.

The objective of this paper is to begin development of chemoelectric and electromechanical transduction models for ionic polymer materials. In this paper we will provide detail on the principles of the energy conversion mechanisms and relate the models to the state of the art in both fields. The common attributes of the transduction models will be identified and the current assumptions regarding the dominant mechanisms discussed. An initial model of both energy conversion mechanisms is presented.

PRINCIPLES OF CHEMOELECTRIC AND ELECTROMECHANICAL ENERGY CONVERSION

The principles of chemoelectric and electromechanical energy conversion are shown in Figure 1. Figure 1 (top) represents the use of ionic polymer membrane as a solid electrolyte for a proton exchange membran(PEM) fuel cell. A PEM fuel cell consists of an ionic polymer membrane coated with a conductive surface on both sides of the material. The conductive surface is constructed by impregnating the surface of the membrane with a platinum-loaded carbon mesh that serves as the charge collector and catalyst for the chemical reaction. Supplying the anode with hydrogen gas produces the chemical reaction

$$2H_2 \rightarrow 4H^+ + 4e^-. \tag{1}$$

The ion selectivity of the polymer allows proton conduction across the membrane but prohibits the conduction of electrons. Connecting the cathode to the anode through an electric circuit produces electron conduction through an external load, thus producing DC electric power. Supplying oxygen gas (or air) to the cathode produces the chemical reaction

$$O_2 + 4H^+ + 4e^- \rightarrow H_2O. \tag{2}$$

Water management in a fuel cell is a critical issue because the ionic conductivity of ion-exchange membranes is strongly dependent on the hydration of the material. Although the chemical reaction at the cathode is a source of water in the system, as shown in equation (1), the water production is typically not sufficient to maintain proper water balance. Humidification of the reactants is required for most fuel cell systems to maintain proper water balance within the membrane. This solution presents another set of difficulties because too much water can reduce the power output of the fuel cell due to 'flooding' of the membrane. As Springer et al. (1991a) states, "... water management within the fuel cell involves walking a tightrope between" dehydration and cathode flooding. Both extremes cause cell failure through membrane degradation.

The fundamental principles of electromechanical transduction in ionic polymer materials are shown in Figure 1 (bottom). In contrast to the fuel cell application, in which chemical reactants produce DC electric power and water, electromechanical transduction involves the coupling between the electric field, electric displacement, the applied stress, and the induced strain.

Although predictive models of electromechanical coupling are still under investigation, it is widely agreed that ionic motion and water transport play a key role in the relationship between applied electric field and mechanical deformation. As in the fuel cell application, hydrating the ionic polymer weakens the bond between the positively-charged cation and the negatively-charged sulfonate group. Application of an electric field produces electrostatic forces between the surface charge and the cations contained within the polymer matrix and results in a net ionic motion towards the cathode. The existence of water bonded to the cations and the electoosmotic drag produced by ionic motion cause migration towards the cathode resulting in a pressure gradient within the material. A moment is produced by the pressure gradient and causes bending about the anode. The pressure gradient relaxes over time due to the reduction in the concentration gradient of water due to ionic motion.

This model of ionic motion and water transport is plausible given some of the experimental observations for ionic polymer actuators. A step change in electric field produces an initial, 'fast' response that peaks in a time on the order of 100 milliseconds to 1 second depending on the geometry of the polymer. A 'slow' relaxation that takes between 10-30 seconds then occurs before the material will reach steady-state. These two phenomena are related to the time scales of the ionic motion and back diffusion of water. The initial response is believed to be related to the initial response of the ions to the application of the electric field, whereas the relaxation to steady state is related to the back diffusion of water through the polymer. For a recent discussion, see the work by Mallavarapu et al. (2001).

Ionic motion is also responsible for the sensing properties of ionic polymer materials. Newbury and Leo (2001) have measured the charge flow of an ionic polymer cantilever when subjected to a mechanical strain and shown that the current is proportional to the rate of motion of the material. Thus, it is plausible to assume that the charge flow is induced by the motion of cations that are weakly bonded to the polymer backbone. The amount of charge flow induced by mechanical deformation is two to three orders of magnitude less than the charge flow that occurs when the same deformation is induced by an electric field. This is explained by the theory that the number of mobile cations that are produced is much smaller under the application of mechanical strain than the application of an electric field.

IONIC POLYMER MODELING

Comparing the principles of chemoelectric and electromechanical energy conversion we see that charge transport and water transport are critical factors for power generation and electromechanical coupling. For this reason our initial modeling efforts have focused on the derivation of transport equations that describe the flux of charged (cationic) and uncharged (water) species through the ionic polymer. Transport models of ionic polymers will be separated into a chemoelectric model and an electromechanical model. The chemoelectric model will address the problem of chemical to electric energy conversion while operating under steady-state conditions, whereas the electromechanical model will address the problem of electrical to mechanical energy conversion under transient and steady-state operation.

Chemoelectric Modeling

Chemoelectric modeling of ionic polymer materials has been performed by the electrochemical research community for the purpose of fuel cell analysis. Although ionic polymers were developed in the late 1960s, interest in fuel cell modeling increased substantially in the late 1980s due to development of ionic polymers with higher ionic conductivity and chemoelectric power density. The complexity of the chemoelectric models can be traced fairly well by reviewing the literature from the early 1990s to the middle of the decade. A model of a fuel cell cathode was presented by Bernardi and Verbrugge (1991) for the purpose of estimating the polarization curve of a polymer electrolyte fuel cell. This work was followed up by a more complete model the following year (Bernardi and Verbrugge, 1992). More sophisticated models of fuel cells were developed by Springer et al. (1991b) and Nguyen and White (1993). This work proceeded into the late 1990s with work by Yi and Nguyen (1998) focusing on the modeling of a complete fuel cell stack.

The primary difference between the different bodies of work are the assumptions about heat transfer, water management, and dimensionality. The early models by Bernardi and Verbrugge (1991) were one-dimensional, constant temperature models of the limiting electrode of the fuel cell. These assumptions were relaxed by future researchers, resulting in a more comprehensive but more.complex model.

In our work the focus is on coupling the chemoelectric energy conversion with electromechanical transduction under conditions in which the heat transfer and water management will not be the dominant issue. For this reason we will concentrate on the one-dimensional, constant temperature models by Bernardi and Verbrugge (1991) as the basis for our initial modeling efforts. A detailed discussion of the derivation is presented in their work. In this paper we will summarize the modeling results and relate the model to the development of a coupled chemolectric-electromechanical system.

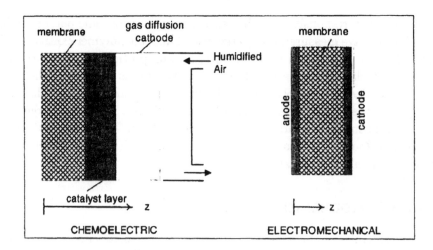

Figure 2: Geometry for chemoelectrical model (left) and electromechanical model (right).

The geometry studied in Bernardi and Verbrugge (1991) is shown in Figure 2. The geometry consists of an ionic polymer membrane bonded to a gas diffusion cathode through a catalyst layer. Oxygen is supplied to the electrode through a chamber that carries pressurized, humidified air to the cathode. The model is separated into three regions:

1. Membrane

2. Catalyst Layer

3. Gas Diffuser Layer

The salient modeling features for each region will be discussed and a summary of the steady-state equations are shown in Table 1.

Model of the Ionic Polymer Membrane

The membrane region of the fuel cell is characterized by the flux of charged and uncharged species. The equations for steady-state operation of a fuel cell have been developed by Bernardi and Verbrugge (1991) and are summarized in Table 1. Flux is modeled with the Nernst-Planck equations that include a convection term,

$$N_i = -z_i \frac{F}{RT} \mathcal{D}_i c_i \nabla \Phi - \mathcal{D}_i \nabla c_i + c_i v. \tag{3}$$

The three terms on the right-hand side of the equation are flux due to migration, diffusion, and convection, respectively. Applying current conservation, mass conservation, and incompressible flow assumptions to the one-dimensional problem, we can derive a relationship between potential and current density, equation (M3), which states that the current and voltage in the membrane are related through Ohm's Law with an additional term due to motion of the charged fluid in the polymer. Application of the continuity and flux equations to the uncharged oxygen in the membrane results in equation M6, whereas conservation of current and incompressible fluid assumptions imply that steady-state velocity in the membrane is constant and the pressure and potential variation through the membrane is linear.

Table 1: Governing equations for steady-state chemoelectric transduction (Bernardi and Verbrugge, 1991)

	Membrane (M)	Catalyst Layer (C)	Gas Diffuser (G)
1	$i = -I$	$\frac{di}{dz} = ai_o e^{\alpha_a f(\phi_{solid} - \phi)} - e^{-\alpha_c f(\phi_{solid} - \phi)}$	-
2	-	$\sigma_c^{eff} \frac{d\phi_{solid}}{dz} = I + i$	$\sigma_d^{eff} \frac{d\phi_{solid}}{dz} = I$
3	$-\mathcal{K}\frac{d\phi}{dz} = i - Fc_f v$	$-\mathcal{K}^{eff}\frac{d\phi}{dz} = i - Fc_f v$	-
4	$\frac{dv}{dz} = 0$	$\rho\frac{dv_s}{dz} = -\frac{s_w}{nF}\frac{di}{dz}$	$\rho\frac{dv_s}{dz} = -\frac{dN_{w,g}}{dz}$
5	-	-	$\frac{p_L}{RT}\frac{D_{w-N_2}^{eff}}{x_{N_2}}\frac{dx_{N_2}}{dz} = \frac{I}{4F}r_{N_2} + N_{w,g}$
6	$\mathcal{D}_{O_2}\frac{d^2 c_{O_2}}{dz^2} = v\frac{dc_{O_2}}{dz}$	$\mathcal{D}_{O_2}^{eff}\frac{d^2 c_{O_2}}{dz^2} = v\frac{dc_{O_2}}{dz} + \frac{di}{dz}\left(\frac{s_{O_2}}{nF} - \frac{s_w}{nF}\frac{c_{O_2}}{\rho}\right)$	-

Catalyst Layer Model

Equations for flux, material balance, and current balance are applied to the catalyst layer in the same manner as those applied to the membrane region. There are two primary differences between the transport model of the membrane and the catalyst layer. First, current must be transferred from the membrane phase of the material to the solid phase of the carbon mesh impregnated in the polymer. Secondly, the chemical reaction occuring in the catalyst layer introduces source terms to the equation that affect the balance of material and the continuity of mass.

The transfer of current from the membrane phase and the solid phase is represented by equations (C1) and (C2). These two equations result from the application of the Butler-Volmer expression for current transfer between the membrane and the solid (C1) and Ohm's Law for the solid portion of the catalyst (C2). Ohm's Law applied to the charged fluid moving through the catalyst layer results in equation (C3) in terms of the effective conductivity of the catalyst layer. Finally, equations (C4) and (C6) are applications of mass continuity and oxygen flux including the source terms associated with the chemical reaction.

Model of the Gas Diffuser Layer

The final component of the chemoelectric model is the gas diffuser region. The defining feature of the gas diffuser region is the diffusion of a multicomponent gas through a porous medium in the presence of water vapor. The fundamental relationship is the Stefan-Maxwell equations that relate the gradient of mole fractions to the species flux through the porous media. Application of Ohm's Law to the solid phase of the material results in equation (G2), and a material balance that incorporates water vapor flux results in equation (G4). Application of the Stefan-Maxwell expressions and relating flux of nitrogen gas with water vapor flux results in equation (G5).

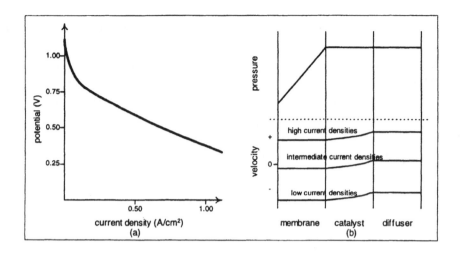

Figure 3: Representative chemoelectric modeling results from Bernardi and Verbrugge (1991).

Discussion of the Chemoelectric Model

The equations summarized in Table 1 are solved to determine the relatonship between potential and current density output of the cell. Bernardi and Verbrugge (1991) present a detailed discussion of the solution of the equations through application of the boundary conditions and the use of a finite difference numerical analysis technique. In addition to solving for the polarization curve of the cell, the model also provides information on concentration gradients of water and oxygen, pressure variation, as well as spatial variation of current density throughout the membrane, catalyst layer, and gas diffuser. The model is parameterized by the material properties of the three regions and the cell geometry.

Representative results that are relevant to the discussion of a combined chemoelectric and electromechanical system are shown in Figure 3. In Figure 3a we see a polarization curve for the power output of the cell. The cell potential falls with increasing current density due to Ohmic losses within the membrane. A typical operating condition for a fuel cell system would be a cell potential of 700 mV and a current density on the order of 500 to 600 mA/cm^2. Figure 3b illustrates the solution for water velocity within the membrane, catalyst layer, and gas diffuser. The solution is a function of the current density in the membrane. The results shown in Bernardi and Verbrugge (1991) demonstrate that at low current densities the direction of the water flow is negative, indicating that water is flowing from the cathode to the anode side of the cell. This is caused by the fact that the pressure gradient term in equation (M3) dominates compared to the term due to potential gradient. At higher cell current densities the situation is reversed. The term due to potential drop across the membrane dominates compared to the pressure gradient term and produces water flow from the anode to the cathode. An interesting balance occurs at intermediate current densities. For values between these two extremes the flow due to potential drop and pressure gradients are approximately equal and water produced by the chemical reaction in the catalyst layer flows out both the anode and the cathode.

Figure 3b is a representative result of the pressure gradient that exists in the polymer for a particular value of current density. Notice that the pressure gradient is linear in the membrane phase of the material and is constant through the catalyst layer and diffuser layer. The pressure gradient will vary with the current density and will couple with the water flow through the relative magnitude of the pressure gradient and potential drop, as discussed above.

Electromechanical Modeling

Charge transport and water transport within the ionic polymer membrane provides the link between the chemoelectric and electromechanical modeling. As discussed in the previous section, pressure gradients and water flow play an important role in chemoelectric modeling because they relate to the hydration state of the polymer electrolyte. The primary focus of chemoelectric modeling is the relationship between potential and current density.

The geometry considered for the transducer model is shown in Figure 2. The system consists of an ion-exchange membrane with conductive electrodes plated on both sides of the material. Without loss of generality we assume that the electrode on the left is the anode and the electrode on the right is the cathode. The z dimension has its origin at the outer surface of the anode and is positive to the right in Figure 2. The gradient of the electric potential is negative, thus, the cation motion is in the positive z direction while current flows from the anode to the cathode.

The equations will be referenced to a small control volume on the interior of the ion-exchange membrane. The flux in the control volume is governed by equation (3), and the current density in the control volume is

$$i = F \sum_i z_i N_i \tag{4}$$

Combining equations (3) and (4) yields

$$i = -\mathcal{K}\nabla\Phi - F\sum_i z_i D_i \nabla c_i + F\left(\sum_i z_i c_i\right)v, \tag{5}$$

where

$$\mathcal{K} = \frac{F^2}{RT}\sum_i z_i^2 D_i c_i. \tag{6}$$

The fluid velocity is governed by Schögl's equation

$$v = \frac{k}{\mu}\left[-\sum_i z_i c_i F\nabla\Phi - \nabla p\right] \tag{7}$$

Applying the equation for material balance

$$\frac{dc_i}{dt} = -\nabla \cdot N_i \tag{8}$$

yields

$$\frac{dc_i}{dt} = z_i\frac{F}{RT}D_i\left(c_i\nabla^2\Phi + \nabla c_i \cdot \nabla\Phi\right) + D_i\nabla^2 c_i - \nabla c_i \cdot v - c_i\nabla \cdot v \tag{9}$$

Under the assumption that the fluid is incompressible,

$$\nabla \cdot v = 0, \tag{10}$$

we can write

$$\nabla \cdot v = 0 = -\sum_i z_i c_i F\nabla^2\Phi - F\sum_i z_i\nabla c_i \cdot \nabla\Phi - \nabla^2 p \tag{11}$$

The five unknowns, c_i, Φ, v, p, and $\mathcal{K}$ can be solved using equations (5), (6), (7), (9), and (11).

One-dimensional Transport Model

Initially we will assume that transport only occurs in the z direction. Under this assumption we can write equations (5), (6), (7), (9), and (11) as functions of z in the following manner

$$\frac{dc_i}{dt} = z_i \frac{D_i}{RT} \left(c_i \frac{d^2\phi}{dz^2} + \frac{dc_i}{dz} \frac{d\phi}{dz} \right) + D_i \frac{d^2 c_i}{dz^2} - v \frac{dc_i}{dz} \tag{12}$$

$$i = -\mathcal{K} \frac{d\phi}{dz} - F \sum_i z_i D_i \frac{dc_i}{dz} + F \sum_i z_i c_i v \tag{13}$$

$$v = -\frac{k}{\mu} \left[F \sum_i z_i c_i \frac{d\phi}{dz} + \frac{dp}{dz} \right] \tag{14}$$

$$0 = -\sum_i z_i c_i F \frac{d^2\phi}{dz^2} - F \sum_i z_i \frac{dc_i}{dz} \frac{d\phi}{dz} - \frac{d^2 p}{dz^2} \tag{15}$$

$$\mathcal{K} = \frac{F^2}{RT} \sum_i z_i^2 D_i c_i \tag{16}$$

Two simplifications to the one-dimensional transport model will be analyzed. The first simplification is the assumption that the fluid velocity within the membrane is zero. We will relax this assumption in the second analysis but assume that the concentration in the membrane is constant with respect to time and space.

Case 1: No Fluid Motion

In the case in which the fluid velocity is assumed to be zero ($v = 0$) within the membrane, equation (14) reduces to

$$\frac{dp}{dz} \propto -\frac{d\phi}{dz} \tag{17}$$

implying that the gradient of the pressure is proportional to the potential gradient across the membrane. For the case in which we apply a step change in the potential gradient to the material, this result indicates that the pressure gradient is independent of the current and the concentration gradient in the membrane.

This result is contradicted by experimental observations. Measurements of mechanical deformation for step changes in the voltage exhibit a fast rise and a slow relaxation to steady-state. The substantial variation in the response contradicts this 'static' model of the pressure gradient within the material. It appears obvious from the measurements of displacement as a function of time that the stress induced by the application of a potential to the polymer produces a time-dependent stress. The time dependence is not modeled by the assumption that there is no fluid motion with the membrane.

Case 2: Steady-state Operation and Zero Concentration Gradient

Steady-state operation and zero concentration gradient implies that

$$\frac{dc_i}{dt} = \frac{d^2 c_i}{dz^2} = \frac{dc_i}{dz} = 0 \tag{18}$$

Substituting these assumptions into equations (12) to (16) yields

$$0 = z_i \frac{D_i}{RT} c_i \frac{d^2\phi}{dz^2} \tag{19}$$

$$i \;=\; -\mathcal{K}\frac{d\phi}{dz} + F\sum_i z_i c_i v \tag{20}$$

$$v \;=\; -\frac{k}{\mu}\left[F\sum_i z_i c_i \frac{d\phi}{dz} + \frac{dp}{dz}\right] \tag{21}$$

$$0 \;=\; -\sum_i z_i c_i F \frac{d^2\phi}{dz^2} - \frac{d^2 p}{dz^2} \tag{22}$$

$$\mathcal{K} \;=\; \frac{F^2}{RT}\sum_i z_i^2 D_i c_i \tag{23}$$

Dividing through by the constant in equation (19) produces

$$\frac{d^2\phi}{dz^2} = 0 \tag{24}$$

Combining this result with equation (22) also yields

$$\frac{d^2 p}{dz^2} = 0 \tag{25}$$

Equations (24) and (25) indicate that the potential gradient and pressure gradients are constant throughout the membrane, and that the potential and pressure variations are linear. Note that these are the same conclusions obtained by Bernardi and Verbrugge (1991) for steady-state analysis of a fuel cell membrane.

Substituting equation (21) into equation (20) results in the expression

$$i = -\left(\mathcal{K} + F^2 \frac{k}{\mu}\sum_i z_i^2 c_i^2\right)\frac{d\phi}{dz} - F\frac{k}{\mu}\sum_i z_i c_i \frac{dp}{dz} \tag{26}$$

This result is combined with equation (21) and written in matrix form

$$\left\{\begin{array}{c} i \\ v \end{array}\right\} = -\left[\begin{array}{cc} \mathcal{K} + F^2\frac{k}{\mu}\sum_i z_i^2 c_i^2 & F\frac{k}{\mu}\sum_i z_i c_i \\ F\frac{k}{\mu}\sum_i z_i c_i & \frac{k}{\mu} \end{array}\right]\left\{\begin{array}{c} \frac{d\phi}{dz} \\ \frac{dp}{dz} \end{array}\right\} \tag{27}$$

This model has the same form as the model proposed by de Gennes et al. (2000) for modeling ionic gels.

Discussion of the Electromechanical Model

Equations (12) through (16) are a set of nonlinear equations for the transient response between current, voltage, and pressure. The simplification studied in Case 2 demonstrates that these equations reduce to a set of linear equations only under the assumptions of a zero concentration gradient and steady-state operation. The next step in the electromechanical coupling analysis is to combine the transport equations with a model of mechanical deformation. The coupling between these two physical phenomena is the pressure distribution caused by charge and water transport through the polymer.

CONCLUSIONS

Modeling chemoelectric and electromechanical coupling in ionic polymer materials requires a model of charge and water transport within the material. Steady-state transport equations combined with the equations for mixing in the diffuser region can be used to model chemoelectric power generation, while the same transport phenomena combined with a pressure coupling term can be used to model electromechanical power generation. One fundamental difference between the two models is the need to model transient events in the case of electromechanical transduction.

The transport equations for electromechanical coupling are a set of coupled nonlinear equations that relate charge, potential, pressure, and the concentration gradient. Assuming zero water flow produces a model which cannot match observed behavior, while assuming steady-state operation and zero concentration gradient produces a linear model that relates charge, potential, water velocity, and pressure. The next step in our research is to combine these results with a mechanics model and correlate numerical simulations with observed behavior.

ACKNOWLEDGEMENTS

This work was funded by a National Science Foundation Career Award, grant number CMS-0093889. The program officer is Dr. Alison Flatau. The authors gratefully acknowledge the support.

NOTATION

$$
\begin{aligned}
i &= \text{current density (A/cm}^2) \\
I &= \text{cell current density (A/cm}^2) \\
F &= \text{Faraday's constant (96,487 C/equivalent)} \\
c &= \text{concentration (mol/cm}^3) \\
v &= \text{water velocity (cm/s)} \\
z &= \text{distance (cm)} \\
\mathcal{D} &= \text{diffusion coefficient (cm}^2\text{/s)} \\
v_s &= \text{superficial water velocity (cm/s)} \\
s &= \text{stoichiometric coefficient} \\
n &= \text{number of components in a gas mixture} \\
N_i &= \text{flux of species } i \text{ (mol/cm}^2 \cdot \text{s)} \\
p &= \text{pressure (atm)} \\
p_i &= \text{partial pressure of species } i \\
R &= \text{universal gas constant (8.3143 J/mol·K)} \\
T &= \text{absolute temperature (K)} \\
x_i &= \text{gas-phase mole fraction of species } i \\
r &= \text{diffusivity ratio} \\
a &= \text{catalyst area per unit volume (cm}^2\text{/cm}^3) \\
f &= \text{defined as } F/RT \text{ (V}^{-1}) \\
k &= \text{hydraulic permeability (cm}^2) \\
z_i &= \text{charge number of species } i
\end{aligned}
$$

Greek Letters

$\mathcal{K}$ = membrane conductivity (mho/cm)

ϕ = potential (V)

σ = electrode conductivity (mho/cm)

ρ = water density (mol/cm^3)

α_a, α_c = anodic and cathodic transfer coefficents

μ = pore-fluid viscosity (g/cm·s)

Subscripts

f = membrane species

c = catalyst layer

w = water

g = gas phase

References

Bernardi, D. M. and Verbrugge, M. W., 1991, "Mathematical Model of a Gas Diffusion Electrode Bonded to a Polymer Electrolyte," AIChE Journal, 37, pp. 1151–1163.

Bernardi, D. M. and Verbrugge, M. W., 1992, "A Mathematical Model of the Solid-Polymer-Electrolyte Fuel Cell," Journal of the Electrochemical Society, 139, pp. 2477–2491.

de Gennes, P., Okumura, K., Shahinpoor, M., and Kim, K. J., 2000, "Mechanoelectric Effects in Ionic Gels," Europhysics Letters, 50, pp. 513–518.

Mallavarapu, K., Newbury, K. M., and Leo, D. J., 2001, "Feedback Control of the Bending Response of Ionic Polymer Metal Composite Actuators," in *Proceedings of the SPIE*, 4329-40.

Newbury, K. M. and Leo, D. J., 2001, "Mechanical Work and Electromechanical Coupling in Ionic Polymer Bender Actuators," in *ASME Adaptive Structures and Materials Symposium, to appear*.

Nguyen, T. V. and White, R. E., 1993, "A Water and Heat Management Model for Proton Exchange Membrane Fuel Cells," Journal of the Electrochemical Society, 140, pp. 2178–2186.

Springer, T., T.A. Zawodzinski, J., and Gottesfeld, S., 1991a, The Electrochemical Society, p. 209.

Springer, T., Zawodzinski, T., and Gottesfeld, S., 1991b, "Polymer Electrolyte Fuel Cell Model," Journal of the Electrochemical Society, 138, pp. 2334–2342.

Yi, J. S. and Nguyen, T. V., 1998, "An Along-the-Channel Model for Proton Exchange Membrane Fuel Cells," Journal of the Electrochemical Society, 145, pp. 1149–1159.

Modeling and Identification

VIBRATION TESTING AND FINITE ELEMENT ANALYSIS OF INFLATABLE STRUCTURES

Marion Sausse, Eric Ruggiero, Gyuhae Park, Daniel J. Inman, and John A. Main

ABSTRACT

This paper presents vibration testing and finite element analysis of an inflated thin-film torus. Inflatable structures show significant promises for future space applications. However, their extremely lightweight, flexible, and high damping properties pose difficult problems in vibration testing and analysis.

In this paper, we experimentally show that smart materials could be used as sensors and actuators for performing vibration tests of an inflated torus. In addition, we developed a predictive model that can be compared with experimental results. A commercial finite element package, ANSYS, is used to model the pre-stressed inflatable structure. Both experimental and finite element results are in good agreement with each other. The predictive model can be used for analyzing the dynamics of inflated structures, and for designing control systems to attenuate vibration in an inflated torus.

INTRODUCTION

Inflated space-based devices have become popular over the past three decades due to their minimal launch-mass and launch-volume [1, 2, 3]. The dynamic behavior is particularly important for satellite structures since they are subjected to a variety of dynamic loadings. However, their extremely lightweight, flexible, and high damping properties pose difficult problems in vibration testing and analysis. These vibration problems are difficult to observe by ground testing. The choice of applicable sensing and actuation systems suitable for use with inflated structure is somewhat limited because of their low stiffness and high flexibility. In addition, excitation methods have to be carefully chosen since the extremely flexible nature causes point excitation to result in only local deformation.

Marion Sausse, Eric Ruggiero, Gyuhae Park, Daniel J. Inman, Center for Intelligent Material Systems and Structures, Virginia Polytechnic Institute and State University, Blacksburg, VA 24061-0261
John A. Main, Department of Mechanical Engineering, University of Kentucky, Lexington, KY 40508

Griffith and Main [4] used a modified impact hammer to excite the global modes of the structure while avoiding local excitation. Lassiter *et al.* [5] tested a torus attached to three struts with a lens in a thermal vacuum chamber. They found significant differences in the response between the structure in ambient and vacuum conditions. Park *et al.* [6] investigated the feasibility of using smart materials, such as PVDF films, to find modal parameters and to attenuate vibration in a flexible inflated structure.

Analytical solutions for free vibrations of toroidal structures without prestress have been studied by many researchers [7, 8, 9]. However, there have been relatively fewer studies on the toroidal shell with prestress. Liepins [10] used a finite difference method to solve governing equations of a prestressed toroidal membrane. Plaut et al. [11] used shell analysis of an inflatable arch with fixed boundary conditions to find the deflections when subjected to snow and wind loadings. Jha and Inman [12] used Sanders linear shell theory to formulate the governing equations and the natural frequencies and mode shapes to compute using Galerkin's method. Recently, commercial finite element packages have been used to model inflatable structures in order to avoid the complexity introduced by conventional analytical solutions of a pre-stressed structure. Briand et al. [13] showed that finite element analysis must investigate an adequate size of meshing, and adequate boundary conditions. Lewis [14] (2000) finite element study found that the aspect ratio of a toroidal inflated structure has a significant impact on its natural frequencies and modes shapes. He also demonstrates that increasing the pressure results in increased natural frequencies.

In this study, we experimentally investigate that smart materials can be used as sensors and actuators for performing vibration tests of an inflated torus. In addition, we have developed a predictive finite element model that can be compared to experimental results in order to understand dynamic behavior of inflatable structures. A commercial finite element package, ANSYS, with a linear thin shell element is used to model the prestressed inflated structure. Finite element results are in good agreement with those of experiments. This paper summarizes the experimental setup, procedures, considerations needed to obtain frequency response functions (FRFs) with high coherence, and some approximations and assumptions made in the finite element modeling.

DESCRIPTION OF THE STRUCTURE

The test structure is an inflatable torus made of Kapton with a 1.8-meter ring diameter and a 0.15-meter tube diameter as shown in figure 1. The torus was made of flat sheets of polyimide film Kapton and fabricated in the Emerging Technology Laboratory at University of Kentucky. Three 120-degree segments were joined together to form the complete torus. The method of fabrication results in many variations in the thickness in the joining regions. In addition, epoxy adds significant mass and stiffness in the bonded joins [4]. The joining region width (flap around the inner and outer diameter) is measured at 5.1 cm and the thickness is about 300μm in this region. A list of the structure's physical properties is given in Table 1

EXPERIMENTAL TESTING

The structure was tested in laboratory conditions in order to find resonant frequencies and mode shapes. The modes of interest are ring modes, which are in plane and out of plane motions of the torus.

The torus was suspended using a rubber wire in order to minimize the effects of the boundary conditions. With the suspended wires, the rigid body modes appears to occur at 1 to 2 Hz, which would be negligible considering that the frequency range of interest is 10-200Hz. The internal pressure of the torus was maintained 0.5 psi using a small aquarium pump. No noticeable effect on the measurement of frequency response functions (FRF) was found from the pump noise and the flow of air into the structure.

Property	Values
Ring Diameter	1.8 m
Tube Diameter	0.15 m
Joining region width	0.051 m
Internal pressure	0.5 psi
Elastic Modulus	2.55e9Pa
Mass Density	1418 kg/m^3
Poisson's Ratio	0.34
Thickness	46e-9m
Joining region	300e-9m

Table 1. Physical parameters of a test structure.

Two difference methods were selected to excite a torus. One is use of a conventional electromagnetic shaker (manufactured by Ling Dynamic Systems) to provide the point force excitation to the structure. A single piece of metal was attached on the arm of the shaker and glued on the torus, as shown in figure 2. The other excitation was realized using the recently developed Macro Fiber Composite (MFC^{TM}) actuator [15]. The MFC offers high performance and flexibility suitable for the use in the inflatable structure. The MFC was bonded to the surface of the torus using double-sided tape, as shown in figure 3. The identified resonant frequencies and mode shapes with the different excitation methods were then compared to each other.

In order to measure the response of the torus, an accelerometer (PCB Model 352C22) and a PVDF sensor were attached to the torus. In order to reduce electromagnetic interference (EMI) effects, a specially designed PVDF sensor (Measurement Specialties SDT1-028K) has been used, as shown in figure 4. This sensor is equipped with a protective coating and a shielded cable that is optimized against the 60 Hz electromagnetic field. These sensors are bonded to the surface of the structure with a doubled-sided tape at sixteen evenly spaced locations.

Excitation was given in two directions. The first was in the out-of-plane direction and the second in the in-plane direction in order to measure the corresponding modes. Force, acceleration and strain data were collected through the DSPT Siglab multi-channel dynamic signal analyzer. The excitation signal was a chirp signal varying from 5-100 Hz. Ten runs were averaged to estimate frequency response functions.

As expected, the extremely flexible nature of the inflatable torus made collecting modal parameter information quite difficult. In particular, local excitation with a shaker as a point input only exaggerated the local properties of the structure. To ensure enough input energy at each frequency from excitation, the chirp input frequency for each test was divided into three sub-ranges (5-35, 35-65, and 65-100 Hz). Furthermore, the time period of excitation was held for as long as possible (51200 samples at each sub-range) so that the excitation ceased and that the corresponding response signal in a data frame was damped out [16]. By focusing the small bandwidth, we were able to provide sufficient input energy to the entire structure, and thereby able to excite the global modes. The careful selection of excitation signal (specific length and bandwidth) was found to be critical for a flexible inflatable torus when trying to obtain reliable vibratory responses with reasonable accuracy.

Figure 1. Inflated test object for dynamic analysis

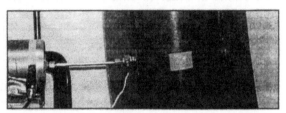

Figure 2: Connection between the torus and the shaker.

Figure 3. The MFC actuator attached to the torus

OVERVIEW ON EXPERIMENTAL RESULTS

After completion of the data measurements, the modal analysis was performed by using the Unified Matrix Polynomial Approach (UMPA) pseudo least square method [17]. This method determines an orthogonal polynomial to curve-fit the measured data, and provides eigenvalues and eigenvectors of the system.

Figure 5 and 6 shows sample frequency response functions of out-of-plane and in-plane motions. The coherence plots, which indicate the correlation between a single input and single output measurement, are also shown to indicate the test results are reasonable. For the out of plane motion, three resonant peaks are identifiable. Frequencies near 13 and 32 Hz are easily characterized as the two first apparent out of plane frequencies. The third resonant peak near 65 Hz is not so easy to define because of a shell mode present around the same frequency. For the in plane motion, the first two resonant peaks are around 16 and 42 Hz. However, as the frequency increases, it is not easy to identify resonant frequencies because of a number of shell modes present.

Experimentally identified mode shapes are expected to have a sequential number of nodal lines. Therefore, mode shapes of each resonant peak are generated in order to determine global in-plane and out-of-plane modes. The identified resonant frequencies are found in table 3 and mode shapes are illustrated in figure 7, as measured with the accelerometer and the MFC excitation.

Figure 4: A PVDF sensor bonded to the torus.

The first conclusion that can be made is that the data measured with the PVDF sensors are consistent with those acquired with an accelerometer. The identified resonant frequencies using the PVDF sensors are almost identical to those measured with the accelerometer. The experimental results presented here validate the usefulness of PVDF sensors for measuring dynamics of inflatable space structures. Further, identified resonant frequencies are almost identical to those found with the shaker and with the MFC excitation. Indeed, the MFC is very lightweight and his mass loading effect is negligible. It is obvious during the tests that the MFC excitation produced less interference with suspension modes of the free-free torus than excitations from the shaker. Without connections to the ground (except for the electrical cable), these actuators and the PVDF sensor can be considered as an integral part of an inflated structure. These combinations could also be used on control devices of an inflatable structure for not only vibration suppression but also for static shape control. A more complete description of the experimental procedure and results can be found in the reference [18].

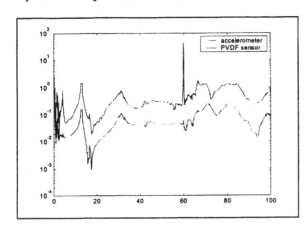

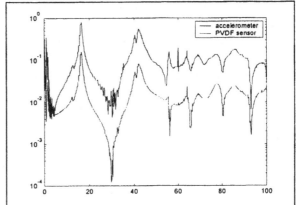

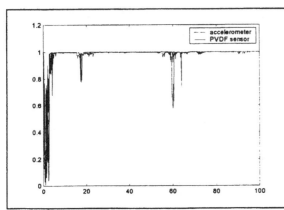

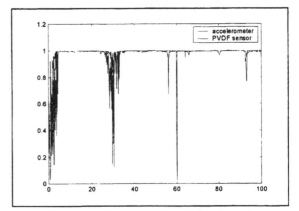

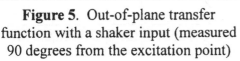

Figure 5. Out-of-plane transfer function with a shaker input (measured 90 degrees from the excitation point)

Figure 6. In-plane transfer function with a shaker input (measured 90 degrees from the excitation point)

FINITE ELEMENT ANALYSIS

Recently, commercial finite element packages have been readily available and their utility has increased with the development of fast computers. The finite element method provides a relatively easy way to model the system. In this section, our experimental results will be compared and validated with the use of commercial finite element analysis. The software package ANSYS was used for the current research effort.

Modeling

Linear thin shell elements were used that have 8 nodes, 4 corners and 4 mid-side nodes. These elements are well suited to model a curved shell. Each node has 6 degrees of freedom, translation and rotation in x, y, and z direction. The model is defined by geometry, based on nodes and elements, the real constants, which define the elements thickness and the material properties. The torus was meshed with 288 elements, 24 elements around the ring, 10 elements around the tube and 1 for the joining regions. Figure 8 shows the mesh configuration of an inflated torus.

Since Lewis [10] showed that the PVDF patches have very limited effect on the vibratory response of low aspect ratio torus (here aspect ratio is 0.083), they were not taken into account in this analysis. Generally the natural frequencies are decreased slightly because of the PVDF patches, as should be expected by adding mass to the system, but there is little influence on the mode shapes [19].

An internal pressure of 0.5 psi was applied to simulate the pressure in the torus. For the static analysis one node had to be restrained to zero displacement to prevent rigid body motion. In the modal analysis, this constraint was removed to solve the problem for free-free response. A prestressed matrix was used to perform the modal analysis. Mass was added to the surrounding elements representing the flaps (resulting from the fabrication) to simulate the presence of epoxy.

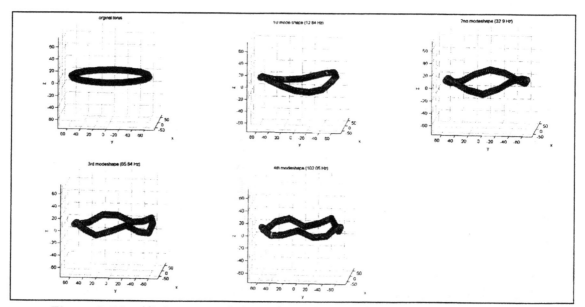

Figure 7. Identified resonant frequencies and mode shapes of the torus with the MFC actuator excitation.

Assumptions

We assumed that the torus was of uniform thickness even in the joining region. We know this is not the case for the torus, but measuring and then modeling all the variations in the thickness would be too complicated. Further, those parameters are found to be not really affecting the analytical results. We also assumed that the pressure remains constant while the torus is vibrating. The material was considered to be linearly elastic.

Finite element model verification

At present there is no reliable published data on modal testing of inflated prestressed toroidal structures. For this reason, we used a previous analytical work to verify the finite element model. Table 2 compares resonant frequencies with the analytical solutions [12] of free vibration analysis of an inflated torus of circular cross section (aspect ratio 0.16). The results show good agreement, even though frequencies found with the software were slightly higher. But this phenomenon is normal and occurs with FEA software because they tend to stiffen the structure.

Discussion of results

The bending modes are orthogonal pairs of ring modes, alternating out of plane and in plane of the torus as represented in figure 14. Table 3 gives the modes frequencies found with the finite element analysis and with the experiments using accelerometer/ PVDF sensors. The results match satisfactorily between FEA and experiments. It is observed that results for in plane modes are closer than for out-of-plane modes. It is also observed that out-of-plane motion modes come before in plane motion modes. There are also other types of motion like for example shell modes. The first shell mode occurs at 55.533 Hz in the finite element analysis represented in the following figure 8, and experimentally identified at 55.6 Hz.

Mode	FEA (Hz)	Analytic analysis(Hz)
1	8.63	7.52
2	11.24	9.32
3	20.02	18.68

Table 2: Comparison between FEA and analytical analysis

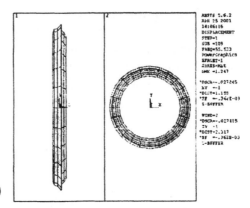

Figure 8: The first shell mode (55.533 Hz)

The test results compare favorable with the analytical model built with Ansys even if some variations appear. These differences can be explained by several, factors such as;

- Non-constant thickness due to method of fabrication and difficult to model in the joining regions.

- The presence of flaps around the torus. Indeed a finite element analysis performed for a torus without the joining region resulting from the method of fabrication showed that the presence of the flaps considerably lower the frequencies of the mode shapes. This confirms the fact that method of fabrication results in adding considerable mass to the structure.
- Presence of epoxy to glue the different parts that also adds mass and stiffness to the model [4].
- Even though air flow in the structure does not affect the measures, the tube through which air is supplied to the structure affects the experimental results.
- Boundary conditions that are not completely for free-free response of the torus due to the rubber wire used to suspend the structure.

Mode	Acceler ometer	PVD F	FEA	Motion
1	12.84	12.81	14.36	Out-of-plane
2	16.37	16.31	15.64	In-plane
3	32.9	31.7	40.61	Out-of-plane
4	40.24	40.24	41.71	In-plane
5	65.64	65.4	62.44	Out-of-plane
6	68.1	67.78	72.77	In-plane
7	102.05	100.9 2	107.16	Out-of-plane
8			109.71	In-plane

Table 3: Comparison between the different results

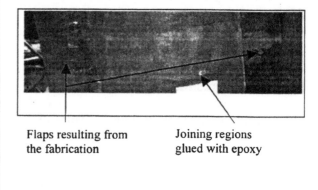

Flaps resulting from   Joining regions
the fabrication   glued with epoxy

Figure 9: One of the joining regions

Impact of the flaps on the structure

A finite analysis was performed for the same torus, but without the joining region (tabs) to see what the impact was on the model. The results show that the impact is considerable. Frequencies are much higher and in-plane motion modes come before out-of-plane motions modes. Moreover the first shell mode occurs for a high frequency that is to say for 216.9 Hz. Results are listed in table 4.

Mode	In plane (Hz)	Out of plane (Hz)
1	31.761	38.747
2	81.192	84.173
3	138.59	139.43

Table 4: Frequencies for a perfect torus.

Figure10: Tube through which air is supplied

DISCUSSION

A finite element analysis was performed using the commercial finite element software, Ansys, to study the dynamics of an inflated torus. Results are in reasonable agreement with the experimental data so we can conclude these previous results are validated. Still there are some variations between theoretical and experimental results but these differences are not of a important scale and are attributed to the assumptions made on the model. The finite element analysis also confirms that the presence of the tabs considerably affects the frequencies and mode shapes by adding mass and stiffness to the structure. Figure 11 shows a global overview on our experimental and simulation results. In conclusion, The predictive model can be used for analyzing the dynamics of inflated structures, and for designing control systems to attenuate vibration in an inflated torus. These tasks are currently under investigation.

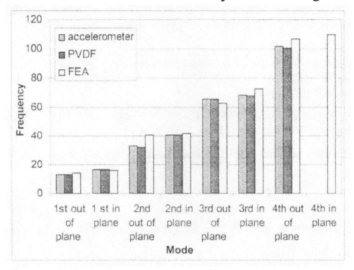

Figure11: Summary of Results

CONCLUSION

Smart materials, such as PVDF sensors and MFC actuators, have been used in vibration testing of an inflated thin-film polyimide torus in order to identify resonant frequencies and associated mode shapes. The experimentally obtained results are in good agreement with the shaker-accelerometer combination. The use of smart materials offers several advantages for the analysis of inflatable structures. These sensors/ actuators are flexible enough to conform to the doubly curved surface of the toroidal shell. Thus, they can be fully integrated in an unobtrusive way into the skin of an inflatable structure. The experimental results also compare favorably with the analytical model built with the finite element software, Ansys. It has been shown that the method of fabrication considerably affects the modes particularly by adding mass and stiffness to the structure.

ACKNOWLEDGEMENT

This work was sponsored partially by the Air Force Office of Scientific Research under grant number F49620-99-1-0231 and AFIT/EN 99-018, and partially by NASA Langley Research Center, grant number LaRC 01-1103 under the direction of Dr. W. Keats Wilkie. The authors gratefully acknowledge the support.

REFERENCES

1 Satter, C.M., and Freeland, R.E., "Inflatable Structures Technology Applications and Requirements," AIAA Paper 95-3737, Presented at AIAA 1995 Space Programs and Technologies Conference, Huntsville, Alabama, September 26-28, 1995

2 Freeland, R.E., "Significance of the Inflatable Antenna Experiment Technology," AIAA Paper 98-2104, pp. 2789-2796, 1998

3 DornHeim, M.A., "Inflatable Structures taking to Flight," Aviation Week & Space Technology, pp. 60-62, 1999

4 Griffith, D.T., and Main, J.A., "Modal Testing of an Inflated Thin Film Polyimide Torus Structures," Proceedings of 18th International Modal Analysis Conference, San Antonio, Texas, February 2000

5 Lassiter, John O., Engberg, Robert, Slade, Kara N., Tinker, Michael L., Comparison of Dynamic Characteristics for an Inflatable Solar Concentrator in Atmospheric and Thermal Vacuum Conditions. Paper 2000-1641, Proceedings of 41st Structures, Structural Dynamics, and Materials Conference and Exhibit, Atlanta, GA 3-8, April, 2000

6 Park, G., Kim, M., Mattias, M., Inman, D.J. "Vibration Control of Inflatable Space Structures using Smart Materials," *Proceedings of 18th ASME Biennial Conference on Mechanical Vibration and Noise*, 9-12 September 2001, Pittsburgh, PA, DETC2001/VIB-21540

7 Kosawada, T., Suzuki, K., Takahashi, S., 1985, "Free Vibration of Toroidal Shells," *Bulletin of the Japan Society of Mechanical Engineers*, Vol. 28, No. 243, pp. 2041-2047.

8 Fang, Z., 1992, "Free Vibration of Fluid-Filled Toroidal Shells," *Journal of Sound and Vibration*, Vol. 155, No. 2, pp. 343-352.

9 Leung, A. Y. T., Kwok, N. T. C., 1994, "Free Vibration Analysis of a Toroidal Shell," *Thin-Walled Structures*, Vol. 18, pp. 317-332.

10 Liepins, A. A., 1965, "Free Vibrations of Prestressed Toroidal Membrane," *AIAA Journal*, Vol. 3, No. 10, pp. 1924-1933.

11 Plaut, R. H., Goh, J. K. S., Kigudde, M., Hammerand, D. C., 2000, "Shell Analysis of an Inflatable Arch Subjected to Snow and Wind Loading," *International Journal of Solids and Structures*, Vol. 37, pp. 4275-4288.

12 Jha, A.K., Inman, D.J., 2001. "Free vibration testing of an inflated torus," *ASME Journal of Vibration and Acoustics*, submitted.

13 Briand, G., Wicks, A.L., and Inman, D.J., "Vibration Testing for Inflated Objects,", Proceeding of 18th International Modal Analysis Conference, San Antonio, Texas, February 2000

14 Lewis, J.A., "Finite Element Modeling and Active Control of an Inflated Torus Using Piezoelectric Devices," Master's Thesis, Virginia Polytechnic Institute and State University. 2000

15 Wilkie, W.K., Bryant, R.G., High, J.W., Fox, R.L., Hellbaum, R.F., Jalink, A., Little, B.D., and Mirick, P.H., "Low-Cost Piezocomposite Actuator for Structural Control Applications," Proceedings of 7th SPIE International Symposium on Smart Structures and Materials, Newport Beach, CA, March 5-9, 2000

16 Guan D.H., Yam, L.H., Mignolet, M.P., and Li, Y.Y.,"Experimental Modal Analysis of Tires," *Experimental Techniques*, December 2000, pp.39-45

17 Allemang R.J., Brown D.L., Fladung W., "Modal Parameter Estimation: Unified Matrix Polynomial Approach," Proceeding of 12th International Modal Analysis Conference pp. 501-514, 1994.

18 Park, G., Ruggiero, E., Sausee, M., Inman, D.J. "Vibration Testing and Analysis of Inflatable Structures using Smart Materials," *Proceedings of 2001 ASME International Mechanical Engineering Congress and Exposition*, November 11-16, 2001, New York, NY, IMECE 2001, to appear

19 Williams R.B., Austin E.M., Inman,D.J., "Limitation of Using Membrane Theory for Modeling PVDF Patches on Inflatable Structures," Journal of Intelligent Material Systems and Structures, to appear, 2000.

FIBER TIP BASED FIBER OPTIC ACOUSTIC SENSORS

Miao Yu and Balakumar Balachandran

ABSTRACT

Recent work conducted on developing a fiber tip based Fabry-Perot (FTFP) sensor system for acoustic measurements is presented in this article. It has been determined that this system can be used to detect acoustic fields in the frequency range of 50 Hz to 7.5 kHz with an optical phase sensitivity of 0.9 rad/Pa. A series of experiments is performed to investigate the possibility and potential use of this sensor system, which is designed to be implemented in a multiplexed architecture for providing response measurements to a structural acoustic control system.

1   INTRODUCTION

In the design of modern transportation vehicles, structural vibration and interior noise have become important problems that need to be addressed. For example, in helicopter systems, control of sound transmission into enclosed spaces is an important problem. Various studies have shown that the predominant frequency components associated with the noise transmission lie in the frequency range of 50 Hz to 5500 Hz [1]. There are various approaches that can be used to control sound fields inside a helicopter cabin [2-4]. Among the different approaches, one approach is based on controlling the radiation (transmission) from (through) a flexible structure by active means, which is referred as active structural acoustics control (ASAC). The following components are important in this approach: a) sensors (error measurements and reference measurements), b) actuators, and c) control scheme. The present efforts are being pursued with the goal of developing a distributed sensor array for control of sound fields inside enclosures and sound radiation from flexible structures.

Fiber optic sensors have the advantages of light weight, high sensitivity, and easy multiplexing. Since the original demonstrations showed that optical fibers could be utilized as acoustic sensors [5, 6], substantial research work has been done in this field [7-14]. Most of these efforts have been directed toward development of hydrophones for the ultrasonic detection. Among those sensors, it has been shown that Bragg grating sensors can be multiplexed by using wavelength division multiplexing (WDM) techniques. However, the limited sensor bandwidth associated with such systems is an issue that needed to be solved. In

Miao Yu and B. Balachadran, Department of Mechanical Engineering, University of Maryland, College Park, MD 20742.

addition, the low sensitivity due to the high Young's modulus of silica results in "small" acoustically induced strains, which also limits the application of these types of sensors. Hence, low-finesse Fabry-Perot sensors have become attractive choices for a high-performance sensing in this area. In such cases, a Fabry-Perot cavity formed between the fiber tip and the object surface is a clear solution. There are two types of modulation schemes used to recover the signal, one being intensity modulation schemes and the other being phase modulation schemes. Intensity modulated sensors offer simplicity of design and ease of implementation, but they suffer from problems of limited sensitivity, low dynamic range, and drift due to intensity fluctuation. Phase modulated sensors are based on detection of the acoustically induced optical phase shift by using an interference technique. They have high sensitivity and do not suffer from optical source and receiver drift problems, but their nonlinear input-output characteristics require careful design of demodulation.

In this paper, a novel fiber tip based Fabry-Perot (FTFP) sensor system is developed for acoustic pressure measurements in the range of 50 Hz to 6 kHz, the typical range of interest for helicopter systems. This sensor system falls in the category of phase modulation schemes discussed above, and it is designed to work in an acoustic control system with multiple sensor inputs. A digital phase demodulation system based on a phase-stepping technique has been developed for this sensor system to decode the optical intensity signal into optical phase signal. The experimental results demonstrate the feasibility of using such sensor system to detect acoustic signals. Details of the sensor modeling, design, and demodulation scheme are discussed in Sections 2 to 4, the experimental arrangement is presented in Section 5, and the experimental results are presented and discussed in Section 6.

2   MODEL DEVELOPMENT

In this section, the interferometers comprising the optical system and the mechanical component response are discussed.

2.1 OPTICAL SYSTEM

The sensing system under investigation is based on a low-finesse Fabry-Perot (FP) cavity. As shown in Figure 1 (a), after the light emerges from the single mode fiber, the electric field components in the multiple-beam interference with Gaussian beam expansion-induced power attenuation can be modeled as [15]

$$E_{1r} = E_0 r_a e^{i\omega t}, \quad E_{2r} = E_0 t_a r_b t_a \sqrt{\alpha} e^{i(\omega t - 2kL_s)}, \text{ and } E_{3r} = E_0 t_a r_b r_a r_b t_a (\sqrt{\alpha})^2 e^{i(\omega t - 4kL_s)}, \tag{1}$$

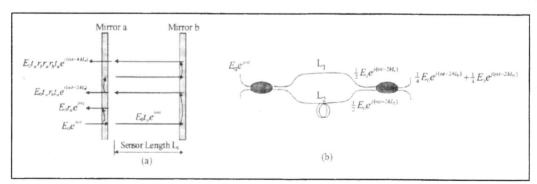

Figure 1. (a) Sensor interferometer and (b) read-out interferometer.

where r_a, r_b are the reflection coefficients of mirror a and b respectively, and t_a is the transmission coefficient of mirror a. The wave number is k, which is equal to $2\pi / \lambda$. The resultant reflected scalar E wave is given by

$$E_r = E_0 e^{i\omega t} \sqrt{R_a} \left[1 - \frac{1-R_a}{R_a} \sum_{m=1}^{\infty} (R_a R_b \alpha)^{\frac{m}{2}} e^{-2imkL_s} \right],$$ (2)

where $r = \sqrt{R}$ and $t = \sqrt{T}$, R and T are reflectivity and transmitivity, respectively. The transfer function H_r of the Fabry-Perot interferometer can be written as

$$H_r^s = \frac{E_r \cdot E_r^*}{E_i \cdot E_i^*} = A_0 - A_1 \cdot \sum_{m=1}^{\infty} (R_a R_b \alpha)^{\frac{m}{2}} \cdot \cos(2mkL_s),$$ (3)

where

$$A_0 = R_a + \frac{(1-R_a)^2 R_b}{(1-R_a R_b)}, \quad \text{and} \quad A_1 = \frac{2(1-R_a - R_b + R_a R_b)}{1-R_a R_b}.$$ (4)

For low finesse Fabry-Perot sensor, the transfer function can be written as

$$H_r^s = A_0 - A_1 \cdot \sqrt{R_a R_b \alpha} \cos(2kL_s).$$ (5)

Here, a path matched differential interferometery (PMDI) system is designed to demodulate the FP sensor. In this system, a read-out interferometer is path-matched to the sensing interferometer. If a read-out interferometer is a Mach-Zehnder interferometer as shown in Figure 1 (b), then the associated transfer function is

$$H_r^r = \frac{E_r \cdot E_r^*}{E_i \cdot E_i^*} = \frac{1}{8}[1 + \cos 2k(L_2 - L_1)] = \frac{1}{8}[1 + \cos 2kL_r],$$ (6)

where L_r is the cavity length of the reference interferometer. When the light passes though the PMDI system, the resulting time dependent intensity function of the sensors, I_T, as detected by the photodetector is given by

$$I_T = \frac{1}{16} \int H_r^r H_r^s i(k) dk,$$ (7)

where H_r^s and H_r^r, which are the transfer functions of a FP sensor interferometer and a Mach-Zehnder read-out interferometer, are given by equations (5) and (6), respectively, and $i(k)$ is the input spectrum of the broadband optical source. After carrying out the integration, equation (7) can be written as

$$I_t \approx \frac{1}{128} I_0 A_0 - \frac{1}{256} I_0 A_1 \sqrt{R_a R_b \alpha} \cos 2k_0 L_s e^{-\left(\frac{2\pi L_s}{L_c}\right)^2} + \frac{1}{128} I_0 A \cos 2k_0 L_r e^{-\left(\frac{2\pi L_r}{L_c}\right)^2} -$$

$$\frac{1}{256} I_0 A_1 \sqrt{R_a R_b \alpha} \left\{ \cos 2k_0(L_s + L_r) e^{-\left[\frac{2\pi(L_s + L_r)}{L_c}\right]^2} + \cos 2k_0(L_s - L_r) e^{-\left[\frac{2\pi(L_s - L_r)}{L_c}\right]^2} \right\},$$ (8)

where L_c is the coherence length of the short coherence light source and $\Delta\lambda$ represents the half-width of the linewidth. When the system is path matched ($L_r \approx L_s$) and L_c is much smaller than L_r and L_s, coherent interference occurs only in the $(L_s - L_r)$ component. Thus equation (8) can be simplified to

$$I_t \approx \tfrac{1}{128} I_0 A_0 - \tfrac{1}{256} I_0 A_1 \sqrt{R_a R_b}\alpha \; \cos 2k_0 (L_s - L_r). \tag{9}$$

2.2 MECHANICAL COMPONENT

Here, the sensor diaphragm is considered as a circular plate with fixed edge. This section is used to determine the relationship between the displacement of the microphone diaphragm and the pressure experienced by the diaphragm. For an isotropic circular plate of radius a and thickness h, the first natural frequency of the diaphragm can be written as

$$f = \frac{10.21}{2\pi a^2}\left[\frac{Eh^2}{12\rho(1-v^2)}\right]^{1/2}. \tag{10}$$

For forced oscillations, the governing equation is of the form

$$D\nabla^4 w + \rho h \frac{\partial^2 w}{\partial t^2} + damping \;\; term = p(t), \tag{11}$$

where $p(t)$ is the dynamic sound pressure to be sensed with amplitude of p, ρ is density of the diaphragm material, v is Poisson ratio, and $D = \dfrac{Eh^3}{12(1-v^2)}$. The solution of equation (11) can be written as

$$w(r,\theta,t) = \sum_{k=0}^{\infty} \eta_k(t)U_{3k}(r,\theta), \tag{12}$$

where η_k are the modal amplitudes and U_{3k} are the natural mode components in the principal directions. Taking advantage of the orthogonality of the modes, for a harmonic loading, equation (11) can be reduced to

$$\ddot{\eta}_k + 2\zeta_k\omega_k\eta + \omega_k^2\eta_k = F_k\, f(t), \tag{13}$$

where ω_k is the natural frequency of the mode of interest and ζ_k is associated the modal damping coefficient; the different coefficients in equation (13) are given by

$$\zeta_k = \frac{\kappa}{2\rho h\omega_k}, \;\; \kappa = \frac{\rho h\omega_k^2}{D}, \;\; F_k = \frac{1}{\rho hN_k}\int_0^a pU_{sk}(r,\theta)2\pi rdr, \;\; and \;\; N_k = \int_0^a 2\pi rU_{sk}^2(r,\theta)dr. \tag{14}$$

For harmonic excitation, the solution of equation (13) can be written as

$$\eta_k = \Lambda_k e^{j(\omega t - \varphi_k)}, \tag{15}$$

where the amplitude function

$$\Lambda_k = \frac{F_k}{\omega_k^2 \sqrt{\left[1 - \left(\frac{\omega}{\omega_k}\right)^2\right]^2 + 4\zeta_k^2 \left(\frac{\omega}{\omega_k}\right)^2}}. \tag{16}$$

Approximating the response given by equation (12) in terms of a single mode, here, the first mode, the response can be written as

$$w(r, \theta) = \Lambda_0 U_{30}(r, \theta), \tag{17}$$

where

$$U_{30}(r, \theta) = A\left[J_0(\kappa r) I_0(\kappa a) - I_0(\kappa r) J_0(\kappa a)\right] \tag{18}$$

From equations (14) to (18), the displacement response is determined to be

$$w(r, \theta) = \frac{2\pi p a}{\rho h \tilde{N}_0 \kappa} \frac{\left[J_1(\kappa a) I_0(\kappa a) - I_1(\kappa a) J_0(\kappa a)\right]\left[J_0(\kappa r) I_0(\kappa a) - I_0(\kappa r) J_0(\kappa a)\right]}{\omega_k \sqrt{\left[1 - \left(\frac{\omega}{\omega_k}\right)^2\right]^2 + 4\zeta_k^2 \left(\frac{\omega}{\omega_k}\right)^2}}, \tag{19}$$

where

$$\tilde{N}_0 = \int_0^a 2\pi r \left[J_0(\kappa r) I_0(\kappa a) - I_0(\kappa r) J_0(\kappa a)\right]^2 dr. \tag{20}$$

For a FTFP sensor, the cavity length change is due to the deflection of the diaphragm center w_0. Hence, the optical phase change $\Delta\varphi$ is related to the sound pressure as

$$\Delta\varphi = \frac{4\pi n}{\lambda} w_0 = \frac{8\pi^2 n p a}{\lambda \rho h \tilde{N}_0 \kappa} \frac{\left[J_1(\kappa a) I_0(\kappa a) - I_1(\kappa a) J_0(\kappa a)\right]\left[I_0(\kappa a) - J_0(\kappa a)\right]}{\omega_k \sqrt{\left[1 - \left(\frac{\omega}{\omega_k}\right)^2\right]^2 + 4\zeta_k^2 \left(\frac{\omega}{\omega_k}\right)^2}} \tag{21}$$

where n is the reflective index of the cavity material and λ is the wavelength of light source.

3   SENSOR SYSTEM DESIGN

The overall sensor system is shown in Figure 2. It consists of a Superluminescent light emitting diodes (SLD), an integrated optical circuit (IOC) phase modulator, a 2×2 optical coupler, 1×N optical switch, FTFP sensor array, a photodetector, and a personal computer based data acquisition system. The advantage of using the optical switch to realize spatial division multiplexing (SDM) is that a larger number of sensors can be detected by using the same base optical system (i.e., source, detectors, and modulators). Furthermore, each sensor can be designed to either sense acoustic field at a particular location of a system or to sense a particular acoustic frequency in the system.

High reliability connector ferrule is used to fabricate each single FTFP sensor. The diaphragm is made of Mylar film with a thickness of 40 μm and a radius of 2.5 mm. Single mode fiber is fixed in the connector ferrule and the distance between the fiber tip and the diaphragm is adjusted to 60 μm, which is half of the imbalance length in the IOC phase

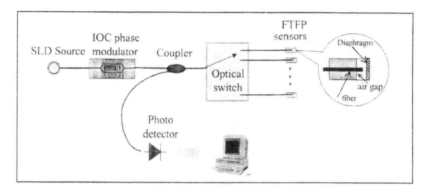

Figure 2. Multiplexed FTFP sensor system.

modulator. A sol-gel process is used to form the TiO_2 mirror on the entire cross-section of the optical fiber so that the reflectivity of the fiber tip can be increased from 4% to 16%.

4   PHASE DEMODULATION SYSTEM

The demodulation system implemented for the current sensor design is a PC-based pseudo-heterodyne scheme based on a four-step phase-stepping algorithm [16]. In this scheme, the optical signal is modulated by an IOC phase modulator instead of a traditional PZT modulator. This technique offers numerous advantages: a) high optical output power, b) large frequency range (up to 3GHz), c) rejection of electrical noise, d) high dynamic range, and e) very high stability. Experiments have been conducted to calibrate corresponding phase output of the IOC phase modulator for different drive voltages as shown in Figure 3.

The modulation signal is a discrete sawtooth wave generated from the digital-to-analog output of the PC. In every period of the modulation signal, four digital voltages are generated and used to drive four step modulated phase values from the IOC phase modulator based on the calibration curve. The modulated phases then are added to the sensor phase change. The combined phase signal is detected by the self-designed high speed photodetector and sent to the analog-to-digital input of the PC. The modulation frequency used here is 100 kHz and the depth of modulation is $3\pi / 2$. The entire modulation and demodulation process is shown in Figure 4.

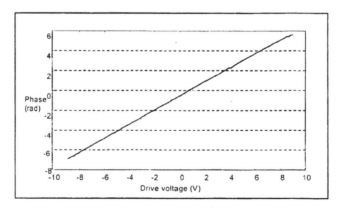

Figure 3. Calibration curve of the IOC phase modulator.

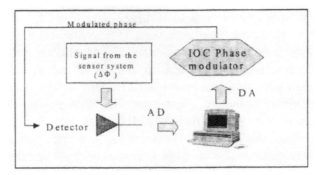

Figure 4. Schematic of PC-based digital phase modulation and demodulation schemes.

The demodulation system works as follows. The optical intensity is sampled four times during each period of the modulation signal. A 12-bit National Instruments digital acquisition board is then triggered to record the intensity every $\pi/2$ radians of the modulation signal. When the depth of modulation is set to $3\pi/2$ and the sampling rate is synchronized with the modulation frequency, the four consecutive optical intensity measurements yield the following:

$$I_0 = A + B\cos(\Delta\phi_s + 0) = A + B\cos(\Delta\phi_s),$$

$$I_1 = A + B\cos(\Delta\phi_s + \frac{\pi}{2}) = A - B\sin(\Delta\phi_s),$$

$$I_2 = A + B\cos(\Delta\phi_s + \pi) = A - B\cos(\Delta\phi_s),$$

$$I_3 = A + B\cos(\Delta\phi_s + \frac{3\pi}{2}) = A + B\sin(\Delta\phi_s). \tag{22}$$

The sensor phase is then determined from these four intensity values by using the following arc-tangent function

$$\Delta\phi_s = \tan^{-1}\left(\frac{I_3 - I_1}{I_0 + I_2}\right). \tag{23}$$

Equation (23) provides a way to determine the phase signal one is trying to detect. However, care has been taken whenever the denominator in equation (23) passes through a zero. Since, the inverse tangent function is multi-valued, the unwrapping algorithm, is written so that this discontinuity is detected and either an addition or subtraction of a phase of π from $\Delta\phi_s$ is carried out to have a continuous phase. The advantage of this algorithm is that the modulation frequency can be much higher than that used in the other techniques and the phase error is also low.

5   EXPERIMENTAL ARRANGEMENT

The prototype sensor system based on Fabry-Perot technique is shown in Figure 5. The system consists of a SLD source, one 2×2 optical coupler, the FTFP sensor, an IOC phase modulator, a photodetector, and a data acquisition personal computer. The Fabry-Perot cavity is produced between the fiber tip and a designed diaphragm structure. The frequency response range of this diaphragm structure extends to 10 kHz. Light from the SLD is sent to the IOC phase modulator first, then via the coupler to the FTFP sensor. The reflected light from the FTFP sensor is then sent to the high-speed detector. The Mach-Zehnder interferometer inside the IOC phase modulator is path-matched to the FTFP sensor to act as a read-out interferometer. The path matching is accomplished by moving the micro-stage to adjust the distance between the fiber tip and the diaphragm. The IOC phase modulator is driven by phase stepping program at a very high frequency (100 kHz). In the experiment, a condenser microphone (Bruel & Kjaer model # 4134) was used as reference sensor for validation. The input acoustic signal was generated by an Altec Lansing computer speaker system (Model No. ACS340). The diaphragm of the FTFP sensor is excited by using this speaker. The vibration changes the distance between the fiber tip and the diaphragm, which is related to the optical phase change. In order to detect this unknown phase change, the phase demodulation algorithm discussed in Section 3 was used. The entire phase modulation and demodulation process was controlled by a PC-based digital signal processing program.

6 RESULTS AND DISCUSSION

The prototype acoustic sensor is studied in a frequency range of approximately 50 Hz to 7.5 kHz by using sinusoidal sound signals. The sensor results have been compared to the results of a Brüel & Kjær 4134 condenser microphone. Representative results from studies performed at 70 Hz, 2.3 kHz and 7.5 kHz are shown in Figures 6, 7, and 8 respectively. The optical sensor data compares well with the condenser microphone in both time domain and frequency domain.

The above results are demonstrative of the applicability of the FTFP sensor system for acoustic measurements. The studies show that the system can be used in the frequency range from 50 Hz to 7.5 kHz. The sensitivity and dynamic range of this sensor are also determined and the experimental and model results are shown in Figure 9. It is found that the experimental sensitivity is lower than the model prediction determined from equation (21). A likely explanation is that the air gap between the diaphragm and the connector ferrule provides additional damping to the vibration of the diaphragm. Based on Figure 9, the sensitivity and dynamic range of this sensor are determined to be 0.9 rad/Pa and 60 dB to 118 dB, respectively.

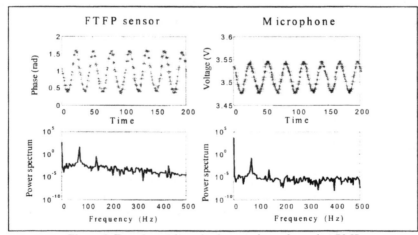

Figure 6. Experimental results for a study performed at 70 Hz.

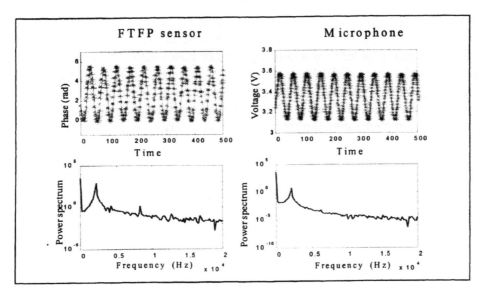

Figure 7. Experimental results for a study performed at 2.3 kHz.

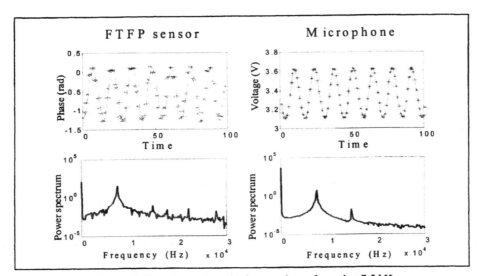

Figure 8. Experimental results for a study performed at 7.5 kHz.

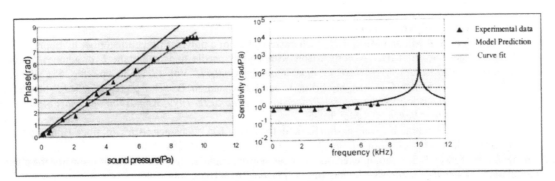

Figure 9. Sensitivity of FTFP sensor.

A novel fiber tip Fabry-Perot (FTFP) sensor has been investigated as a possible element of a fiber optic sensor system for acoustic measurements. A PC-based pseudo-heterodyne scheme is developed in order to achieve large sensor bandwidth. The studies were performed at frequencies in the range of 50 Hz to 7.5 kHz. Experimental results showed that the sensor was able to capture the acoustic field with an acceptable accuracy and confirm the model predictions. Further steps will aim at developing the multiplexed sensor system by using the optical switch and implementing the sensor system in acoustic control applications.

ACKNOWLEDGMENTS

Partial support received for this work, the U.S. Army Research Office (ARO) under contract No. DAAH 049610334 is gratefully acknowledged. Current support being received from the National Science Foundation through Award No. CMS-0123222 is also gratefully acknowledged.

REFERENCES

1. Nashif, A.D, Jones, D.I.G., and Henderson, J.P., "Vibration damping," John Wiley &Sons, Inc., New York, 1985.
2. Nelson, P.A. and Elliott, S.J., "Active control of sound," Academic, New York, 1994.
3. Balachandran, B., Sampath, A., and Park, J. "Active control of interior noise in a three-dimensional enclosure," *Smart Materials and Structures*, 5 (1), pp. 89-97, 1996.
4. Fuller, C.R., Elliott, S.J., and Nelson, P.A., " Active control of vibration," Academic Press, New York, 1996.
5. Bucaro, J.A., Dardy, H.D., and Carome, E.F., "Fiber optic hydrophone," *Journal of Acoustical Society of America*, 62, pp. 1302-1304, 1977.
6. Cole, J.H., Johnson, R.L., and Buta, P.G., "Fiber Optic detection of sound," *Journal of Acoustical Society of America*, 62, pp. 1136-1138, 1977.
7. Hess, C.F., " Optical microphone for the detection of hidden helicopters," *AIAA Journal*, 30(11), pp. 2626-2631 Nov. 1992
8. He, G. and Cuomo F., " The analysis of noises in a fiber optic microphone," *Journal of Acoustical Society of America*, 92(5), pp. 2521-2526, Nov 1992.
9. Zhou, C.H., Letcher, S.V., and Shukla, A., " Fiberoptic microphone based on a combination of Fabry-Perot interferometry and intensity modulation." *Journal of the Acoustical Society of America*, 98 (2), pp. 1042-1046, Aug. 1995.
10. Li, D.L. and Zhang, S.Q., "The ring-type all-fiber Fabry-Perot interferometer hydrophone system," *Journal of the Acoustical Society of America*, Vol 104, No.5, pp. 2798-2806, 1998.
11. Koch, C., " Measurement of ultrasonic pressure by heterodyne interferometry with a fiber-tip sensor," *Applied Optics*, Vol. 36, No.13, 1999.
12. Beard, P.C., Hurrell, A.M., and Mills, T.N, " Characterization of a polymer film optical fiber hydrophone for use in the range 1 to 20 MHz: A comparison with PVDF needle and membrane hydrophones," *IEEE Transactions on Ultrasonics, Ferroelectrics, and Frequency Control*," Vol. 47, No. 1, pp. 256-265, 2000.
13. Baldwin, C., Yu, M., Miller, C., Chen, S., Sirkis, J., and Balachandran, B., "Bragg grating based Fabry-Perot sensor system for acoustic measurements," *Proceedings of the SPIE 1999 Symposium on Smart Structures and Materials, Newport Beach, CA*, March 1-5.
14. Takahashi, N., Yoshimura, K., Takahashi, S., and Imamura, K., " Characteristics of fiber Bragg grating hydrophone," *IEIEC TRANS. ELECTRON.*, Vol. E83-C, No. 3, pp. 275-281, 2000.
15. Chang, C. and Sirkis, J. S. " Multiplexed optical fiber sensors using a single Fabry-Perot resonator for phase modulation," *Journal of Lightwave Technology*, 14(7), pp. 1653-1663, July 1996
16. Gasvik, K. " Optical Metrology," Wiley, New York, 1995.

MODELING OF PLATE STRUCTURES EQUIPPED WITH CURRENT-DRIVEN ELECTROSTRICTIVE ACTUATORS FOR ACTIVE VIBRATION CONTROL

Frederic Pablo, Daniel Osmont, and Roger Ohayon

ABSTRACT

The study here reported is focused on active vibration control applications performed on plate host structures equipped with electrostrictive patches. In such applications the design of controllers requires to simulate the behavior of the coupled structure. These simulations can then be performed through a Finite Element Method which implies the elaboration of electrostrictive finite elements. The purpose of the present paper is to present the elaboration of a plate electrostrictive finite element with a priori plate assumptions. We underlined in a previous paper that using current as driving input for these actuators would imply simplifications in active vibration control. We thus set out here finite elements for both current driven, and voltage driven actuators, with the view of highlighting theses simplifications. The elements here developped are characterized by the reduction of the initial electromechanically coupled problem to a purely mechanical one. Comparisons of numerical simulations of the control of the vibration of an elastic beam using different models of the system is moreover presented in this article.

INTRODUCTION

With the view of improving micro-vibrations absorption, more powerful actuators have been searched for. In the last few years, electrostrictive ceramics characterized by important strains when subjected to an electric field have been elaborated and widely studied with the aim of supplying this need. These materials make it possible to design very attractive actuators for active vibrations control. Nevertheless they are characterized by a non-linear electromechanical behavior and a thermal sensitivity which considerably complicate their use.

Frederic Pablo, ONERA, 29 ave. de la div. Leclerc, BP 72, 92322 Châtillon cedex, France.
Daniel Osmont, ONERA, 29 ave. de la div. Leclerc, BP 72, 92322 Châtillon cedex, France.
Roger Ohayon, CNAM / LMSSC, 2, rue Conté, 75003 Paris, France.

255

Studies focused on understanding the behavior of electrostrictive ceramics such as 0.9PMN-0.1PT under cycling electric field showed important sensitivity of this behavior to operating parameters. Moreover, recent papers [1,2] underlined a significant heating of electrostrictive patches under cycling electric field and showed that the sensitivity of electrostrictive behavior to operating parameters could in fact be reduced to a sensitivity to the ceramic own temperature. From these observations some models of electrostrictive behavior giving good accuracy have been elaborated [1–4].

In active vibration control applications the design of controllers requires simulation models of the structure. These simulations can be performed through a Finite Element Method which implied the elaboration of some three-dimensional [5,6] and two-dimensional electrostrictive finite elements [7].

The study here reported is focused on active vibration control applications performed on plate host structures equipped with electrostrictive patches. The use of plate finite elements to simulate the behavior of such structures thus seems to be more suitable. As no plate electrostrictive finite element, with direct a priori plate assumptions, has been up to now presented (to our knowledge) the purpose of this article is to briefly set out the elaboration of such a finite element for electrostrictive patches used as actuators.

The finite element formulations, presented in a previous paper [2] and recalled here, are based on electromechanical constitutive equations derived in an earlier paper [1], mechanical and electrical considerations and direct a priori plate assumptions. It is shown that the electromechanical problem can be reduced to an equivalent mechanical one. The electrical phenomena are then taken into account through prescribed stresses and non usual modifications of the elastic constitutive laws. This method considerably simplifies the resolution of the problem since classical finite elements for laminated plates can be used to model the electrostrictive plate.

We moreover showed, in a previous paper [2], that using current instead of voltage as the driving input of actuators may simplify active vibrations control with electrostrictive patches. Based on these considerations, two thin plate electrostrictive finite elements are described depending on the driving input used [2].

Comparisons of numerical simulations of the control of the vibration of an elastic beam using different models of the system will be presented. These simulations are related to current or voltage driving input and finite element models of the beam with and without the electrostrictive patch. From these comparisons we will determine conditions for which the simplest control process, i.e. a current driven control without any modeling of the actuators can be efficient.

PROBLEM STATEMENT

The present paper aims at presenting the elaboration of thin plate electrostrictive finite elements for current driven, and voltage driven actuators (patches) with the view of integrating those elements in control loops.

This section is focused on establishing the basic assumptions of a plate theory. As electrostrictive ceramics are characterized by an electromechanical behavior, the problem here to solve is a fully electromechanically coupled one. This coupling then induces the need for simultaneously solving mechanical and electrical relations presented in Ref. [2,8], taking into account the non-linear electromechanical constitutive laws derived in Ref. [9,1].

We underline that in Ref. [9,1] we developed electromechanical constitutive equations for current driven actuators. Moreover, a similar developpement can be used to derive voltage driven constitutive equations.

Material Assumptions

This study is focused on a particular class of materials which will be called mechanically and electrically orthotropic. These assumptions can then be translated as follows:

1. there is no coupling between transverse distorsions and stresses on the one hand, and, plane strains and stresses on the other hand, through the constitutive law,

2. there is no coupling between transverse electric field and displacement on the one hand, and, plane electric fields and displacements on the other hand, through the constitutive law,

3. electrostriction couples transverse electric displacement and field, and plane stresses and strains on the one hand, and, in-plane electric displacements and field, and transverse stresses and strains on the other hand.

A Priori Plate Assumptions

We will suppose afterwards that the structures we are interested in are thin plates. We thus suppose that the ratio of the thickness h and the in-plane characteristic dimension L is small compared to 1. Given motion equations, transverse stresses T_{iz} $(i = x, y, z)$ are negligible compared to in-plane stresses $T_{\alpha\beta}$ $(\alpha, \beta = x, y)$.

From these assumptions, one can express displacements as linear functions of z (transverse variable):

$$U_x(x, y, z, t) = u(x, y, t) - z\, w_{,x}(x, y, t) \tag{1}$$
$$U_y(x, y, z, t) = v(x, y, t) - z\, w_{,y}(x, y, t) \tag{2}$$
$$U_z(x, y, z, t) = w(x, y, t) \tag{3}$$

Given the strain-displacement relationship, strains associated with these displacements can be written as follows :

$$S_p = S^0 - z\,R^1 = \begin{pmatrix} v_{x,x} \\ v_{y,y} \\ v_{x,y} + v_{y,x} \end{pmatrix} - z \begin{pmatrix} w_{,xx} \\ w_{,yy} \\ 2w_{,xy}, \end{pmatrix} \quad \text{and} \quad S_{iz} \sim 0 \quad (i = x, y, z) \tag{4}$$

where S^0 and R^1 are respectively in-plane strains deriving from membrane and the curvature. This equations are the expression of the well-known Kirchhoff-Love assumptions.

PLATE THEORY

The plate theory here presented is based on a priori assumptions. We proved in an earlier paper [2] that for metallized patches for which in plane electric field may be neglected, in-plane electric displacements are negligible compare to transverse electric displacement. The Gauss law can thus be reduced to : $\boldsymbol{D}_{z,z} = q^d$.

One is then able to integrate this relation, whatever the driving input used. The solution of the Gauss law will then make it possible to eliminate the unknown electrical variables in favour of the mechanical displacements and to obtain a purely mechanical problem with equivalent electric forces.

Current driven actuator

In this operating configuration a current is imposed through the electrostrictive patch. This current then induces charges on the upper and lower ceramic surfaces which can be associated with electric displacements. The transverse electric displacements ($\boldsymbol{D}_z$) is thus known for each time t on the upper and lower patch surfaces.

Given the electric boundary conditions (electric displacement D_z^d is imposed) the integration of the Gauss law leads to a constant transverse electric displacement through the thickness of the plate:

$$\boldsymbol{D}_z = \boldsymbol{D}_z^d(x,y) \tag{5}$$

Assuming plane stresses and introducing this solution in the stress constitutive law, the plane stresses-strains relationship can be read:

$$\boldsymbol{T}_p = \widetilde{C}_{pp}^D \boldsymbol{S}_p + \overline{Q}_{pz}^{\sigma D}(D_z^d(x,y))^2, \tag{6}$$

where $\widetilde{C}_{pp}^D$ are the isolated elastic constants.

We thus have eliminated the electric displacement from the unknowns of the electromechanical problem. The problem now to solve is then a purely mechanical problem with electrically induced prescribed stresses. Nevertheless, the current density have to be measured at each point of the faces of the plate if we want to establish these stresses.

This approximation will be satisfied if the transverse electric displacement is almost constant when the faces of the patches are assumed to be voltage equipotentials. If not, this more critical case need further investigations.

Voltage driven actuator

In this operating configuration a transverse voltage is imposed through the electrostrictive patch. As patches are equipped with electrodes on the upper and lower surfaces, electric potentials (V) are supposed to be known and uniform on these faces.

When electrostrictive patches are voltage driven, the electromechanical problem is strongly non-linear and difficult to solve. However, this problem can be simplified through a linearization around an operating point. Any variable v can then be read $v = v^P + \overline{v}$, where v^P and $\bar{v}$ are respectively the known value at the operating point and an increment around this point for this variable. We then have to establish the value of increments.

The constitutive law, we have to use for voltage driven actuators, can also be read as a relation expressing stresses as a function of strains and electric displacement, where $D = D(E)$. Differentiating these constitutive laws with respect to strains S and electric field E, and, applying materials and plane stresses assumptions, one can read:

$$\overline{T}_p = \widetilde{\mathbb{C}}_{pp}^{D} \overline{S}_p + \widetilde{\mathsf{d}}_{pz}^{\sigma D} \overline{D}_z, \tag{7}$$

$$\overline{D}_z = \widetilde{\mathsf{d}}_{zp}^{D\varepsilon} \overline{S}_p + \widetilde{\mathbb{h}}_{zz}^{E} \overline{E}_z, \tag{8}$$

where submatrices $\widetilde{\mathbb{C}}$, $\widetilde{\mathsf{d}}$ and $\widetilde{\mathbb{h}}$ are calculated at the operating point.

These "incremental" constitutive laws can be compared to linear piezoelectric constitutive laws. One can then use similar methods as for current driven actuators to eliminate the electric field from the problem unknowns. Considering Gauss law for the incremental problem, one can read:

$$D_{z,z} = D_{z,z}^{P} + \overline{D}_{z,z} = q^d. \tag{9}$$

Integrating this relationship and taking into account the incremental constitutive law (8), one can establish the following value for the transverse electric displacement increment:

$$\overline{D}_z = \Phi_0^D \overline{S}^0 - \Phi_1^D \overline{R}^1 + \overline{D}_z^{elec}, \tag{10}$$

where Φ_i^D and $\overline{D}_z^{elec}$ are submatrices depending on $\widetilde{\mathbb{C}}$, $\widetilde{\mathsf{d}}$, $\widetilde{\mathbb{h}}$ and the electrical boundary conditions, and, are thus fully established.

Introducing this relationship in the incremental constitutive law (7) then implies:

$$\overline{T}_p = \Phi_0^\sigma \overline{S}^0 - \Phi_1^\sigma \overline{R}^1 + \overline{T}_p^{elec}, \tag{11}$$

where Φ_i^σ and $\overline{T}_p^{elec}$ are submatrices depending on $\widetilde{\mathbb{C}}$, $\widetilde{\mathsf{d}}$, $\widetilde{\mathbb{h}}$ and the electrical boundary conditions, and, are thus fully established.

In the particular case of homogeneous materials and null electrical volumic charges, incremental laws (10) and (11) :

$$\begin{cases} \overline{T}_P = \left(\widetilde{\mathbb{C}}_{PP}^{D} + \widetilde{\mathsf{d}}_{Pz}^{\sigma D} \widetilde{\mathsf{d}}_{zP}^{D\varepsilon} \right) \overline{S}^0 - z \widetilde{\mathbb{C}}_{PP}^{D} \overline{R}^1 - \widetilde{\mathsf{d}}_{Pz}^{\sigma D} \widetilde{\mathbb{h}}_{zz}^{E} \dfrac{\overline{V}^+ - \overline{V}_-}{h} \\[2mm] \overline{D}_z = \widetilde{\mathsf{d}}_{zP}^{D\varepsilon} \overline{S}^0 - \widetilde{\mathbb{h}}_{zz}^{E} \dfrac{\overline{V}^+ - \overline{V}_-}{h}, \end{cases} \tag{12}$$

where $\widetilde{\mathbb{C}}_{PP}^{D} + \widetilde{\mathsf{d}}_{Pz}^{\sigma D} \widetilde{\mathsf{d}}_{zP}^{D\varepsilon} = \widetilde{\mathbb{C}}_{PP}^{E}$ and $\overline{V}^+$ and $\overline{V}_-$ are electric potentials increments imposed on the electrodes. Thus, the elastic constants for membrane stresses are the short-circuited ones and elastic constants for bending stresses are the isoled ones.

We thus have eliminated the electric displacement from the unknowns of the electromechanical problem. The problem now to solve is then a purely mechanical problem with electrically induced prescribed stresses (known) and modification of the mechanical constitutive law.

Partial conclusions

We thus proved that whatever the driving input used, the non-linear electromechanical problem can be reduced to a purely mechanical problem where electrical phenomena are taken into account by additional prescribed stresses and consistent modification of the mechanical constitutive law. Moreover, the simplification of the problem when current is used instead of voltage as the driving input is highlighted by the previous developments.

CONTROL OF A CANTILEVERED BEAM

Driving input considerations

We will here apply the theory proposed above on the active vibration control of a 40*220*1 mm³ cantilevered beam, for its first mode. We will thus study the time evolution of the mechanical displacements of a released beam. The beam is moreover equipped with a piezoelectric or electrostrictive actuator placed at 45 mm from the clamping. The whole structure can then been modelled as presented in Figure 1(a):

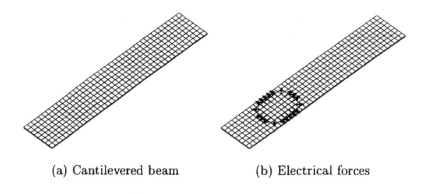

(a) Cantilevered beam   (b) Electrical forces

Figure 1. Modelled structure

The first frequency of the beam respectively equipped with a short-circuited and isolated piezoelectric patch, obtained by a Finite Element Method, are about 17.95 Hz and 18.02 Hz. The electric state of the patch thus does not significantly modify the first frequency.

Let's now suppose that a static voltage is applied to the actuator. The electrical forces applied to the beam through the patch are then plane forces and are distributed as presented in Figure 1(b)

If we now observe the transverse electric displacement D_z distribution over the patch, obtained through the constitutive laws, one can observe that mechanically induced electrical displacements, and thus total electric displacements are quasi uniform (Cf. Figure 2). Indeed, from numerical results, the variation of mechanically induced electric displacement is less than 5% compared to its mean value.

The mean value of the transverse electric displacement can then supposed to be uniform on the patch, and it is thus possible to use the current as the driving input given that $D_z = I/S$ (if I and S are respectively the imposed current and the patch area).

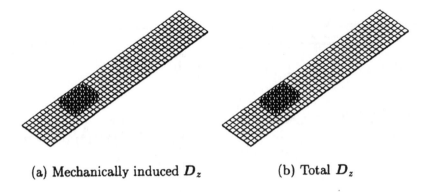

(a) Mechanically induced D_z (b) Total D_z

Figure 2. Transverse electric displacement distribution over the patch

Active vibration control

Let's now suppose a classical electrostrictive behavior as presented in Figure 3.

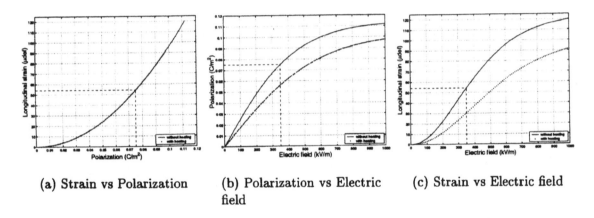

(a) Strain vs Polarization   (b) Polarization vs Electric field   (c) Strain vs Electric field

Figure 3. Modeled electrostrictive behavior ((-): without heating, (.): with heating)

Let us polarize the patch with an electric field of 350 V/mm, an electric displacement of 7.45e-2 C/m^2 and use the patch around the neighbourhood of this state of polarisation.

A linear and a non-linear control proportional to the speed have then been elaborated with the of view of reducing the vibration induced by a release of the cantilever beam. The linear control has been obtained by assuming that the behavior of the electrostrictive ceramic is linear around the operating point. As for the non-linear control, it is related to the non-linear behavior. Applying these two controls on the non-linear ceramic with a gain of 0.08, one obtains the results presented in Figure 4, whatever the driving input used (voltage or current).

One can then observe that the time evolution of the displacement of the beam (Figure 4(a)) is not affected by the control, but that the driving input is slightly modified (for the case considered here).

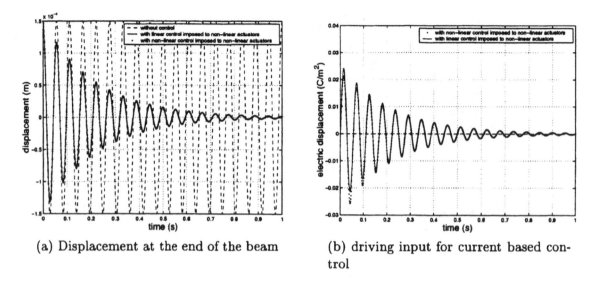

(a) Displacement at the end of the beam   (b) driving input for current based control

Figure 4. Active vibration control of the cantilever beam with an electrostrictive patch

Let's now compare the electric energy supplied to the actuator as a function of the control, and of the driving input used. One then obtain curves presented in Figure 5

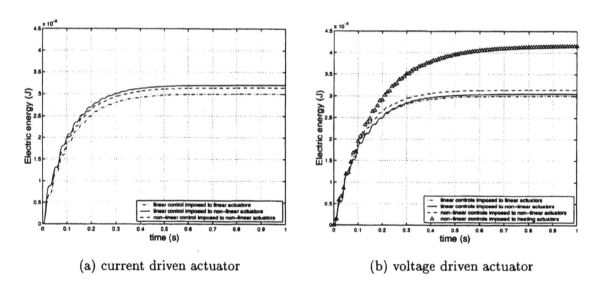

(a) current driven actuator   (b) voltage driven actuator

Figure 5. Electric energy supplied to the electrostrictive patch

One observe that the electric energy supplied to the patch is similar for the two driving inputs. But, the patch is subjected to a heating under cycling operating conditions which modifies the voltage driven behavior. Not taking into account this heating leads to modifications in the vibration absorption (Cf. Figure 6) and to a greater electric energy to be supplied to the actuator.

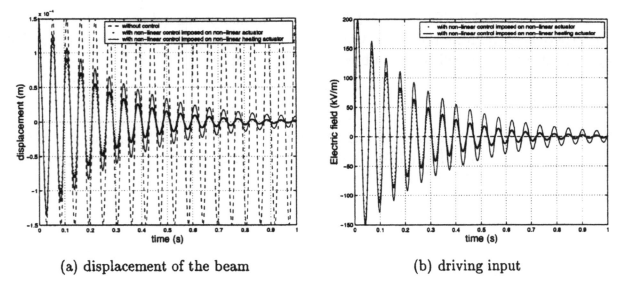

(a) displacement of the beam   (b) driving input

Figure 6. Active vibration control with a heating electrictive ceramic

An active vibration control using current driven electrostrictive actuators thus seems more reliable than such a control using similar voltage driven patches. We moreover underline here that the modelized behavior used in these simulations do not take into account the hysteresis. As these hysteresis are more important in the strain-electric field behavior than in the strain-polarization one, the electric energy to be supplied to the ceramic is more important for a voltage driven electrostrictive patch than for a current driven one. This point has been experimentally confirmed.

CONCLUSION

The study reported in this paper aims at performing active vibration control of plate structures, using electrostrictive patches.

We have shown in this paper that using a priori plate assumptions and a priori material assumption for a metallized electrostrictive patch, the electromechanical problem can be reduced to a purely mechanical problem. The electrical phenomena are then taken into account through the modification of elastic constants and through prescribed stresses calculated from electrical inputs. Classical FEM elements for stratified plates can then be used to simulate the behavior of the coupled structure.

Using current instead of voltage as the driving input is very interesting for electrostrictive ceramics since it avoid strain saturations and lower the hysteresis.

We then proved that using current as the driving input is consistent since the transverse electric displacement is almost uniform over the electrodes of the actuator. One can thus design controllers with current as driving input similar to the classical "voltage controllers".

The experimental verification of the proposed method of control are in progress. The first experiences confirme the interest of current driving the electrostrictive patch. A full report of these results will be the subject of a future paper.

References

[1] F. Pablo and B. Petitjean. Characterization of 0.9PMN-0.1PT Patches for Active Vibrations Control of Plate Host Structures. *Journal of Intelligent Materials, Systems and Structures*, 11(11):857–867, November 2000.

[2] F. Pablo, D. Osmont, and R. Ohayon. A thin plate electrostrictive element for active vibration control. In *Smart Structures 2001 Conferences and SPIE's International Symposium on NDE for Health Monitoring and Diagnostics*, Newport Beach, USA, March 04-08 2001.

[3] C. L. Hom and N. Shankar. A Fully Coupled Constitutive Model for Electrostrictive Ceramic Materials. *Journal of Intelligent Materials, Systems and Structures*, 5:795–801, 1994.

[4] M. Fripp. Distributed Structural Actuation and Control with Electrostrictors. Master's thesis, Massachusetts Institute of Technology, 1995. Thesis Number: 1995-256.

[5] C. L. Hom and N. Shankar. A Finite Element Method for Electrostrictive Ceramic Devices. *International Journal of Solids and Structures*, 33(12):1757–1779, 1996.

[6] K. Ghandi and N. W. Hagood. A Hybrid Element Model for Phase Transitions in Nonlinear Electro-Mechanically Coupled Material. *SPIE's 4th Annual Symposium on Smart Structures and Materials : Smart Structures and Materials*, 3039:97–112, 1997.

[7] J. C. Debus, B. Dubus, and J. Coutte. Finite Element Modeling of Lead Magnesium Niobate Electrostrictive Materials: Static Analysis. *Journal of the Acoustical Society of America*, 103(6):3336–3343, June 1998.

[8] F. Pablo, D. Osmont, and R. Ohayon. A thin plate electrostrictive element. part one : modeling and variational formulations. Submitted to "Journal of Intelligent Materials, Structures and Systems".

[9] F. Pablo and B. Petitjean. Electrostrictive Patches for Active Vibration Control of Thin Plate Host Structures. In Norman M. Wereley, editor, *SPIE's 7th Annual Symposium on Smart Structures and Materials : Smart Structures and Integrated Systems*, volume 3985, pages 818–829, 2000.

ELECTRO-MECHANICAL COUPLED MODELING AND OPTIMIZATION OF PASSIVELY DAMPED ADAPTIVE COMPOSITE STRUCTURES

Robert P. Thornburgh and Aditi Chattopadhyay

ABSTRACT

A smart structural model has been developed to analytically determine the response of arbitrary structures with piezoelectric materials and attached electrical circuitry. The theory simultaneously models both the structural and the electrical components, and the complex state of strain that may exist in the piezoelectric patches, thereby providing accurate mechanical and electrical response. The model is used to optimally determine both the placement of piezoelectric actuators and parameters describing associated electrical components in a passively damped structure. A robust multi-objective optimization procedure is developed to design the passive system for simultaneous damping of several critical modes of interest. The influence of stacking sequence in augmenting passive damping can also be examined by including ply orientations as design variables. Since the optimization problem now involves both continuous and discrete design variables, a hybrid optimization technique is used that allows the inclusion of both types of design variables. Also, since multiple design objectives are introduced, the Kreisselmeier-Steinhauser function approach is used allowing the multiple and conflicting design objectives and constraints to be combined into a single unconstrained function.

INTRODUCTION

Piezoelectric actuators are often used to control the dynamic response of structures, however accurate description of the interaction between the structural characteristics and the associated electrical circuitry is an often overlooked aspect in the design of smart structural systems. Piezoelectric materials (PZT) add additional stiffness to the structure, but this stiffness is dependent upon the electrical circuitry attached to each PZT. Electrical components can contain, store and dissipate both potential and kinetic energy, and piezoelectric material allows for transformation between mechanical and electrical energy. For efficient performance of a smart structural system, it is necessary to address all of theses issues associated with the entire system to accurately capture the response to a set of arbitrary external conditions. By simultaneously modeling both the structure and the electrical components, accurate calculation of mechanical and electrical response is possible.

A common application of the electrical interaction with the structural deformation is in the design of passive electrical damping systems[1-9]. Passive damping circuits have the ability to convert mechanical strain energy into electrical energy in the PZT and then dissipate this

Arizona State University, Dept. of Mechanical and Aerospace Engineering, Tempe, AZ 85282-6106.

energy as heat in resistors. However, to be most effective, these circuits must be tuned to damp out particular modes. Hagood et al. [1,2] conducted significant investigation in the areas of passive electrical damping and self-sensing actuators using coupled equations similar to those used in this work. Wu et al. [4-6] has developed several techniques for passive damping circuits and has demonstrated methods for damping out multiple vibrational modes. Wang et al. [7-9] demonstrated how passive damping circuits and active control can be combined in active-passive hybrid piezoelectric networks to enhance damping capability. Their research also made use of optimization techniques to choose electrical parameters to maximize control authority [8]. However, in all of these works, although significant effort was made in addressing the electrical aspects of the damping systems, very simple structural models, that do not take into account the complex state of strain that may exist in the piezoelectric patches, were used. In addition, the placement of the piezoelectric patch has not been considered concurrently with the design of the electrical system. Much of the work available in the literature addressed the vibration of cantilevered beams and plates where a piezoelectric patch located near the root can be effective in controlling all of the lower order modes. Structures with other boundary, where optimal location varies for each mode and are not intuitively obvious have not been examined.

The objective of this work is to demonstrate how multidisciplinary optimization (MDO) techniques can be utilized to optimize both structural and electrical aspects of an adaptive structural system. To simultaneously optimize multiple performance requirements, the Kreisselmeier-Steinhauser (K-S) function approach [10-12] is used. The K-S technique is a multiobjective optimization procedure that combines all the objective functions and the constraints to form a single unconstrained composite function to be minimized. Then an unconstrained solver is used to locate the minimum of the composite function. The advantage of this method is that it does not rely on arbitrary weight factors to combine multiple objective functions although they can be used in cases where a designer wishes to emphasize specific design criteria [12].

Optimization of an integrated structural and electrical system involves both discrete and continuous variables. Gradient based methods are generally ineffective in the optimization of discrete variables and nongradient based techniques such as genetic algorithms (GA) and simulated annealing (SA) can be computationally very expensive [13]. A hybrid method [14] has been proposed that uses a discrete search combined with a gradient based technique for the continuous design variables. The use of this technique provides significant improvement in computational efficiency over traditional discrete searches by using a combinatorial search algorithm that allows the use of gradients for the continuous variables, while using a discrete search technique for the discrete design variables.

COUPLED PIEZOELECTRIC-MECHANICAL FORMULATION

A recently developed two-way coupled piezoelectric-mechanical theory [15] is used to model composite plates with piezoelectric actuators. The coupled equations are formulated in terms of the mechanical strain and the electric displacement as opposed to the strain and electric field.

$$\sigma = \left(\mathbf{Q} + \mathbf{P}\mathbf{B}^{-1}\mathbf{P}^{\mathrm{T}}\right)\varepsilon - \left(\mathbf{P}\mathbf{B}^{-1}\right)\mathbf{D} \tag{1}$$

$$E = -\left(B^{-1}P^{T}\right)\varepsilon + B^{-1}D \tag{2}$$

Using this formulation, the electric displacement (**D**) can be taken as constant through the thickness of the PZT, thus ensuring conservation of charge on each of the electrodes. The equations of motion are derived using a variational approach and Hamilton's Principle. Using a finite element formulation, they can be expressed in matrix form as follows

$$\begin{bmatrix} M_u & 0 \\ 0 & 0 \end{bmatrix} \begin{Bmatrix} \ddot{u}_e \\ \ddot{D} \end{Bmatrix} + \begin{bmatrix} C_u & 0 \\ 0 & 0 \end{bmatrix} \begin{Bmatrix} \dot{u}_e \\ \dot{D} \end{Bmatrix} + \begin{bmatrix} K_{uu} & K_{uD} \\ K_{Du} & K_{DD} \end{bmatrix} \begin{Bmatrix} u_e \\ D \end{Bmatrix} = \begin{Bmatrix} F_u \\ F_D \end{Bmatrix} \tag{3}$$

where u_e is the nodal displacement vector and **D** is the vector of nodal electrical displacements. The stiffness matrices are further defined in Ref [12].

The absence of any electrical inertia or damping terms in Eq. (3) is a result of only considering the mechanical aspects of the system. In modeling an integrated smart structural system, it is necessary to include additional terms associated with the electrical components. For a simple LRC circuit, the linear equations of motion can be written as follows

$$M_q \ddot{q}_e + C_q \dot{q}_e + K_q q_e = F_q \tag{4}$$

then these equations can be directly combined with Eq. (3). This combination results in a completely coupled electrical-mechanical system of the following form

$$\begin{bmatrix} M_u & 0 \\ 0 & A_q^T M_q A_q \end{bmatrix} \begin{Bmatrix} \ddot{u}_e \\ \ddot{D}_e \end{Bmatrix} + \begin{bmatrix} C_u & 0 \\ 0 & A_q^T C_q A_q \end{bmatrix} \begin{Bmatrix} \dot{u}_e \\ \dot{D}_e \end{Bmatrix} + \begin{bmatrix} K_{uu} & K_{uD} \\ K_{Du} & K_{DD} + A_q^T K_q A_q \end{bmatrix} \begin{Bmatrix} u_e \\ D_e \end{Bmatrix} = \begin{Bmatrix} F_u \\ F_D \end{Bmatrix} \tag{5}$$

where A_q is an area matrix, used to convert the electric displacements into total charge.

As shown in Ref. [15] it is possible to reduce a system using the mode shapes rather than work with the full finite element model. The system can be partially reduced using the structural mode shapes if the assumption is made that the mode shapes for the coupled system can be expressed in terms of the open circuit mode shapes. This assumption is generally quite reasonable since changes in the elastic stiffness of PZTs due to open circuiting or short circuiting cause only modest shifts in natural frequency and very small alterations in the mode shapes. The structural stiffness introduced by the PZTs is generally small compared to the stiffness of the host structure and the difference between open circuit and short circuit stiffness is only around twenty percent. This combined with the fact that PZTs typically cover a fraction of the structural surface indicates that the differences in mode shapes between the open circuited condition and the actual structure will be generally small and localized.

First, the coupled system is reduced so that the open circuit eigenvalue problem can be solved for the desired number of eigenvalues.

$$K_{uu}\varphi = \omega_{oc}^2 M_u \varphi \tag{6}$$

Then using Eq. (6) and

$$\Phi^T K_{uu} \Phi = \mathrm{diag}\left[\omega_{oc1}^2 \quad \omega_{oc2}^2 \quad \cdots \quad \omega_{ocm}^2\right] \tag{7}$$

$$\Phi^T M_u \Phi = I_m \tag{8}$$

the coupled system of Eq. (8) can be reduced to

$$\begin{bmatrix} I_m & 0 \\ 0 & M_q \end{bmatrix} \begin{Bmatrix} \ddot{r} \\ \ddot{q}_e \end{Bmatrix} + \begin{bmatrix} \mathrm{diag}[\mu_m] & 0 \\ 0 & C_q \end{bmatrix} \begin{Bmatrix} \dot{r} \\ \dot{q}_e \end{Bmatrix} + \begin{bmatrix} \mathrm{diag}[\omega_m^2] & \Phi^T K_{uq} \\ K_{qu}\Phi & K_{qq}^* \end{bmatrix} \begin{Bmatrix} r \\ q_e \end{Bmatrix} = \begin{Bmatrix} \Phi^T F_u \\ F_q \end{Bmatrix} \tag{9}$$

where

$$\mathbf{r} = \mathbf{\Phi}^{-1}\mathbf{u}_e$$

(10)

Thus, the problem has been reduced to a small system composed of only the electrical degrees of freedom and the chosen number of mode shapes. This drastically reduces the number of degrees of freedom and allows for much faster computations once the eigen system is solved for.

OPTIMIZATION TECHNIQUE

Using the developed model, the objective now is to formulate an optimization problem to minimize the vibrational response of a composite plate under a steady state vibrational load. The developed formulation uses a passive electrical damping circuit to control vibration. The circuit is assumed to be composed of linear inductors, resistors and capacitors with values that are determined during the optimization process. Other nonlinear electrical components could be used and optimized, but a corresponding increase in computational effort would be expected. The optimization procedure must also take into account location and orientation of the piezoelectric patches on the plate and determine the location for maximum damping.

The mathematical optimization problem is stated as follows

$$\text{Min } f(\phi)$$

(11)

where

$$\phi = \begin{bmatrix} XC_h \\ YC_h \\ ANG_h \\ R_i \\ L_j \\ C_k \end{bmatrix} \quad \begin{array}{l} h = 1 \ldots N_a \\ \\ i = 1 \ldots N_r \\ j = 1 \ldots N_i \\ k = 1 \ldots N_c \end{array}$$

(12)

subject to the constraints

$$\begin{array}{c} R_i \geq 0 \\ L_j \geq 0 \\ C_k \geq 0 \\ \phi_L \leq \phi \leq \phi_U \end{array}$$

(13)

where $f(\phi)$ is the objective function, ϕ is the vector of design variables, N_a is the number of actuators and N_r, N_i, and N_c are the numbers of resistors, inductors and capacitors to be optimized. The design variables include the x- coordinate (XC), the y- coordinate (YC) and the orientation angle (ANG) for each actuator and the resistance (R), inductance (L) and the capacitance values for each electrical component. It is also possible to include ply orientation angles as design variables. Geometric constraints are imposed so that the electrical components maintain positive values in order to represent physical hardware.

It is important to note that the structural design variables (XC, YC, ANG) are all discrete, while the electrical design variables (R, L, C) are all continuous. The x- and y- coordinates of the piezoelectric patch are chosen as discrete variables based on the finite element model of the plate structure. The plate is meshed with a uniform grid of elements and the piezoelectric patches are required to have locations that align the PZT with the mesh. This procedure is used to avoid remeshing of the plate during every optimization iteration. The small numerical variations that result from changing the mesh in a finite element analysis can introduce errors that may lead to sub-optimal designs. Thus, it is more efficient to use a uniform mesh and restrict the PZTs to discrete locations. The analysis can start with a relatively coarse mesh to determine the general coordinates of the optimum location, and then if further accuracy is desired, the process can be repeated using a more refined mesh and the previous optimum solution as a starting point. The orientation angles are included to allow modeling of inter-digitated electrode (IDE) piezoelectric patches. These actuators are very powerful since they make use of d_{11} actuation and exert an actuation force in a preferential direction.

The optimization formulation is based on the Kreisselmeier-Steinhauser (K-S) function approach. In this technique the multiple objective functions are first transformed into reduced objective functions, which can be expressed as follows

$$\hat{f}_i = \left(F_i(\phi)/F_{i0}\right) - 1 - g_{max} \leq 0 \quad i = 1\ldots \text{number of objective functions} \tag{14}$$

where F_{i0} represents the original value of the ith objective function, F_i is its value based on the current design variables and g_{max} is the largest value of the original constraint vector. These normalized objective functions are now analogous to constraints and are combined into a single vector, $f_m(\phi)$; $m=1,2,\ldots,M$, with M being the sum of the number of constraints and the number of objective functions. The constraints and objective functions are then combined into a single composite objective function, $F_{KS}(\phi)$, defined as

$$F_{KS}(\phi) = f_{max} + \frac{1}{\rho}\ln\sum_{m=1}^{M} e^{\rho(f_m(\phi)-f_{max})} \tag{15}$$

where f_{max} is the largest constraint corresponding to the new constraint vector $f_m(\phi)$. The parameter ρ acts as a draw-down factor controlling the distance from the surface of the K-S envelope to the surface of the maximum constraint function.

The optimization procedure consists of a hybrid scheme that uses simulated annealing to optimize the discrete variables and a continuous search procedure to optimize the continuous variables. Because the K-S function approach reduces the problem to an unconstrained one, any unconstrained gradient based technique can be used for the continuous variables. A modified Newton method was used for the unconstrained minimization and gave generally satisfactorily performance. The modified Newton method used a finite difference scheme to calculate the gradient and the Hessian matrix in order to determine the search direction. The computational cost associated with the second-order search is often overcome by convergence in significantly fewer iterations compared to other first and zero order search techniques.

The optimization algorithm begins with the simulated annealing procedure and a user defined initial point. The simulated annealing procedure controls the values of the discrete variables only, and attempts to minimize the objective function using a probabilistic approach. Whenever the value of the objective function is requested by the simulated annealing algorithm, the current values of the discrete variables are used to calculate the open circuit eigen values and eigen vectors of the plate. Then an unconstrained search procedure is invoked to minimize the objective function at that point by varying the continuous design variables. Since the continuous variables are the electrical components, the procedure can be described as determining the optimum passive damping circuit for a particular set of PZT locations and orientations. The objective function value returned to the simulated annealing search is the minimum achievable K-S function value determined by the unconstrained search.

When the value of the objective function is required by the unconstrained search, the current values for the continuous variables are used to construct the linear electrical matrices. These are then combined with the reduced structural matrices based on the eigen values and vectors. These equations are then solved to obtain the system response for any disturbance frequency. Next, the frequency response curves are calculated and the maximum response peak is computed at each mode to be minimized. These parameters are then used to calculate the K-S function value.

RESULTS

The developed optimization procedure is demonstrated first through the integrated design of a cantilevered plate with a single actuator connected to a passive shunt circuit. The plate is assumed to be made of 3.17mm thick aluminum and the detailed dimensions are shown in Fig. 1. A single piezoelectric actuator is placed on one side to induce vibration and a single actuator is placed on the opposite side and is connected to the shunt circuit. The plate is

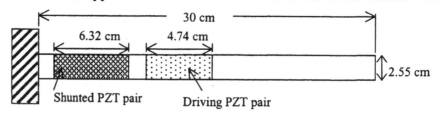

Figure 1. Configuration for the cantilevered plate prior to optimization.

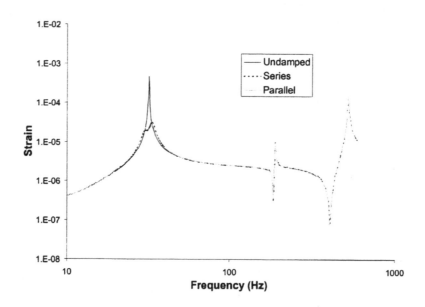

Figure 2. Frequency response curves for a cantilever plate with and without passive damping circuits.

modeled with a 19x3 element finite element mesh and the system is reduced using the first twelve modes. Optimization is performed to determine the optimum location of the piezoelectric patch and the electrical parameters governing the inductor and the resistor in the passive shunt to minimize the vibratory response associated with the first mode. The initial location for the patch is at the center of the plate. Results are presented for two configurations, one once with the inductor and resistor in series and the other in parallel.

The frequency response curves before and after optimization are shown in Fig. 2. It can be seen that the optimization algorithm is able to reduce the peak response of the first mode by over an order of magnitude in both cases. The optimum location for the piezoelectric patch is at the root of the plate. This is expected since the root is the location of maximum stress in a cantilevered plate. The optimum values calculated for the inductor and resistor in series are 422.032 henries and 14406 ohms, respectively. For the case with the inductor and resistor in parallel, the corresponding values are 417.927 henries and 489630 ohms.

Next the optimum location and electrical parameters are calculated for a composite plate simply supported on all four sides. This example provides a more interesting design challenge since the mode shapes each provide different locations and orientations of maximum strain. The plate is a 32cm square plate, 1.6mm thick, made of carbon fiber-epoxy laminae. The initial lay-up for the plate is eight laminae in $[0,90]_{2s}$ configuration. Also, the piezoelectric actuators used in this example are Active-Fiber Composite (AFC) actuators to allow optimization of orientation angle as well. Vibration is induced by as 2cmx2cm piezoelectric actuator on the backside of the plate. The plate is modeled with a 16x16 element mesh with the actuator in the center of the plate as an initial design.

First a single 4cmx4cm patch with a parallel shunt circuit is optimized for reducing the response of the first vibrational mode. The frequency response curves associated with the initial and the optimum configurations are presented in Fig. 3. The optimized location of the piezoelectric patch is away from the center of the plate, towards the corner at a point 12cm from two sides, as illustrated in Fig. 4. The optimized orientation angle for the AFC patch is 45° and the inductor and resistor has values of 50.884 henries and 520270 ohms, respectively.

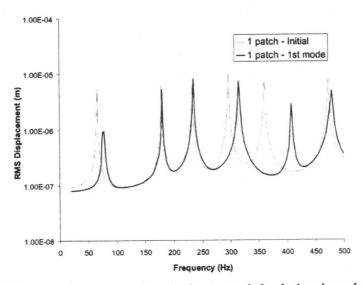

Figure 3. Frequency response curves for a simply supported plate both undamped and with a parallel damping circuit tuned for the first mode.

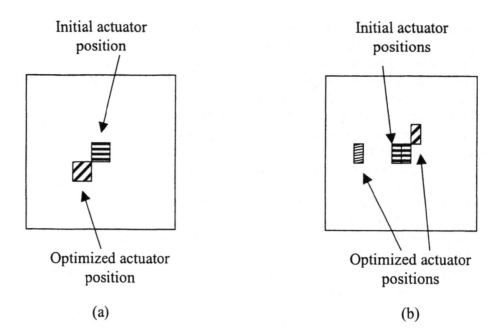

Figure 4. Position of the actuators on the simply supported plate before and after optimization for (a) one actuator, first mode and (b) two actuators, four modes.

It can be seen that optimization results in significant reduction in the response associated with the first mode.   These results demonstrate the benefit of including both structural and electrical parameters simultaneously in the optimization procedure. The procedure not only takes into account the effect of the passive shunt circuit, the results also demonstrate the influence of the mechanical stiffness of the piezoelectric patch.   The stiffness of the piezoelectric patch not only affects the first mode, it also reduces the static response of the system. It must be noted that the static response can be included as an additional objective function for minimization using the K-S function approach.

Next the same plate is optimized with two 2cmx4cm AFC actuators. Each actuator is connected to a single parallel shunt circuit. In this example the objective is the minimization of the first four vibrational modes. This method is very different from the multi-mode damping circuit proposed by Wu [6] which uses four different circuits to damp the four modes. The optimized results of the circuit are shown in Fig. 5. The piezoelectric patches move to locations at 7cmx16cm and 19cmx18cm, with orientations of 15° and 45° respectively, as seen in Fig. 4. The inductor and resistor have values of 11.645 henries and 195790 ohms for the first actuator and 103.497 henries and 140420 ohms for the second actuator. It must be noted that, as with any optimization algorithm, the use of simulated annealing does not guarantee convergence to a global minimum. As seen in Fig. 5, the first three modes are damped by only two circuits, although no reduction in the fourth mode is achieved.

Although the examples shown may be considered relatively simple, they demonstrate the potential of using MDO techniques for designing integrated adaptive structural systems. For passive damping circuits, the optimization determines best possible values of both structural and electrical components and can be used to design systems to damp multiple modes. But greater value lies in the potential of designing systems with synergistic characteristics, where

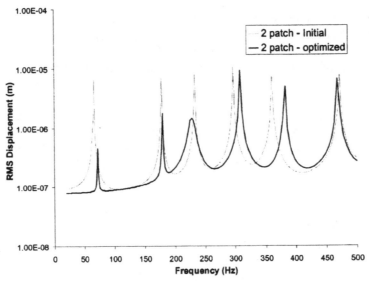

Figure 5. Frequency response curves for a simply supported plate both undamped and with two actuators.

careful combinations of actuators and circuits provide significant damping with little or no power consumption.

CONCLUDING REMARKS

A smart structural model has been developed to analytically determine the response of arbitrary structures with piezoelectric materials and attached electrical circuitry. The model simultaneously includes both the structural and the electrical components and can be used to optimally determine both the placement of piezoelectric actuators and parameters describing associated electrical components in a passively damped structure. A robust multi-objective optimization procedure was developed to design the passive system for simultaneous damping of several critical modes of interest. The ability to minimize the frequency response of multiple modes was accomplished by using the Kreisselmeier-Steinhauser function approach. The Kreisselmeier-Steinhauser function approach allows the multiple and conflicting design objectives and constraints to be combined into a single unconstrained function. Since the optimization problem involves both continuous design variables, such as electrical component values, and discrete design variables, such ash piezoelectric actuator placement and orientation, a hybrid optimization technique was used. Results show how this technique can be effective in optimizing both structural and electrical aspects of a smart structural system.

ACKNOWLEDGEMENTS

The authors would like to thank the Air Force Office of Scientific Research and technical monitor Dan Segalman whose support made this research possible.

REFERENCES

1. Hagood, N. W. and A. Von Flotow. 1991. "Damping of Structural Vibrations With Piezoelectric Materials and Passive Electrical Networks," *J. of Sound and Vibration*, 146(2): 243-268.
2. Hagood, N. W. and E. F. Crawley 1991. "Experimental Investigation of Passive Enhancement of Damping for Space Structures," *J. of Guidance, Control and Dynamics*, 14(6): 1100-09.
3. Agnes, G. S. 1994 "Active/Passive Piezoelectric Vibration Suppression," *Proc. of the Int. Soc. for Optical Eng.*, 2193: 24-34.
4. Wu, S. Y. 1996. "Piezoelectric Shunts with a Parallel R-L Circuit for Structural Damping and Vibration Control," *Proc. of the Int. Soc. for Optical Eng.*, 2720: 259-69.
5. Wu, S. Y. and A. S. Bicos. 1997. "Structural Vibration Damping Experiments Using Improved Piezoelectric Shunts," *Proc. of the Int. Soc. for Optical Eng.*, 3045: 40-50.
6. Wu, S. Y. 1998. "Method for Multiple Mode Shunt Damping of Structural Vibration Using a Single PZT Transducer," *Proc. of the Int. Soc. for Optical Eng.*, 3327: 112-22.
7. Kahn, S. P. and K. W. Wang. 1994. "Structural Vibration Controls Via Piezoelectric Materials With Active-Passive Hybrid Networks," *Proc. of ASME IMECE*, DE-75: 187-194.
8. Tsai, M. S. and K. W. Wang. 1996. "Control of a Ring Structure With Multiple Active-Passive Hybrid Networks," *J. of Smart Materials and Structures*, 5: 695-703.
9. Tsai, M. S. and K. W. Wang. 1999. "On the Structural Damping Characteristics of Active Piezoelectric Actuators With Passive Shunt," *J. of Sound and Vibration*, 221(1): 1-22.
10. Wrenn, G. A. 1989. "An Indirect Method For Numerical Optimization Using the Kreisselmeier-Steinhauser Function," *NASA Contractor Report 4220*.
11. Sethi, S. S. and A. G. Striz. 1997. "On Using the Kreisselmeier-Steinhauser Function in Simultaneous Analysis and Design," *Proc. of the 38th AIAA/ASME/ASCE Structures, Structural Dynamics and Materials Conf.*, 1357-1365.
12. Rajadas, J. N., R. A. Jury and A. Chattopadhyay. 2000. "Enhanced Multiobjective Optimization Technique For Multidisciplinary Design," *Eng. Opt.*, 33: 113-133.
13. Belegundu, A. D. and T. R. Chandrupatla. Optimization Concepts and Applications in Engineering. 1999. Prentice Hall. Upper Saddle River, New Jersey.
14. Seeley, C. E., A. Chattopadhyay and D. Brei. 1996. "Development of a Polymeric Piezoelectric C-Block Actuator Using a Hybrid Optimization Procedure," *AIAA J.*, 34(1): 123-128.
15. Thornburgh, R. P. and A. Chattopadhyay. 2001. "Electrical-Mechanical Coupling Effects on The Dynamic Response of Smart Composite Structures," *Proc. of the Int. Soc. for Optical Eng.*, 4327:413-424.

INTERROGATION OF BEAM AND PLATE STRUCTURES USING PHASED ARRAY CONCEPTS

Ashish S. Purekar and Darryll J. Pines

ABSTRACT

The structural dynamics of beams and plates can be interpreted as structural waves traveling through a structure. The presence of a discontinuity such as a crack or a hole in the structure cause reflections from incident waves. Phased arrays, which are typically used in a acoustic and electromagnetic wave applications, are used on a beam and plate to determine the direction of incoming waves. The movement of structural waves on a beam is seen by using a phased array. The presence/location of a discontinuity can also be inferred based on timing of reflected waves. The phased array is also shown to distinguish the direction that incident wave arrives on a plate. A phased array is shown to be a useful tool in determining the direction that wave is moving on a beam and plate structure.

INTRODUCTION

Recently there has been considerable interest in Health and Usage Monitoring Systems (HUMS) for various aerospace applications such as rotorcraft, fixed wing aircraft, and space structures. Damage in structures can occur from a variety of conditions such as excessive vibratory loads causing fatigue damage, interlaminar stresses in composite structures caused by off-design loading, or even simple phenomenon such as dropping of tools on panels. Typically damage critical structures are replaced periodically regardless of whether there is damage and this can lead to needless disposal of good parts. Normally, damage detection is done visually or the part is removed from the system and inspected independently which causes a loss of productivity of the system. A desirable HUMS system would determine the integrity of the part while it is still in the system and provide real-time status of the part. An online system would save both time and resources and allow the system to function more efficiently.

Damage detection for beams and plates is an area of active research since many components of aerospace systems can be modeled as beams or plates. One approach which can be taken for damage detection relies on a modal analysis approach to the structural

[1] Department of Aerospace Engineering, University of Maryland, College Park, MD, 20742, U.S.A.

dynamics [1–5]. Sensors placed on different locations can determine the modes and natural frequencies of the structure. Damage in a structure would result in shifts in natural frequencies and changes in the modes. Various damage detection techniques are available which allows the mass and stiffness matrices to be updated based on these changes in natural frequencies and modes. However, one drawback of this approach is that the natural frequencies and modes do not shift significantly due to small incipient damage cases. In order to get better resolution, the order of the system is increased by adding more elements which can result in a high computation cost. Another method of damage detection is to use ultrasonic techniques to examine specific locations on the structure. This method is powerful in that small damage cases can be detected; however, the ultrasonic techniques need to be local to the damage and much time and effort would go into scanning the entire structure.

A wave mechanics approach strikes a balance between modal methods, which are global, and ultrasonic methods, which are typically local as seen in Figure 1. In the wave mechanics approach, the structural dynamics are expressed in terms of waves moving through the structure. Natural frequencies and modeshapes found by modal analysis can be determined using wave mechanics as well as the interaction between an incident wave and a discontinuity which is a method commonly used in ultrasonic techniques.

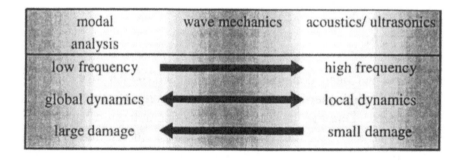

Figure 1. Wave mechanics in relation to conventional models.

Phased arrays are used in acoustics and antenna theory in order to actively shift the direction that an array listens or transmits. For an array which is mobile, the array can be physically steered to look in specific directions. Arrays which cannot be physically steered can be still be pointed to listen or transmit in specific directions using the appropriate signal processing. The combination of phased arrays along with wave mechanics models can be used in a damage detection scheme to actively interrogate a structure to search for discontinuities. This can be done by a simple pulse-echo scanning approach where the array sends out a plane wave in a particular direction and senses in the same direction for any reflections. The presence of a reflection would then indicate a discontinuity and the position of the discontinuity can be determined.

WAVE MECHANICS

Beam Dynamics

Wave propagation analysis has been traditionally been used as a way to model the dynamics of the structure and in the control the vibration of beams [6–9]. Additionally, some promising work has been done using wave mechanics for damage detection in beams [10–13].

The wave approach to structural dynamics work well for one-dimensional structures. The governing partial differential equation for a simple Euler-Bernoulli beam:

$$EI\frac{\partial^4 w(x,t)}{\partial x^4} + \rho A\frac{\partial^2 w(x,t)}{\partial t^2} = 0 \tag{1}$$

is solved for when the response is assumed to be harmonic. The solution can be thought of as structural waves traveling up and down the structure.

$$w(x,\omega) = w_{rp}e^{-ikx} + w_{lp}e^{ikx} + w_{re}e^{-kx} + w_{le}e^{kx} \tag{2}$$

where EI is the bending stiffness, ρA is the mass per unit length, and $k = \sqrt[4]{\omega^2\rho A/EI}$ is the wavenumber. In Equation 2, the four solutions correspond to rightward propagating, leftward propagating, rightward evanescent, and leftward evanescent waves, respectively. A diagram of the different waves are shown in Figure 2. The evanescent components die out spatially and are typically not considered when the application is based in the higher frequency region.

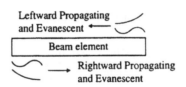

Figure 2. Structural waves present in beam

A spatial transition matrix can be used to describe a wave as it moves down the beam. Similarly, a scattering matrix is used to describe the creation of reflected waves from incident waves at a discontinuity. For example, a set of incident propagating and evanescent waves approaching a clamped boundary will reflect back as another set of reflected propagating and evanescent waves:

$$\begin{bmatrix} w_{prop} \\ w_{evan} \end{bmatrix}_{ref} = S_{clamped}\begin{bmatrix} w_{prop} \\ w_{evan} \end{bmatrix}_{inc} \tag{3}$$

where $S_{clamped} = \begin{bmatrix} -i & -1-i \\ -1+i & i \end{bmatrix}$ for the clamped boundary. Different boundary conditions have different scattering matrices: $S_{free} = \begin{bmatrix} -i & 1+i \\ 1-i & i \end{bmatrix}$ and $S_{pinned} =$

$\begin{bmatrix} -1 & 0 \\ 0 & -1 \end{bmatrix}$. The scattering matrix for a discontinuity such as a crack becomes more complicated because simple quantities such as displacement or slope are not prescribed but rather the properties of the crack section have to be addressed. A simple first order model of the crack would have a different EI and ρA than an undamaged region, Figure 3. The scattering matrix for a crack would depend on the depth of the crack. A relatively small crack would not reflect waves because there is little difference in the structural properties whereas a crack which cuts through most of the thickness of the beam has stronger reflections.

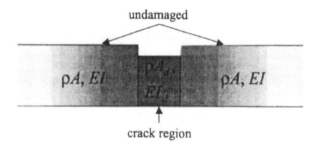

undamaged

$\rho A, EI$ $\rho A_d,$ EI_d $\rho A, EI$

crack region

Figure 3. Diagram of crack region

Plate Dynamics

The two dimensional analog of an Euler-Bernoulli beam leads to the differential of a simple plate:

$$D\left(\frac{\partial^4 w}{\partial x^4} + 2\frac{\partial^4 w}{\partial x^2 \partial y^2} + \frac{\partial^4 w}{\partial y^4}\right) + m\frac{\partial^2 w}{\partial t^2} = 0 \tag{4}$$

For the current application, the differential equation in polar coordinates is more applicable. Since the assumption is that the waves are symmetric with θ, the $\partial/\partial\theta$ terms will be ignored resulting in:

$$D\left(\frac{\partial^4 w}{\partial r^4} + 2\frac{1}{r}\frac{\partial^3 w}{\partial r^3} + \frac{1}{r^2}\frac{\partial^2 w}{\partial r^2}\right) + m\frac{\partial^2 w}{\partial t^2} = 0 \tag{5}$$

The solution to the differential equation for a plate in polar coordinates can be found by assuming the response is harmonic and is expressed as

$$w = c_1 H_0^{(1)}(\beta r) + c_2 H_0^{(2)}(\beta r) + c_3 I_0(\beta r) + c_4 K_0(\beta r) \tag{6}$$

where $\beta = \sqrt[4]{\omega^2 m/D}$. The solution is composed of a combination of Bessel functions which are similar to the beam solutions, given in Equation 2. The $H_0^{(1)}(\beta r)$ and $H_0^{(2)}(\beta r)$ terms correspond to the e^{ikx} and e^{-ikx} beam propagating solutions respectively and the $I_0(\beta r)$ and $K_0(\beta r)$ terms correspond to the e^{kx} and e^{-kx} beam near field solutions respectively. A major difference between the plate propagating solutions and the beam propagating solutions is that the beam propagating solutions maintain their magnitude as the the wave moves down the beam whereas the magnitude of the plate propagating solutions decrease as the wave moves away from the origin. As in the beam case, the near field terms can be ignored when interested in the higher frequency region.

PHASED ARRAYS

Phased arrays are used in acoustic and antenna theory to listen to a source at a given position [14]. An array is normally positioned so that it is facing the source; however, if an array cannot be moved, there is a signal processing technique which allows the array to listen in a particular direction just as if it was physically rotated. The signal from a particular direction can be given a higher gain than a signal from any other direction. In this way, the array can selectively listen in a particular direction while reducing the effect of signals from other directions. A similar approach can be used to sense the direction of waves traveling in a structure. Damage in the structure produces reflected and scattered energy which could be used in a damage detection scenario using concepts from acoustics [15,16].

The general equation for the response from an array, as shown in Figure 4 is given by:

$$\psi(\vec{r}) = \int_{V_0} X\left(\vec{r}_0, \omega\right) A\left(\vec{r}_0, \omega\right) g\left(|\vec{r} - \vec{r}_0|, \omega\right) dV_0 \qquad (7)$$

where X is a user specified gain applied to the sensors, A is a function which describes the distribution of the array, and g is the response from $\vec{r}$ to $\vec{r}_0$. For the sensing case, the source term is located at position $\vec{r}$ and the array is along $\vec{r}_0$ and the g term is the response at $\vec{r}_0$ from a disturbance at $\vec{r}$.

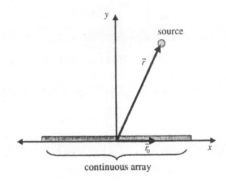

Figure 4. Diagram of Array along x-axis.

Typically, the g function can be approximated as $g = Ce^{-i\beta\hat{r}\cdot\vec{r}_0}$ where β is the wavenumber of interest, $\hat{r}$ is the unit vector pointing to the source, and $\vec{r}_0$ is the vector to each point along the array. This approximation can be used for plate dynamics where C is frequency dependent and includes leading constants. In the formulation of the phased array, the near field terms in both the plate and beam cases are ignored because the high frequency range is of interest. The expression $\vec{\beta} = \beta\hat{r}$ is interpreted as a wavenumber vector which not only describes the scalar properties of the wave but also the direction that the wave is traveling. Using $\beta = 2\pi/\lambda$, the wavenumber vector can be expressed as $\vec{\beta} = (2\pi a/\lambda)\,\vec{i}_x + (2\pi b/\lambda)\,\vec{i}_y$ where a and b are the direction cosines of $\hat{r}$ and $\vec{i}_x$ and $\vec{i}_y$ are the unit vectors in the x and y axes respectively.

For a simple array where the weights are uniform along the array and A defines a continuous array of length L along the x-axis, the response of the array is found by:

$$\psi = \int_{L/2}^{L/2} e^{-i\beta_x x_0}\,dx_0 \tag{8}$$

where $\beta_x = 2\pi a/\lambda$. The solution to the integral is a *sinc* function

$$\psi = L\frac{\sin\left(\pi\frac{L}{\lambda}a\right)}{\pi\frac{L}{\lambda}a} \tag{9}$$

Since L and λ are constants, the quantity of importance is the direction cosine, a. The plot in Figure 5(a) shows the normalized response of the array as a function of the direction cosine, a, which describes the direction that the wave is coming from. The response of the array is the greatest when $a = 0$ and when a moves in either direction, the response of the array drops. This shows that the array acts as a directional filter and in this case, the response of the array is greatest when $a = 0$.

Steering of the array is accomplished by introducing complex weights to the elements of the array in order to shift the main peak in Figure 5(a) to a different location along the x-axis, as shown in Figure 5(b).

$$\psi = \int_{L/2}^{L/2} e^{-i(\beta_x - \beta'_x)x_0}\,dx_0 \tag{10}$$

where β'_x are the complex weights. The integral can be evaluated to be

$$\psi = L\frac{\sin\left(\pi\frac{L}{\lambda}(a - a')\right)}{\pi\frac{L}{\lambda}(a - a')} \tag{11}$$

where a' is the direction that the array is to be steered toward. An alternate way to look at the steered array is to map the direction cosine, a, to the θ domain in a polar plot. By doing this, a lobe pattern is produced which shows where the array is steered toward. Figure 6(a) shows a typical lobe pattern for an unsteered array. The maximum response of the array occurs at 90° which indicates the array is most sensitive to a signal coming from directly in front. As the angle moves away from 90°, the response of the array drops indicating that signals from that direction are not seen by the array. When the array is steered, the main lobe moves toward that direction, as seen in Figure 6(b) where the array is steered toward 60° and the main lobe moves accordingly.

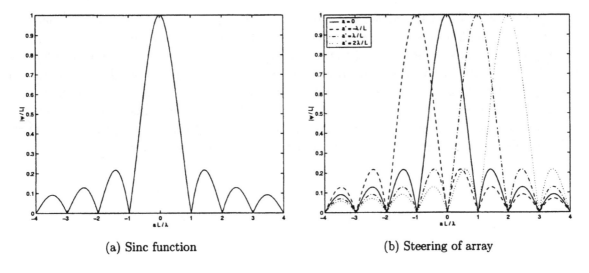

(a) Sinc function (b) Steering of array

Figure 5. Response of array

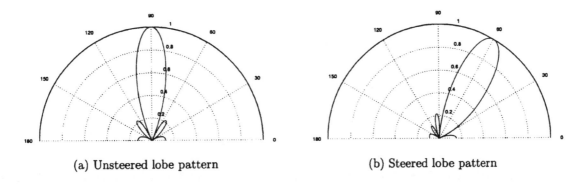

(a) Unsteered lobe pattern (b) Steered lobe pattern

Figure 6. Lobe patterns for unsteered and steered array

Discrete Arrays

Typically, arrays are composed of individual elements and in structures, an array could be made up of strain gages, accelerometers, or even piezoelectric sensors to measure properties of the deflection of the plate. Since the array is discrete, the integral in Equation 8 becomes a summation over each element. The response of the array then becomes

$$\psi = \frac{1}{N} \sum_N e^{-i\beta_x x_n} \tag{12}$$

where N is the total number of the elements located at x_n. For an odd number of elements with sensor spacing dx_0, the summation can be evaluated explicitly as

$$\psi = \frac{1}{N} \frac{\sin\left(N\pi \frac{a}{\lambda} dx_0\right)}{\sin\left(\pi \frac{a}{\lambda} dx_0\right)} \tag{13}$$

A typical response of the array, for a given N and dx_0, is shown in Figure 7, where the x-axis is a function of a since λ is held constant. There are now three peaks visible in the range of the x-axis. This means that there are three possible locations where the response of the array will be greatest. In order to ensure that array has a single maximum, the array must be designed so that only one peak will appear. This can be accomplished by satisfying the Nyquist criterion in space. In order to avoid the other peaks, the spacing of the sensors, dx_0, needs to be less than half the wavelength at the frequency of interest. Once this is satisfied in the design of the array, the other lobes will not exist.

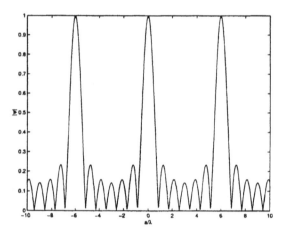

Figure 7. Response of discrete array.

EXPERIMENTAL SETUP AND RESULTS OF PHASED ARRAYS WITH PLATE AND BEAM STRUCTURES

Phased Array on Beam

A phased array on a beam would only need to look in two directions because the structural waves are only traveling in one of two directions. If near field components are neglected, the phased array should be able to distinguish between leftward and rightward traveling waves as shown in Figure 8. By steering the array to the left and the right, the rightward and leftward waves, respectively, can be seen.

A 1/16" thick, 1.5" wide, 6' long aluminum beam was instrumented with 7 strain gages bonded at the midspan of the beam and placed in cantilever boundary conditions. An impulse was placed on the beam and the array, composed of the strain gage elements, was used to determine the presence of the leftward and rightward waves. Figure 9 describes how the processing of the raw vibration data is accomplished. A short duration window function is used to capture high frequency components of the signal. Figure 10(a) shows the response of the array when the impulse was placed on the tip of the beam. The plot indicates that initially a leftward wave is seen by the array which is then followed by a rightward wave produced by a reflection of the leftward wave off of the fixed end. After

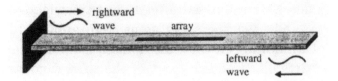

Figure 8. Diagram of Array Along Beam.

the rightward wave, another leftward wave is seen which is the reflection of the rightward wave off of the free end. This pattern of leftward and rightward waves following one another is repeated in the time domain. Figure 10(b) shows the response of the array when an impulse was placed near the root of the beam. In this case, a rightward wave is seen first followed by a leftward wave. For Figures 10(a) and 10(b), the frequency of interest is 1500 Hz. Both cases show that the phased array discriminates between the leftward and rightward waves. Figures 11(a) and 11(b) show the response of the array for the case of 2000 Hz. In all these figures, the x-axis is the time domain and the relative time increments are of importance rather than the absolute time information because the starting time depends on the triggering.

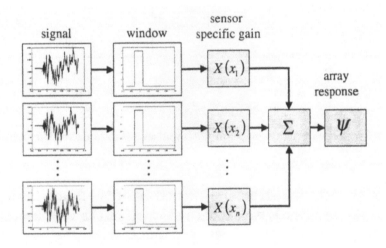

Figure 9. Processing of data.

The responses between Figures 10 and 11 are very similar except that the waves are seen to move faster for the 2000 Hz case. This is due to the dispersive nature of the beam dynamics. In the rod case, the speed of a wave at each frequency is constant; however, for the beam case, the speed of the wave is dependent on the frequency. The phase speed, which is the speed of a single wave of frequency ω, is $c = \omega/k$ where k is the wavenumber found earlier. Since k is proportional to $\sqrt{\omega}$, the speed of a single wave component varies with frequency. The group speed, which describes the collective speed of a group of waves, is found by $c_g = d\omega/dk$ and is related to the phase velocity

c by $c_g = 2c$. The signals were processed after they were windowed, similar to a short time Fourier transform, and then the array gains were applied to construct the array response. The effect of windowing in the time domain is a convolution in the frequency domain. Therefore, a particular frequency out of the time windowed spectrum actually represents a smearing in and around that frequency of the true frequency spectrum. For the responses of the arrays shown in Figures 10 and 11, the waves that are seen are moving at the group speed. This can be verified because the time it takes for the wave to move the distance from the middle of the array to either end and back again corresponds to the group velocity at that frequency.

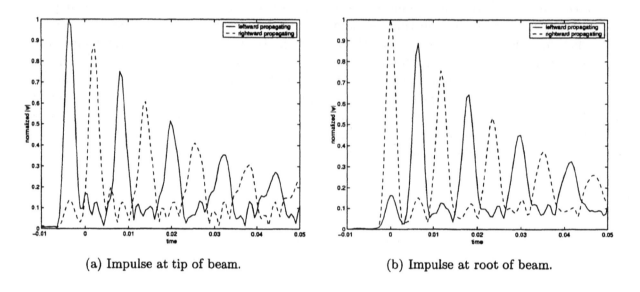

(a) Impulse at tip of beam.   (b) Impulse at root of beam.

Figure 10. Response of array on beam for impulses placed at the two ends (1500 Hz).

In the case of a crack, the structural properties of the beam have changed local to the discontinuity. An incident wave should reflect off of the discontinuity and reach the array quicker than the case where the wave continues down the beam and reflects off of a boundary condition. In terms of damage detection, a wave which appears sooner than expected would indicate the presence of damage. To see the effect that a discontinuity would have on the waves and the timing, a discontinuity was introduced on the beam by attaching a mass between the array and the fixed end and excited with an impulse at the free end, shown in Figure 12. The array should see the leftward wave coming from the impulse and then see the rightward wave reflect off of the fixed end or the discontinuity. Figure 13 shows the rightward waves that pass through the array for the cases where there is no discontinuity and the case where there is an added mass. The results show that the array sees a rightward wave off of the discontinuity appear before the reflection off of the fixed end. This would indicate the presence of some sort of discontinuity between the array and the fixed end.

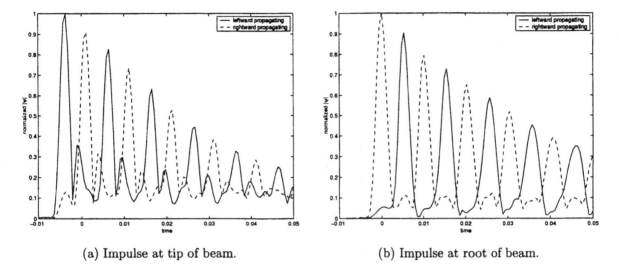

(a) Impulse at tip of beam.   (b) Impulse at root of beam.

Figure 11. Response of array on beam for impulses placed at the two ends (2000 Hz).

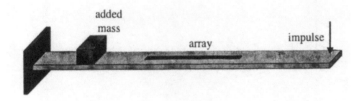

Figure 12. Diagram of beam with discontinuity.

Phased Array on Plate

An array, made up of piezoelectric sensor elements, was set up on a .02" thick, $2' \times 3'$ plate and two opposite edges were held fixed, as in Figure 14. For damage detection applications, a hole in a plate would reflect incoming waves. In a sense, the hole can be considered a source and if an phased array can detect the presence and location of a source, then the location of damage can be inferred for a uniform plate. For the plate experiment, an impulse was placed at various locations on the plate to excite the structure and the phased array was used to determine the direction, θ, of the incoming waves.

The normalized response of the array, ψ, is shown as a function of steering angle in the plots in Figure 15. Figure 15(a) shows the response when the location of the impulse was placed at 30° relative to the array as shown in Figure 14. The magnitude of the response is greatest at about 30°. Figures 15(b), 15(c), and 15(d) show that the response of the array is greatest when steered toward the direction of impulse. One concern in the experiments was the effect of reflections off of the boundaries on the array response. Since only an initial segment of the time record of each sensor was used in producing the array response, the reflections were not seen because they arrive at a later time.

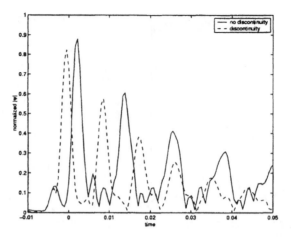

Figure 13. Rightward waves with and without discontinuity.

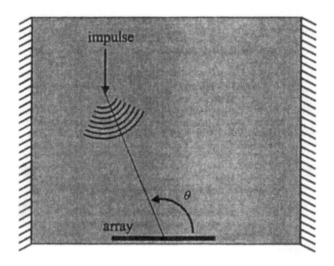

Figure 14. Diagram of Array on Plate.

CONCLUSIONS

Phased arrays are applicable in the use of structural dynamics to determine the direction of wave movement. From the beam results, the presence and timing of leftward and rightward waves is determined based on steering the array to look in either direction. The timing of these arrays points to the distance from the ends the array is located. A discontinuity in the beam can be determined based on the timing of the reflected waves. Although the presence of discontinuity was determined based on an added mass, the same concepts apply if the discontinuity is a crack. In this way, a damage detection scheme is possible based on looking at reflected waves. Another important characteristic seen by the array in the beam experiments is the decrease in the wave amplitude with time. In an ideal case where there is no damping, the amplitude of the waves should be constant; however experimentally there is damping present so the magnitude of the waves decrease

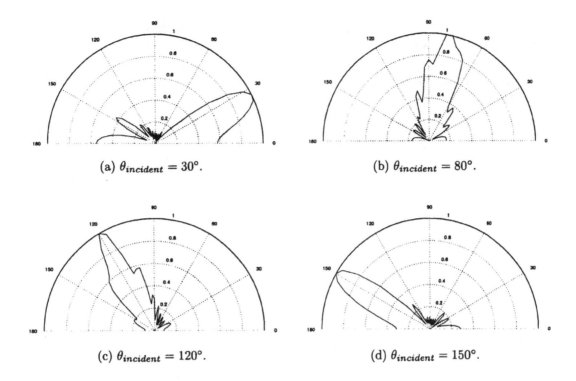

(a) $\theta_{incident} = 30°$.   (b) $\theta_{incident} = 80°$.

(c) $\theta_{incident} = 120°$.   (d) $\theta_{incident} = 150°$.

Figure 15. Response of array to impulse placed at various positions.

with time. The phased array on the plate showed that the direction of the incoming wave can be determined by steering the array from one extreme to the other.

REFERENCES

1. C. Farhat and F. Hemez, "Updating finite element dynamic models using an element-by-element sensitivity methodology," *AIAA Journal* **31**, pp. 1702–1711, 1993.

2. T. Kashangaki, S. W. Smith, and T. W. Lim, "Underlying modal data issues for detecting damage in truss structures," in *Proceedings of the 33rd AIAA/ASME/AHS/ASC Structures, Structural Dynamics, and Materials Conference*, pp. 1437–1446, 1992.

3. H. Luo and S. Hanagud, "An integral equation for changes in the structural dynamics characteristics of damaged structures," *International Journal of Solids and Structures* **34**, pp. 4557–4579, 1997.

4. T. Y. Kam and T. Y. Lee, "Crack size identification using an expanded mode method," *International Journal of Solids and Structures* **31**, pp. 925–940, 1994.

5. D. C. Zimmerman and M. Kaouk, "Structural damage detection using a subspace rotation algorithm," in *Proceedings of the 33rd AIAA/ASME/AHS/ASC Structures, Structural Dynamics, and Materials Conference*, pp. 2341–2350, 1992.

6. A. H. von Flotow and B. Schafer, "Wave-absorbing controllers for a flexible beam," *AIAA Journal of Guidance* **9**, pp. 673–680, 1986.

7. D. J. Pines and A. H. von Flotow, "Active control of bending wave propagation at acoustic frequencies," *Journal of Sound and Vibration* **142**, pp. 391–412, 1990.

8. D. W. Miller and S. R. Hall, "Experimental results using active control of traveling wave power flow," *AIAA Journal of Guidance* **14**, pp. 350–359, 1991.

9. H. Fujii, T. Ohtsuka, and T. Murayama, "Wave-absorbing control for flexible structures with noncollocated sensors and actuators," *AIAA Journal of Guidance, Control, and Dynamics* **15**, pp. 431–439, 1992.

10. J. F. Doyle, "Determining the size and location of transverse cracks in beams," *Journal of Experimental Mechanics* **35**, pp. 272–280, 1995.

11. K. A. Lakshmanan and D. J. Pines, "Damage identification of chordwise crack size and location in uncoupled composite rotorcraft flexbeams," *Journal of Intelligent Materials Systems and Structures* **9**, pp. 146–155, 1998.

12. A. S. Purekar and D. J. Pines, "Detecting damage in non-uniform beams using the dereverberated transfer function response," *Smart Materials and Structures* **9**, pp. 429–444, 2000.

13. J. Ma and D. J. Pines, "Dereverberation and its application to damage detection in one-dimensional structures," *AIAA Journal* **39**, pp. 902–918, 2001.

14. L. J. Ziomek, *Fundamentals of Acoustic Field Theory and Space-Time Signal Processing*, CRC Press, 1995.

15. S. P. Joshi, "Feasibility study on phased array of interdigital mesoscale transducers for health monitoring of smart structures," in *Proceedings of SPIE*, vol. 2443, pp. 248–257, 1995.

16. P. Agrawal, "Health monitoring of smart structures by method of wave scattering techniques," Master's thesis, The University of Texas at Arlington, 1996.

MODELIZATION AND NUMERICAL APPROXIMATION OF ACTIVE THIN SHELLS

Michel Bernadou and Christophe Haenel

ABSTRACT

This work is concerned with the numerical analysis of a general active thin shell structure which is constituted by (i) a thin shell made of a "classical" material and having a general shape; (ii) some piezoelectric patches which are bonded upon the external surfaces of the shell and which work as actuators. In this way, a convenient 2D modelization is obtained, an existence and uniqueness result of a solution is proved, an approximation by accurate conforming finite elements is obtained, the associate convergence is studied and a numerical experiment proves the effectiveness of the method.

1. INTRODUCTION

Generally, an active thin shell structure comprises

i) a thin structure made of a "classical" material which has a purely mechanical behaviour;

ii) some piezoelectric patches which are bonded upon the external surfaces of the structure and which work as sensors or actuators.

In this paper, we consider the three following aspects of the numerical analysis of active thin shell structures:

(i) *Derivation of a 2D modelization*: We start by giving the intrinsic three-dimensional equations (mechanical + electrical equations), we consider the associate variational formulation and we prove an existence and uniqueness theorem in a convenient functional space.

Next, by introducing appropriate hypotheses on the behaviour of main unknowns through the thickness, we obtain a two-dimensional variational formulation expressed upon the middle surface of the shell ; some additional simplifications allow to drop terms which are small with respect to the energy. Thus, the main unknowns are

* the three components of the displacement;

* the three terms of the quadratic development of the potential with respect to the thickness variable.

Pôle Universitaire Léonard de Vinci, F-92916 Paris La Défense Cedex

Here again, for the 2D formulation, we prove an existence and uniqueness theorem of a solution in a convenient functional space.

From this basic active thin shell structure, it is possible without any major difficulty to extend the study to more general situations like a "classical" shell activated by many different patches bonded upon the structure or equipped with piezoelectric thin films.

(ii) *Approximation by an accurate conforming finite element method*: The problem

∗ is of order two with respect to the membrane components of the displacement and with respect to the three terms of the quadratic development of the potential: these five unknowns are approximated by a $P_4 - C^0$ finite element (Ganev-Dimitrov's triangle [1]);

∗ is of order four with respect to the bending component of the displacement: this unknown is approximated by a $P_5 - C^1$ finite element (Argyris-Fried and Scharpf's triangle [2]).

We obtain convergence results of the approximated solution to the continuous solution and good criteria on the choice of numerical integration schemes.

(iii) *Implementation and a numerical experiment*: This 2d approximation method has been implemented [3]. After the assemblage, a condensation method allows to express the potential terms as a linear combination of the displacement degrees of freedom. Thus, the substitution of these potential degrees of freedom leads to a symmetric linear system which can be easily solved. A numerical experiment on a thin active cylindrical structure illustrates the effectiveness of the method.

2. THE TWO-DIMENSIONAL VARIATIONAL FORMULATION

Let us start from the 3d basic equations and show how to obtain a convenient set of 2d approximated equations.

2.1. THE 3D EQUATIONS

The generic case of a piezoelectric patch bonded upon a "classical" structure is shown in Figure 1. For clarity, we recall the equations of the 3d-problem in intrinsic form [4]:

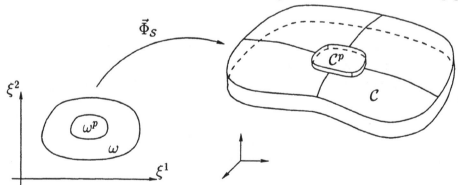

Figure 1: Shell C, patch C^p and their representations using plane reference domains ω and ω^p

i) *Mechanical and electrical field equations*

$$-div\underline{\sigma} = \vec{p} \quad \text{in } C \cup C^p, \quad div\vec{D} = 0 \quad \text{in } C^p$$

where $\underline{\sigma}$ and $\vec{D}$ denote respectively the stress tensor and the electrical displacement vector. By $\vec{p}$, we mean the density of a distributed loading.

ii) *Mechanical boundary conditions*

The boundary of $C \cup C^p$ is subdivided into a clamped part $\{\partial(C \cup C^p)\}_0^M$ and a loaded part $\{\partial(C \cup C^p)\}_1^M$ so that

$$\vec{U} = \vec{0} \quad \text{on} \quad \{\partial(C \cup C^p)\}_0^M \quad \text{and} \quad \underline{\sigma} \cdot \vec{n} = \vec{q} \quad \text{on} \quad \{\partial(C \cup C^p)\}_1^M.$$

iii) *Electrical boundary conditions*

Likewise, the boundary of the patch C^p is subdivided into an electrically loaded part $\{\partial C^p\}_0^E$ and a free part $\{\partial C^p\}_1^E$ so that

$$\varphi = \varphi_d \quad \text{on} \quad \{\partial C^p\}_0^E \quad \text{and} \quad \vec{D} \cdot \vec{n} = 0 \quad \text{on} \quad \{\partial C^p\}_1^E,$$

where φ denotes the electric potential.

iv) *Transmission conditions*

$$\vec{U} \quad \text{and} \quad \underline{\sigma} \cdot \vec{n} \text{ are continuous on the interface } C \cap C^p.$$

v) *Behaviour laws*

$$\underline{\sigma} = \underline{C}^{el} \cdot \underline{\varepsilon} \quad \text{in} \quad C$$

and

$$\underline{\sigma} = \underline{C}^p \cdot \underline{\varepsilon} - {}_pe^p \cdot \vec{E} \quad \text{and} \quad \vec{D} = {}_p^t e^p \cdot \underline{\varepsilon} + \underline{d}^p \cdot \vec{E} \quad \text{in} \quad C^p,$$

where $\underline{C}^{el}$ and $\underline{C}^p$ mean the elasticity tensors of the elastic and piezoelectric bodies, $\underline{\varepsilon}$ denotes the strain tensor, ${}_pe^p$ and $\underline{d}^p$ denote the piezoelectric and dielectric coefficients while $\vec{E}$ denotes the electrical vector field which is such that

$$\vec{E} = -\nabla\varphi.$$

Main unknowns: These equations can be entirely expressed in terms of the main unknowns of the problem: the three dimensional displacement vector $\vec{U}$ and the potential φ.

Existence and uniqueness: With these 3d equations, we associate the variational equation:

$$\boxed{\begin{array}{c} \text{Find } (\vec{U}, \varphi) \in \vec{V}_M(C \cup C^p) \times V_{E_{\varphi_d}}(C^p) \text{ such that} \\ \overset{*}{a}\,[(\vec{U}, \varphi), (\vec{V}, \psi)] = \overset{*}{f}\,(\vec{V}), \quad \forall (\vec{V}, \psi) \in \vec{V}_M(C \cup C^p) \times V_{E_0}(C^p) \end{array}} \tag{1}$$

where

$$\overset{*}{a}\,[(\vec{U}, \varphi), (\vec{V}, \psi)] = \int_{C \cup C^p} \underline{\sigma}(\vec{U}, \varphi) : \underline{\varepsilon}(\vec{V}) \, dC + \int_{C^p} \vec{D}(\vec{U}, \varphi) \cdot \vec{E}(\psi) \, dC, \tag{2}$$

$$\overset{*}{f}\,(\vec{V}) = \int_{C \cup C^p} \vec{p}\vec{V} \, dC + \int_{\{\partial(C \cup C^p)\}_1^M} \vec{q}\vec{V} \, d(\partial C), \tag{3}$$

$$\vec{V}_M(C \cup C^p) = \{\vec{V} \in (H^1(C \cup C^p))^3 \ ; \ \vec{V} = \vec{0} \text{ on } \{\partial(C \cup C^p)\}_0^M\},$$

$$V_{E_{\varphi_d}}(C^p) = \{\psi \in H^1(C^p) \ ; \ \psi = \varphi_d \text{ on } \{\partial(C^p)\}_0^E\}.$$

Theorem 1 *For sufficiently regular data, problem (1) has one and only one solution.*
Proof (see [5] and [6] for more details): The substitution of the behaviour laws into the bilinear form (2) gives

$$\overset{*}{a}\,[(\vec{U},\varphi),(\vec{V},\psi)] = \int_{\mathcal{C}} \underline{C}^{el} \cdot \underline{\varepsilon}(\vec{U}) : \underline{\varepsilon}(\vec{V})d\mathcal{C} + \int_{\mathcal{C}^p} (\underline{C}^p \cdot \underline{\varepsilon}(\vec{U}) - {}_pe^p \cdot \vec{E}(\varphi)) : \underline{\varepsilon}(\vec{V})d\mathcal{C}$$
$$+ \int_{\mathcal{C}^p} ({}_p^t e^p \cdot \underline{\varepsilon}(\vec{U}) + \underline{d}^p \cdot \vec{E}(\varphi)) \cdot \vec{E}(\psi)d\mathcal{C}.$$

This is a bilinear and continuous form on the space $\vec{V}_M(\mathcal{C} \cup \mathcal{C}^p) \times V_{E_0}(\mathcal{C}^p)$; moreover, $\overset{*}{a}\,[\cdot,\cdot]$ is elliptic since the symmetry of ${}_pe^p$ involves

$$\overset{*}{a}\,[(\vec{V},\psi),(\vec{V},\psi)] = \int_{\mathcal{C}} (\underline{C}^{el} \cdot \underline{\varepsilon}(\vec{V})) : \underline{\varepsilon}(\vec{V})d\mathcal{C} + \int_{\mathcal{C}^p} [(\underline{C}^p \cdot \varepsilon(\vec{V})) : \underline{\varepsilon}(\vec{V}) + (\underline{d}^p \cdot \vec{E}(\psi)) \cdot \vec{E}(\psi)]d\mathcal{C},$$

and then, the result is a consequence of the elliptic properties of tensors $\underline{C}^{el}$, $\underline{C}^p$ and $\underline{d}^p$ and of Korn's inequality [7]. Moreover, the application f^* defined by (3) is a linear and continuous form and, for concluding, it remains to apply a variant of the Lax-Milgram lemma [8] [9] valid for affine spaces. ∎

2.2. THE ASSOCIATE 2D EQUATIONS

Such a derivation is very technical and extensively described in [5], [10] and particularly in [6]. Hereunder, we report the main results. In order to be able to consider thin shells with general shapes, we use a representation of the shell by a set of curvilinear coordinates: more precisely, the middle surface S of the shell $\mathcal{C}$ is assumed to be the image through a mapping $\vec{\Phi}_S$ of a plane reference domain ω, i.e., with Figure 1,

$$\mathcal{C} = \{M \in \mathbb{R}^3, \overrightarrow{OM} = \vec{\Phi}_{\mathcal{C}}(\xi) = \vec{\Phi}_S(\xi^1,\xi^2) + \xi^3 \vec{a}_3, (\xi^1,\xi^2) \in \omega, -\frac{1}{2}e < \xi^3 < \frac{1}{2}e\}$$

while the patch $\mathcal{C}^p$ is defined by

$$\mathcal{C}^p = \{M \in \mathbb{R}^3, \overrightarrow{OM} = \vec{\Phi}_{\mathcal{C}}(\xi) = \vec{\Phi}_S(\xi^1,\xi^2) + \xi^3 \vec{a}_3, (\xi^1,\xi^2) \in \omega^p, +\frac{1}{2}e < \xi^3 < \frac{1}{2}e + e^p\}.$$

We use the local covariant and contravariant bases ($\vec{a}_\alpha = \vec{\Phi}_{S,\alpha}$; $\vec{a}_3 = \vec{a}_1 \times \vec{a}_2/|\vec{a}_1 \times \vec{a}_2|$) and ($\vec{a}^i$) such that $\vec{a}^i \cdot \vec{a}_j = \delta^i_j$ and we make the following assumptions:
  i) Stresses are approximatively plane and parallel to the tangent plane to the middle surface.
  ii) Conservation of the normals during the deformation: the displacement $\vec{U}$ of the shell $\mathcal{C} \cup \mathcal{C}^p$ can be approximated with the help of the displacement field $\vec{u}$ of the points located upon the middle surface S by the relation

$$\vec{U}(\xi^\alpha,\xi^3) = \vec{u}(\xi^\alpha) - \xi^3(u_{3,\mu} + b^\lambda_\mu u_\lambda)\vec{a}^\mu.$$

  iii) The electrical potential $\varphi(\xi^\alpha,\xi^3)$ can be approximated by a polynomial of degree two with respect to the thickness variable ξ^3, i.e.,

$$\varphi(\xi^\alpha,\xi^3) = \varphi_0(\xi^\alpha) + \left(\xi^3 - \frac{e+e^p}{2}\right)\varphi_1(\xi^\alpha) + \frac{1}{2}\left(\xi^3 - \frac{e+e^p}{2}\right)^2 \varphi_2(\xi^\alpha).$$

Then, after

i) integration of the 3D-equations through the thickness,

ii) simplification of the resulting equations by neglecting some terms which are of order $0(\frac{e}{R})$ with respect to the leading terms (R means the minimum of the curvature radius),

we obtain the following statement for the final 2D variational formulation on the plane reference domain:

$$
\begin{array}{c}
\text{Find } (\vec{u}, \varphi_0, \varphi_1, \varphi_2) \in \vec{V}_{KL}(\omega) \times V_{P_{\varphi_d}}(\omega^p) \text{ such that} \\
a_{e\ell}(\vec{u}, \vec{v}) + a_p[(\vec{u}, (\varphi_0, \varphi_1, \varphi_2)), (\vec{v}, (\psi_0, \psi_1, \psi_2))] = f(\vec{v}) \\
\forall (\vec{v}, (\psi_0, \psi_1, \psi_2)) \in \vec{V}_{KL}(\omega) \times V_{P_0}(\omega^p).
\end{array}
\tag{4}
$$

In this formulation, we note

$$
a_{e\ell}(\vec{u}, \vec{v}) = \int_\omega e\bar{C}^{\alpha\beta\lambda\mu,e\ell}(\gamma_{\lambda\mu}(\vec{u})\gamma_{\alpha\beta}(\vec{v}) + \frac{e^2}{12}\bar{\rho}_{\alpha\beta}(\vec{u})\bar{\rho}_{\lambda\mu}(\vec{v}))\sqrt{a}\,d\xi^1 d\xi^2;
$$

$$
a_p[(\vec{u}, (\varphi_0, \varphi_1, \varphi_2)), (\vec{v}, (\psi_0, \psi_1, \psi_2))]
$$

$$
= \int_{\omega^p} \{\bar{C}^{\alpha\beta\lambda\mu,p}[I_0^{0,p}\gamma_{\alpha\beta}(\vec{u})\gamma_{\lambda\mu}(\vec{v}) + I_1^{0,p}(\gamma_{\alpha\beta}(\vec{u})\bar{\rho}_{\lambda\mu}(\vec{v}) + \bar{\rho}_{\alpha\beta}(\vec{u})\gamma_{\lambda\mu}(\vec{v})) + I_2^{0,p}\bar{\rho}_{\alpha\beta}(\vec{u})\bar{\rho}_{\lambda\mu}(\vec{v})]
$$

$$
+ {}_p\bar{e}^{\alpha\lambda\mu,p}[(I_0^{0,p}\varphi_{0,\alpha} + \frac{1}{2}I_0^{2,p}\varphi_{2,\alpha})\gamma_{\lambda\mu}(\vec{v}) + (I_1^{0,p}\varphi_{0,\alpha} + I_1^{1,p}\varphi_{1,\alpha} + \frac{1}{2}I_1^{2,p}\varphi_{2,\alpha})\bar{\rho}_{\lambda\mu}(\vec{v})]
$$

$$
+ {}_p\bar{e}^{3\lambda\mu,p}[I_0^{0,p}\varphi_1\gamma_{\lambda\mu}(\vec{v}) + (I_1^{0,p}\varphi_1 + I_1^{1,p}\varphi_2)\bar{\rho}_{\lambda\mu}(\vec{v})]
$$

$$
- {}_p\bar{e}^{\alpha\lambda\mu,p}[(I_0^{0,p}\gamma_{\lambda\mu}(\vec{u}) + I_1^{0,p}\bar{\rho}_{\lambda\mu}(\vec{u}))\psi_{0,\alpha} + I_1^{1,p}\bar{\rho}_{\lambda\mu}(\vec{u})\psi_{1,\alpha} + \frac{1}{2}(I_0^{2,p}\gamma_{\lambda\mu}(\vec{u}) + I_1^{2,p}\bar{\rho}_{\lambda\mu}(\vec{u}))\psi_{2,\alpha}]
$$

$$
- {}_p\bar{e}^{3\lambda\mu,p}[(I_0^{0,p}\gamma_{\lambda\mu}(\vec{u}) + I_1^{0,p}\bar{\rho}_{\lambda\mu}(\vec{u}))\psi_1 + I_1^{1,p}\bar{\rho}_{\lambda\mu}(\vec{u})\psi_2]
$$

$$
+ \bar{d}^{\alpha\lambda,p}[I_0^{0,p}\varphi_{0,\alpha}\psi_{0,\lambda} + I_0^{2,p}(\frac{1}{2}\varphi_{0,\alpha}\psi_{2,\lambda} + \varphi_{1,\alpha}\psi_{1,\lambda} + \frac{1}{2}\varphi_{2,\alpha}\psi_{0,\lambda}) + \frac{1}{4}I_0^{4,p}\varphi_{2,\alpha}\psi_{2,\lambda}]
$$

$$
+ \bar{d}^{\alpha3,p}[I_0^{0,p}\varphi_1\psi_{0,\lambda} + I_0^{2,p}(\frac{1}{2}\varphi_1\psi_{2,\lambda} + \varphi_2\psi_{1,\lambda})]
$$

$$
+ \bar{d}^{3\lambda,p}[(I_0^{0,p}\varphi_{0,\lambda} + \frac{1}{2}I_0^{2,p}\varphi_{2,\lambda})\psi_1 + I_0^{2,p}\varphi_{1,\lambda}\psi_2]
$$

$$
+ \bar{d}^{33,p}[I_0^{0,p}\varphi_1\psi_1 + I_0^{2,p}\varphi_2\psi_2]\}\sqrt{a}\,d\xi^1 d\xi^2
$$

with

$$
I_n^{m,p} = \int_{\frac{e}{2}}^{\frac{e}{2}+e^p} (\xi^3)^n \left(\xi^3 - \frac{e+e^p}{2}\right)^m d\xi^3,
$$

the explicit values of which are given in Table 1 for the useful values of m and n. Likewise, the expression of $f(\vec{v})$ presents the same kind of technicallity so that we refer to [6] for the details. The spaces $\vec{V}_{KL}(\omega)$, $V_{P_{\varphi_d}}(\omega^p)$ and $V_{P_0}(\omega^p)$ take into account the mechanical and electrical boundary conditions. Thus, by denoting $\vec{v} = v_i\vec{a}^i$,

$$
\vec{V}_{KL}(\omega) = \{\vec{v} = (v_1, v_2, v_3) \in (H^1(\omega))^2 \times H^2(\omega) \,;\, \vec{v} = \vec{0} \text{ and } v_{3,n} = 0 \text{ on } \gamma_0^M\}
$$

and, in the case of a patch excited through electrodes located on its upper and lower surfaces,

n \ m	0	1	2	4
0	e^p	0	$\frac{1}{12}(e^p)^3$	$\frac{1}{80}(e^p)^5$
1	$\frac{1}{2}e^p(e+e^p)$	$\frac{1}{12}(e^p)^3$	$\frac{1}{24}(e^p)^3(e+e^p)$	–
2	$e^p(\frac{1}{4}e^2+\frac{1}{2}ee^p+\frac{1}{3}(e^p)^2)$	–	–	–

Table 1: Values of $I_n^{m,p}$

$$V_{P_{\theta_d}} = \{(H^1(\omega^p))^3 \cup \{(\theta_0, \theta_1, \theta_2) \text{ such that}$$

$$\theta_0 + \frac{(e^p)^2}{8}\theta_2 = \frac{\theta_d\left(\frac{e}{2}+e^p\right) + \theta_d\left(\frac{e}{2}\right)}{2} \text{ and } \theta_1 = \frac{\theta_d\left(\frac{e}{2}+e\right) - \theta_d\left(\frac{e}{2}\right)}{e^p}\},$$

where θ_d is a given constant on each electrode. Of course, the space $V_{P_0}(\omega^p)$ corresponds to the particular value $\theta_d = 0$.

Here again, we obtain [6]:

Theorem 2 *For sufficiently smooth data, the variational formulation (4) has a unique solution.* ∎

Remark 1: In [6], we consider more general situations:

    i) the number and the location of the patches can be changed without difficulty;

    ii) for each patch, the number and the location of the electrodes can be changed as well. ∎

Remark 2: In section 2.1, we have adopted the intrinsic presentation. From such a 3d intrinsic presentation, it is possible to derive intrinsic models of piezoelectric shells. For instance, in [11], we have developed 2d modelizations of the behaviour of a piezoelectric thin shell. ∎

3. APPROXIMATION BY A CONFORMING FINITE ELEMENT METHOD

We realize a conforming finite element approximation of the solution $(\vec{u}, (\varphi_0, \varphi_1, \varphi_2))$ of problem (4). In this way, we use Ganev and Dimitrov's triangle for the approximation of $(u_1, u_2, \varphi_0, \varphi_1, \varphi_2)$ and Argyris-Fried and Scharpf's triangle for the approximation of u_3 (Figure 2). These approximations are combined with the use of numerical integration techniques.

We can prove [6] [12]:

Theorem 3 *Assume that* $(\vec{u}, (\varphi_0, \varphi_1, \varphi_2)) \in (H^5(\omega))^2 \times H^6(\omega) \times (H^5(\omega))^3$, $\mathcal{P}^i \in W^{4,q}(\omega)$, $q \geq 2$, $\vec{\Phi}_S$ *is sufficiently smooth, the numerical integration schemes are exact for polynomials of degree 6.*

Then, there exist constants C and h_1, independent of h such that, for any $h < h_1$, we have

$$\|\vec{u} - \vec{u}_h\|_{V_{KL}(\omega)} + \|(\varphi_0, \varphi_1, \varphi_2) - (\varphi_{0_h}, \varphi_{1_h}, \varphi_{2_h})\|_{V_{P_0}(\omega)}$$

$$\leq Ch^4\left\{\left(\sum_\alpha \|u_\alpha\|_{5,\omega}^2 + \|u_3\|_{6,\omega}^2 + \sum_{i=0}^2 \|\varphi_i\|_{5,\omega}^2\right)^{1/2} + \left(\sum_{i=1}^3 \|\mathcal{P}^i\|_{4,q,\omega}^q\right)^{1/q}\right\}.$$ ∎

Implementation

The implementation uses Modulef library (see [13] and http://www-rocq.inria.fr/modulef). In particular, it is convenient to use a condensation method to eliminate the three terms $\varphi_0, \varphi_1, \varphi_2$ of the development of the potential φ.

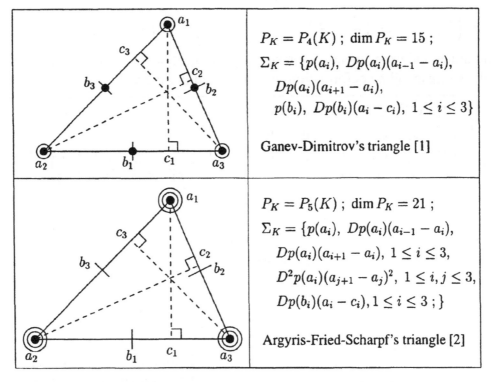

$$P_K = P_4(K) \; ; \; \dim P_K = 15 \; ;$$
$$\Sigma_K = \{p(a_i), \; Dp(a_i)(a_{i-1} - a_i),$$
$$Dp(a_i)(a_{i+1} - a_i),$$
$$p(b_i), \; Dp(b_i)(a_i - c_i), \; 1 \leq i \leq 3\}$$

Ganev-Dimitrov's triangle [1]

$$P_K = P_5(K) \; ; \; \dim P_K = 21 \; ;$$
$$\Sigma_K = \{p(a_i), \; Dp(a_i)(a_{i-1} - a_i),$$
$$Dp(a_i)(a_{i+1} - a_i), \; 1 \leq i \leq 3,$$
$$D^2 p(a_i)(a_{j+1} - a_j)^2, \; 1 \leq i, j \leq 3,$$
$$Dp(b_i)(a_i - c_i), 1 \leq i \leq 3 \; ;\}$$

Argyris-Fried-Scharpf's triangle [2]

Figure 2: Finite elements used to approximate the 2d-problem.

4. A NUMERICAL EXPERIMENT

Let us consider a half cylinder of length ℓ, radius R and thickness e. Its middle surface is the image of the plane rectangular domain $] - \frac{\pi R}{2}, \frac{\pi R}{2} [\times]0, \ell[$ through the mapping $\vec{\Phi}_S$ displayed in Figure 3.

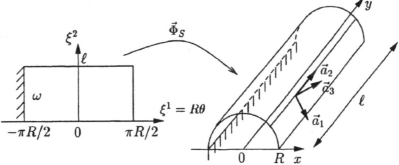

Figure 3: Geometrical definition of the half cylinder: $\vec{\Phi}_S(\xi^1, \xi^2) = (R\sin(\xi^1 R), \xi^2, R\cos(\xi^1 R))$

This shell is clamped along the side $\xi^1 = -\frac{\pi R}{2}$ and free on the other boundaries. It is made of an elastic material (graphite/epoxy) and its mechanical characteristics are displayed in Table 2. On the upper and lower faces of the cylinder we bond slices of piezoelectric patches the location of which in the reference domain is shown on Figure 4. Each patch has a thickness e^p and its mechanical, piezoelectric and dielectric coefficients are given in Table 3. The upper and lower faces of these patches are covered by electrodes and the following potentials are applied

$$\varphi^+\left(\frac{e}{2}\right) = 0 \ ; \ \varphi^+\left(\frac{e}{2}+e^p\right) = V \ ; \ \varphi^-\left(-\frac{e}{2}-e^p\right) = V \ ; \ \varphi^-\left(-\frac{e}{2}\right) = 0.$$

For the numerical experiments, we take $\ell = 10cm$, $R = 20cm$, $e = 0.5cm$, $e^p = 0.2cm$, $V = 100V$. Corresponding deformations are presented in Figure 5.

$$[\underline{C}^{e\ell}] = \begin{pmatrix} C_{11}^{e\ell} & C_{12}^{e\ell} & C_{13}^{e\ell} & 0 & 0 & 0 \\ C_{12}^{e\ell} & C_{11}^{e\ell} & C_{13}^{e\ell} & 0 & 0 & 0 \\ C_{13}^{e\ell} & C_{13}^{e\ell} & C_{33}^{e\ell} & 0 & 0 & 0 \\ 0 & 0 & 0 & C_{44}^{e\ell} & 0 & 0 \\ 0 & 0 & 0 & 0 & C_{44}^{e\ell} & 0 \\ 0 & 0 & 0 & 0 & 0 & C_{66}^{e\ell} \end{pmatrix} \ ; $$

$C_{11}^{e\ell}$	$C_{12}^{e\ell}$	$C_{13}^{e\ell}$	$C_{33}^{e\ell}$	$C_{44}^{e\ell}$	$C_{66}^{e\ell}$
14.41	5.18	7.16	134.9	5.6	3.6

Table 2: Mechanical characteristics of the material which constitutes the circular cylinder

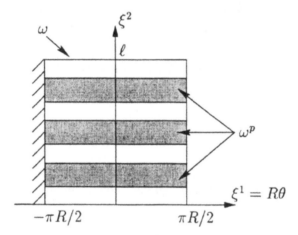

Subdomains ω^p are associated with the orthogonal projection on the middle surface of the upper and lower piezoelectric patches.

Figure 4: Location of the piezoelectric patches in the reference domain ω

$$[\underline{C}] = \begin{pmatrix} C_{11} & C_{12} & C_{13} & 0 & 0 & 0 \\ C_{12} & C_{11} & C_{13} & 0 & 0 & 0 \\ C_{13} & C_{13} & C_{33} & 0 & 0 & 0 \\ 0 & 0 & 0 & C_{44} & 0 & 0 \\ 0 & 0 & 0 & 0 & C_{44} & 0 \\ 0 & 0 & 0 & 0 & 0 & C_{66} \end{pmatrix} \ ; \ [\underline{pe}] = \begin{pmatrix} 0 & 0 & {}_pe_{31} \\ 0 & 0 & {}_pe_{31} \\ 0 & 0 & {}_pe_{33} \\ 0 & {}_pe_{15} & 0 \\ {}_pe_{15} & 0 & 0 \\ 0 & 0 & 0 \end{pmatrix} \ ;$$

$$[\underline{d}] = \begin{pmatrix} d_{11} & 0 & 0 \\ 0 & d_{11} & 0 \\ 0 & 0 & d_{33} \end{pmatrix}$$

GPa						$C \cdot m^{-2}$			$F \cdot m^{-1}$	
C_{11}	C_{12}	C_{13}	C_{33}	C_{44}	C_{66}	${}_pe_{31}$	${}_pe_{33}$	${}_pe_{15}$	d_{11}	d_{33}
126	79.5	84.1	117	23	23.25	-6.5	23.3	17.0	1.503	1.3

Table 3: Characteristics of the piezoelectric material

Figure 5: Deformation of the cylinder and its undeformed position

REFERENCES

1. Ganev, H.G. and Tch.T. Dimitrov. 1980. "Calculation of arch dams as a shell using IBM-370 Computer and curved finite elements", in *Theory of Shells*, Amsterdam, North-Holland, 691–696.
2. Argyris, J.H., I. Fried and D.W. Scharpf. 1968. "The TUBA family of plate elements for the matrix displacement method", *Aero. J. Royal Aeronaut. Soc.*, 72, 701–709.
3. Haenel, C. 2000. *Modélisation, Analyse et Simulation Numérique de Coques Piézoélectriques*, Thèse de l'Université Pierre et Marie Curie, Paris, France.
4. Ikeda, T. 1990. *Fundamental of Piezoelectricity*, Oxford Univ. Press, Oxford.
5. Bernadou, M. and C. Haenel. (to appear). "Modelization and numerical approximation of piezoelectric thin shells. Part 1: The continuous problem".
6. Bernadou, M. and C. Haenel. (to appear). "Modelization and numerical approximation of piezoelectric thin shells. Part 3: From the patches to the active structures".
7. Ciarlet, P.G. 1978. *Mathematical Elasticity, Vol. III: Theory of Shells*, North Holland, Amsterdam.
8. Ciarlet, P.G. 1978. *The Finite Element Method for Elliptic Problems*, North-Holland, Amsterdam.
9. Delfour, M. 2000. Communication personnelle.
10. Bernadou, M. and C. Haenel. (to appear). "Modelization and numerical approximation of piezoelectric thin shells. Part 2: Approximation by finite element methods and numerical experiments".
11. Bernadou, M. and M. Delfour. 2000. "Intrinsic models of piezoelectric shells", Proceedings of ECCOMAS 2000, Barcelona, 11–14, September.
12. Bernadou, M. 1996. *Finite Element Methods for Thin Shell Problems*, J. Wiley and Sons, Chichester.
13. Bernadou, M., P.L. George, A. Hassim, P. Joly, P. Laug, B. Muller, A. Perronnet, E. Saltel, D. Steer, G. Vanderbork and M. Vidrascu. 1988. *MODULEF: Une Bibliothèque Modulaire d'Eléments Finis*, Editions INRIA, Rocquencourt, Deuxième édition.

DEVELOPMENT OF A PVDF PIEZOPOLYMER SENSOR FOR UNCONSTRAINED IN-SLEEP CARDIORESPIRATORY MONITORING

Feng Wang, Mami Tanaka, and Seiji Chonan

ABSTRACT

This paper reports the development of an unconstrained sensing technique for monitoring the respiration and heartbeats during sleep using a PVDF piezopolymer film sensor with the aim that the sensor can be used on the ordinary bed together with the condition that the use of the sensor does not interfere the daily sleep of the patient under measurement. A polyvinylidene fluoride (PVDF) film is used as the sensory material in the sensor system. The film is placed under the sheet at the location of the thorax to pick up the fluctuation of the pressure on the bed caused by the respiratory movement and heartbeats. Wavelet multiresolution decomposition analysis is used for the detection of respiration and heartbeat from the sensor output. It is shown that the respiration and heartbeats can simultaneously be detected by the sensor with the use of the wavelet multiresolution decomposition analysis.

INTRODUCTION

Measurement of the physiological information such as the respiration and heart rate during sleep is of great importance for public health care, especially for the early diagnosis of cardiorespiratory sleep disorders [1]. In sleep laboratories, in-sleep cardiorespiratory monitoring is generally carried out by using the polysomnography (PSG) that attaches several sensors to the body of a patient. However, using of PSG confronts the following problems: a patient might unconsciously remove or even destroy the sensor(s) and thus interrupt the measurement during sleep; and the attachment of sensors to a patient body brings about considerable physical as well as psychological burden to the patient, thus making his/her sleep differ from his/her daily sleep. To overcome these problems, new techniques of unconstrained sensing of physiological information are necessary to be developed.

Over the past years, numerous attempts have been made on the development of unconstrained sensing techniques for in-sleep monitoring of physiological information. For example, Alihanka et al and Salmi et al proposed a static charge sensitive bed for unconstrained monitoring of respiration and heart rate during sleep [2, 3]. Nishida et al embedded 221 pressure sensors to a bed to monitor the in-sleep respiration and body movement during sleep [4]. Tanaka and Watanabe et al developed pressure sensor embedded air mattresses for monitoring the in-sleep respiration and heart rate [5, 6]. However, in these methods, special beds or mattresses are necessary, which may again lead to the problem that the difference in feeling of beds affects the patient sleep and thus makes the patient sleep under measurement different from his daily

Feng Wang, Mami Tanaka, Seiji Chonan, Tohoku University, Graduate School of Engineering, Department of Mechatronics and Precision Engineering, 04 Aoba-yama, Aoba-ku, Sendai, 980-8579, Japan.

298

sleep. With the progress of computer image processing techniques, new unconstrained respiration monitoring method of capturing and monitoring the movement of the quilt using a video camera was also proposed [7]. But the high cost of the equipment and the psychological oppressive feeling of being watched by a camera make it hard to call it an ideal method.

In this study, considering the above respects, an unconstrained sensing technique for in-sleep cardiorespiratory monitoring using a PVDF piezopolymer film sensor is proposed. The aim of the research is to develop an unconstrained monitor system that can be used on the ordinary bed together with the condition that the use of system does not affect the daily sleep of patient under the measurement condition so that the system can be used in both hospitals and homes. This paper reports the development of the sensor system and the signal processing method for detection of the respiratory movement and heartbeat from the sensor output using wavelet multiresolution decomposition technique.

PRICIPLE OF DETECTION

A human maintains the respiration by alternative contracting and relaxing of the respiratory muscles. The movement of the respiratory muscles causes the motion of the internal organs. As a result, it causes the fluctuation of the center of gravity of the human trunk. When a human is lying on a bed, the fluctuation of the center of gravity of his trunk results in the fluctuation of the pressing force applied to the bed through his body. Meanwhile, at each heartbeat, the ejection of blood into arteries by the heartbeat causes the slight vibration of his body surface, known generally as the ballistocardiography (BCG). The BCG also brings about the slight fluctuation of the pressing force applied to the bed when he is lying on a bed. It is expected in this case that the respiration and heartbeats can be monitored by detecting the fluctuation of the pressing force caused by the respiratory movement and BCG.

STRUCTURE OF SENSOR SYSTEM

The polyvinylidene fluoride (PVDF) is a kind of piezoelectric polymer. It is formed to a very thin and flexible film. Moreover, the voltage output of a PVDF film is characterized by the temporal differential property, which means that it only responds to the dynamic change of pressure, not to the static pressure [8, 9]. These properties make it an ideal candidate of sensory material for the detection of the respiratory movement and heartbeats during sleep. Taking advantage of the characteristics of the PVDF film, a sensor system for unconstrained measurement of the respiration and heartbeats during sleep is developed.

Figure 1 shows the schematic of the sensor system. The system is consisting of a PVDF film, a charge amplifier and a notch filter. A PVDF film of the size of 140mm in width, 200mm in length and 28μm in thickness is used as the sensory material. The PVDF film is placed under the sheet at the location of the thorax to pick up the fluctuation of the applied pressing force to the bed. The PVDF film is wrapped with a towel to protect the film, and further to enhance the sensitivity of the sensor. The total sensor has a thickness of only 2mm and is flexible as well, thus the existence of the sensor is hardly perceptible. It may not affect the sleep of the patient under measurement. The shielded cable leads the charge output of PVDF film to the charge amplifier. Figure 2 shows the charge amplifier and 50Hz notch filter. With the field effective transistor (FET) input structure, the charge amplifier transfers the input charge into the voltage

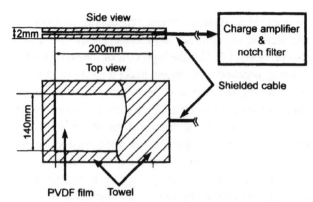

Figure 1. Schematic of PVDF sensor system

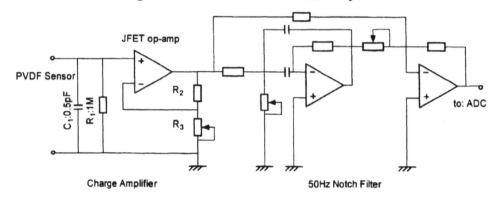

Figure 2. Charge amplifier and notch filter

signal and further amplifies it. The input resistance R_1 and capacitor C_1 of the charge amplifier are selected as $1M\Omega$ and $0.5pF$ empirically by experiments for best performance. Gain of the amplifier is

$$G = 1 + \frac{R_2}{R_3},\qquad(1)$$

which is adjustable by the variable resistor R_3. The notch filter cuts off the 50Hz line noise induced from the human body and surrounding environment.

MEASUREMENT EXPERIMENT

In order to test the functions of the sensor system, the experimental measurement was conducted on healthy subjects on deferent types of bed. The schematic of the experimental setup is shown in Fig. 3. The PVDF film is placed under the sheet at the location of the thorax to pick up the fluctuation of the applied pressing force to the bed. Output of the sensor system was digitized via an AD converter at the sampling rate of 100Hz and then it was recorded to a personal computer. The output of the traditional band-type plethysmography respiration monitor and ECG were also recorded together with the output of the sensor system. They are used as the reference of the respiration and the heartbeat respectively. It should be noticed that no restriction on the sleeping posture was applied to the subjects during the experiments.

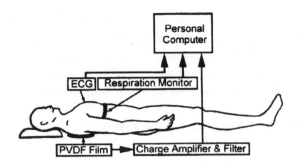

Figure 3. Experimental setup for functional verification of PVDF sensor

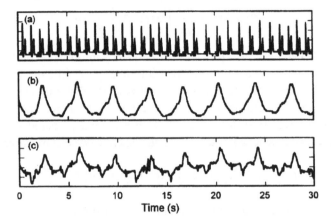

Figure 4. Example of experimental records: (a) ECG, (b) output of respiration monitor and (c) output of proposed PVDF film sensor.

As an example of the experimental result, 30-sencond records of ECG, output of traditional plethysmography respiration monitor and output of proposed PVDF sensor are shown in Fig. 4. As shown in the figure, in the output of the sensor system, large waves synchronizing with the output of traditional respiration monitor, which is used as respiration reference, are obviously noticeable. Also can be noticed are the localized small pulses in the sensor output synchronized with the QRS complexes of ECG, which are the high spikes in the ECG signal shown in Fig. 4 (a), and indicate the beginning of the excitation of the ventricular muscles. These facts suggest that the signals corresponding to the respiratory movement and the heartbeats can simultaneously be detected from the sensor output by the signal decomposition such as digital filters. Besides, similar results are obtained from different subjects and beds, confirming that the sensor output is independent of bed types. Furthermore, subjects of the experiments confirm that the existence of the PVDF sensor under the sheet on the bed is not perceptible.

DETECTION OF RESPIRATION AND HEARTBEAT

As shown in Fig. 4, the respiratory movement can be easily detected from the output of PVDF sensor by a low-pass filter. However, the heartbeat signal component is of the nature of localized ripples rather than sinusoidal, which means that most of its energy concentrates on the high frequency harmonies rather than base frequency—frequency of heartbeat. In addition, inter-individual and intra-individual differences of the heart rate are significant. All these make it difficult to design and implement the traditional linear filters for the separation of respiration and heartbeat components in this application. On the other hand, the wavelet analysis has been known as an effective tool in dealing with this kind of localized signals [10]. Therefore, we tried to detect the respiration and heartbeat from the PVDF sensor output using wavelet analysis.

Multiresolution decomposition using wavelet analysis

Continuous wavelet transform of a signal $x(t)$ using mother wavelet $\psi(t)$ is defined as

$$W(a,b) = \frac{1}{\sqrt{|a|}} \int_{-\infty}^{\infty} x(t)\psi*(\frac{t-b}{a})dt, \qquad (2)$$

where a is the scale parameter that provides the frequency information of the signal and b is the shift parameter that provides the time position information of different frequency component. By applying wavelet transform to a signal, time-frequency distribution of it can be obtained. However, continuous wavelet transform is a tedious procedure, and much of redundant information is generated. If scales a and shifts b are choused based on powers of two as $(2^j, 2^{-j}k)$ —so called dyadic scales and shifts, the discrete wavelet transform can be defined as

$$DWT_k^{(j)} = 2^j \int_{-\infty}^{\infty} x(t)\psi*(2^j t - k)dt, \qquad (3)$$

which is much more calculating-efficient and just as accurate. The inverse transform of discrete wavelet transform of $DWT_k^{(j)}$ Eq. (3) is

$$x(t) = \sum_{j}\sum_{k} DWT_k^{(j)}\psi(2^j t - k). \qquad (4)$$

If the inside sum in Eq. (4) be rewritten as

$$h_j(t) = \sum_{k} DWT_k^{(j)}\psi(2^j t - k), \qquad (5)$$

and write $l_j(t)$ as

$$l_j(t) = h_{j-1}(t) + h_{j-2}(t) + h_{j-3}(t) + \cdots \qquad (6)$$

and further consider the signal $x(t)$ as $l_0(t)$, Eq. (4) can be rewritten as

$$\begin{aligned} x(t) &= l_0(t) \\ &= h_{-1}(t) + l_{-1}(t) \\ &= h_{-1}(t) + h_{-2}(t) + l_{-2}(t) \\ &= \cdots \end{aligned} \qquad (7)$$

In this way, by applying discrete wavelet analysis to a signal $x(t)$, the signal is split into a high frequency component $h_{-1}(t)$ and a low frequency component $l_{-1}(t)$. Generally, the high frequency component $h_{-1}(t)$ is also called as first-level detail d_1, and the low frequency component $l_{-1}(t)$ is called as first-level approximation a_1. The first-level approximation a_1 is

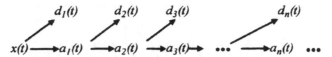

Figure 5. Wavelet multiresolution decomposition

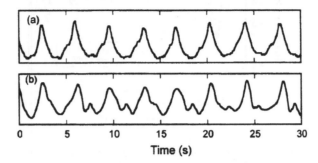

Figure 6. Comparison of (a) output of traditional plethysmography respiration monitor and (b) a_6 component of the PVDF sensor output

again split into d_2 and a_2, second-level detail and approximation. And a_2 is further split into d_3 and a_3, and so on. This procedure is generally known as wavelet multiresolution decomposition, as shown in Fig. 5.

Applying the wavelet multiresolution decomposition analysis to the output of the sensor to 7-level, it is found that the a_6 and d_1 components of the sensor output are respectively suitable for the discrimination of the respiration and the heartbeats components.

Detection of respiration

Figure 6 shows the output of the traditional plethysmography respiration monitor, which is used as respiration reference, and the a_6 component of the PVDF sensor output. In Fig. 6 (b), it is noticed that compared with the PVDF sensor output shown in Fig. 4 (c), in the a_6 component of it, the components originated from the heartbeats in the PVDF sensor output are completely removed. Thus respiration can be easily monitored by detection the peaks that are higher than a threshold in the a_6 component. For example, in Fig. 6, the number of peaks of the a_6 component and the number of peaks of the respiration reference are exactly the same, 8 in 30 seconds. It means that 8 times of respiration are detected with both the proposed PVDF sensor and the traditional plethysmography respiration monitor.

However, as one can expect, a human does not keep his sleeping posture and position steady over long time, in stead, he might turn his body now and then during sleep. As a result, the pressure on the bed varies with the posture of the human body. And so does the height of the peaks in the a_6 component during measurement. Therefore, the peak detection cannot be simply carried out using a determined threshold. Inspired from the automatic controlled gain circuit (AGC) that is widely used in electronic appliances, by analysis of the experimental data, a time-varying adaptive threshold is defined as:

$$Threshold_i = \sum_{i-k}^{i-1} \beta_j H_j, \tag{8}$$

where H_j is the height of the jth peak of the a_6 component and β_j is the weight applied to it.

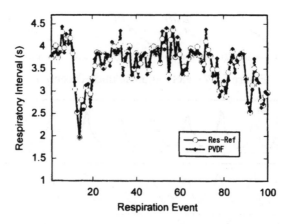

Figure 7. Comparison of instantaneous respiration intervals measured using
plethysmography respiration monitor used as respiration reference and PVDF sensor

Using this adaptive threshold, an automatic detecting program was developed and tested on
the experimental data. From the a_6 component of PVDF sensor output, 989 peaks are detected
in one hour, which means the average respiratory rate is 16.48 times-per-minute. At the same
time, 989 times of respiration are detected using the plethysmography respiration monitor. The
results using the proposed sensor and traditional respiration monitor agree perfectly.
Furthermore, by visual examination, neither false positive (FP) nor false negative (FN) is found.
Here FN means that the automatic detecting program failed to detect a peak of respiration from
the PVDF sensor output while respiration is detected using traditional respiration monitor and to
the opposite, FP refers to a peak of respiration was detected from the PVDF sensor output while
it did not actually exist.

In clinical practice, instantaneous respiratory rate/interval is more meaningful than
average respiratory rate. From Fig 6., it is noticed that the position of peaks of the a_6 component
of PVDF sensor output and peaks of the plethysmography respiration monitor also agree well.
This suggests that instantaneous respiratory interval can also be measured from the PVDF
sensor by detecting the position of peaks in the a_6 component of PVDF sensor output and
measuring the time between successive peaks. Figure 7 shows the comparison of instantaneous
respiration intervals of 100 respiratory events measured using the plethysmography respiration
monitor and the PVDF sensor. As shown in the figure, the two results agree very well.

From the above results, it is verified that the respiration during sleep can be successfully
detected using the proposed PVDF sensor.

Detection of heartbeat

Figure 8 (a) shows the record of ECG, which is used as heartbeat reference, while the d_1
component of the PVDF sensor output is shown in Fig. 8 (b). Physiologically, the high spikes in
the ECG, known as QRS complexes, indicate the beginning of the excitation of the ventricular
muscles. The QRS complex is generally used as indication of heartbeat in heart rate monitors.
From the figure, it is noticed that synchronizing with the QRS complexes in the ECG shown in
Fig. 8 (a), there are grouped vibrating waves in the d_1 component of the PVDF sensor output as
shown in Fig. 8 (b). Therefore, each group of vibrations can be considered related to a heartbeat.

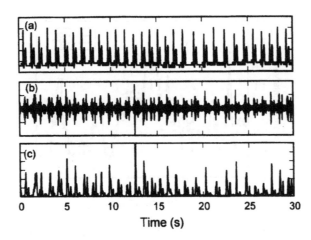

Figure 8. Comparison of (a) ECG, (b) d_l component of PVDF sensor
output, and (c) smoothed-rectified d_l

Hence it is expected that the heartbeat detection can be carried out by detecting these groups in the d_l component of the PVDF sensor output.

To detect the vibrations in the d_l component of the PVDF sensor output, the d_l component is first squared to rectify it into unipolar, and then the envelope of the rectified signal is calculated using a moving average smoothing algorithm. The result, being named as SRD1 (smoothed-rectified d_l), is shown in Fig. 8 (c). From Fig. 8 (c), it is known that the number and positions of peaks of SRD1 agree with the number and positions of QRS complexes approximately. However, the pressure fluctuation caused by the heartbeat on the bed is very weak, and the shape of it is not always steady as shown in Fig. 4. These make the detection of heartbeat much more difficult than the detection of respiration. For example, in Fig. 8 (c), in 25-27 second, very low peaks of SRD1 are found. Therefore, like the detection of respiration described in the previous section, a time-varying adaptive threshold is also need. Here, unlike the adaptive threshold using the height of previous peaks in the detection of respiration, by analysis of experimental data, standard deviation in three-second interval is choused as adaptive threshold for SRD1 peak detection empirically. Besides, as shown in Fig. 8 (c), at about 7 second, it is noticed that corresponding to one heartbeat, two peaks of SRD1 of similar height exist. To deal with this kind of problem, like most existing QRS detecting method, an intelligent judgment method is introduced. That is, all peaks higher than the threshold are first choused as candidates of heartbeat, and if two peaks are detected in 0.25 second and the interval between them is very short compared with successive interval around it, one of the candidates is considered false positive detection. In this case, the higher peak is considered as the real heartbeat and the lower one is discarded.

Applied the above method to experimental data lasting of one hour, 3821 heartbeats are detected from the PVDF sensor output, while 3827 heartbeats are detected from the ECG. Thus the average heart rates measured from the PVDF sensor and from the ECG are 63.68 and 63.78 beat-per-minute respectively. By visual examination, it is found that among 3821 detected heartbeats, 21 of them are FP detections. That is, from the PVDF sensor output, 21 heartbeats are detected while they do not exactly exist. Thus the real heartbeat number detected from the PVDF sensor is 3800. Compared with 3827 heartbeats detected from ECG, which is considered

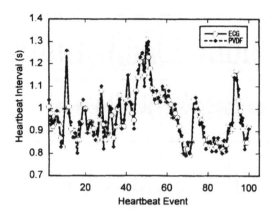

Figure 9. Comparison of instantaneous heartbeat intervals measured from
ECG and PVDF sensor.

actual number of heartbeat in the hour, the above-mentioned method fails to detect 27 heartbeats
from the PVDF sensor output. This means that 27 FN detections are found. If the error rate of
heartbeat detection be defined as

$$\text{error rate} = \frac{\text{FP+FN}}{\text{actual heartbeat number}} \times 100\%, \qquad (9)$$

form the above data of FP, FN and actual heartbeat number detected from the ECG, the error
rate in this case is 1.25%.

Again, measurement of instantaneous heart rate or heartbeat intervals is more meaningful
in clinical monitoring. Intervals between successive peaks in the SRD1 are plotted in Fig. 9
together with corresponding QRS intervals measured from the ECG, which is clinically used as
instantaneous heartbeat intervals. As shown in the figure, the two intervals agree well,
suggesting that the interval between two successive peaks in the SRD1 can satisfactorily be used
as instantaneous heartbeat interval.

The above results confirm that heartbeats during sleep can also be successfully detected
using the proposed PVDF sensor.

CONCLUSION

In this paper, an unconstrained sensor system for the in-sleep cardiorespiratory monitoring
using a PVDF piezopolymer film as the sensory material has been presented. The system was
tested on different types of bed. Experiments show that the system can be used on different
types of bed and using of the sensor does not interfere with the daily sleep of the subjects under
measurement. Wavelet multiresolution decomposition and time-varying adaptive threshold
peak-detecting algorithm is developed for detection of respiration and heartbeat from the sensor
output. Accurate simultaneous detection of both the respiration and heartbeats has been verified
experimentally.

With the further improvements on the sensor for the enhancement of the accuracy of the
system, and with the realization of real-time detection algorithm and diagnosis-assistance
algorithm, it is expected that the system can be used in early diagnosis of cardiorespiratory sleep
disorders. Furthermore, by combining the system with the network communication techniques,

important data can be sent to medical centers in real time if a patient is suffering such a disorder. In this way, home health-care and home telemedicine applications of the system can be expected.

REFERENCES

1. Saunders, N. and C. Sullivan. 1994. *Sleep and Breathing*, New York, NY: Marcel Dekker, Inc.
2. Alihanka, J., K. Vaahtoranta, and I. Saarikivi. 1981. "A New Method for Long-Term Monitoring of the Ballistocardiogram, Heart Rate, and Respiration," *Am. J. Physio.*, 240: 384-392.
3. Salmi, T. and L. Leinonen. 1986. "Automatic analysis of sleep records with static charge sensitive bed," *Electroenceph. Clin. Neurophysio.*, 64:84-87.
4. Nishida, Y., M. Takeda, T. Mori, H. Mizoguchi and T. Sato. 1998. "Unrestrained and Non-Invasive Monitoring of Human's Respiration and Posture in Sleep Using Pressure Sensor," *J. Robot. Soc. Jpn.*, 16(5): 705-711.
5. Tanaka, S. 2000. "Unconstrained and Noninvasive Automatic Measurement of Respiration and Heart Rates Using a Strain Gauge," *Trans. SICE*, 36 (3): 227-233.
6. Watanabe, H. and K. Watanabe. 2000. "Study on the non-restrictive vital bio-measurement by the air mattress methods," *Trans. SICE*, 36 (11): 894-900.
7. Nakajima, K., A. Osa, S. Kasaoka and K. Nakashima: "Detection of physiological parameters without any physical constrains in bed using sequential image processing," *Jpn. J. Appl. Phy.*, 35 (2B): L269-L272.
8. Harsányi, G. 1995. *Polymer Films in Sensor Applications*, Lancaster, PA: Technomic Publishing Co., Inc.
9. Dargahi, J. 1998. "Piezoelectric and pyroelectric transient signal analysis for detections of the temperature of a contact object for robotic tactile sensing," *Sensors & Actuators (A)*, 71: 89-97.
10. Strang, G. and T. Nguyen. 1997. *Wavelet and Filter Banks*, Wellesley, MA: Wellesley-Cambridge Press.

SIMULTANEOUS DESIGN OF STRUCTURAL TOPOLOGY AND CONTROL

Ye Zhu, Jinhao Qiu, Hejun Du, Junji Tani, and Masanori Murai

ABSTRACT

Simultaneous optimization with respect to the structural topology, actuator locations and control parameters of an actively controlled plate structure is investigated in this paper. The system consists of a clamped-free plate, a H_2 controller and four surface-bonded piezoelectric actuators utilized to suppress the bending and torsional vibrations induced by external disturbances. The plate is represented by a rectangular design domain which is discretized by a regular finite element mesh and for each element the thickness ratio is used as a design variable. A nested solving approach is adopted in which Ricatti-based control syntheses are considered as sub processes and nested into the main optimization process using the method of moving asymptotes to optimize the structural topology and actuator locations. To obtain clear and manufacturable topology, the popular techniques in the topology optimization area including penalization, filtering and perimeter restriction are also employed. Numerical example shows that the approach used in this paper can produce systems with clear structural topology and high control performance.

INTRODUCTION

Since the second half of 1980s, a number of studies [1, 2] on simultaneous structural -control optimizations have been presented. Although it is difficult to categorize the existing studies due to their variousness, as for the structural design, most of them share an approach in which the structural topology is predetermined and only a few dimensional parameters of the structural components are chosen as design variables.

In this paper, simultaneous optimization with respect to the structural topology, actuator locations and control parameters is investigated. The main purpose is to develop a procedure which extends the simultaneous structural-control optimization to include the structural topology design and demonstrate the potentiality of utilizing the new method to enhance the performance.

The example system considered in this paper consists of a clamped-free plate structure, a H_2 controller and four surface-bonded piezoelectric actuators utilized to suppress the bending

Ye Zhu, Jinhao Qiu, Junji Tani and Masanori Murai, Institute of Fluid of Science, Tohoku University, 2-1-1 Katahira, Aoba-ku, Sendai, Japan 980-8577.
Hejun Du, School of Mechanical and Production Engineering, Nanyang Technological University, Nanyang Avenue, Singapore 639789.

and torsional vibrations induced by external disturbances. The structural topology, actuator placement and control parameters are dealt with simultaneously to minimize a mixed objective function considering both control performance and structural stiffness.

Without any a priori decision on its connectivity, the plate is represented by a rectangular design domain which is discretized by a regular finite element mesh and for each element the parameter indicating the presence or absence of material is used as a design variable. Due to the unavailability of large-scale 0-1 optimization algorithms, the binary parameters of the original topology design problem are relaxed to take all values between 0 and 1. The popular techniques in the topology optimization area including penalization, filtering and perimeter restriction are used to suppress numerical problems such as intermediate thickness, checkerboards, and mesh dependence. Moreover, since it is not efficient to treat the structural and control design variables equally within the same framework, a nested solving approach is adopted in which Ricatti-based [3] control syntheses are considered as sub processes and nested into the main optimization process using the method of moving asymptotes (MMA) [4] to optimize the structural topology and actuator locations.

The numerical results show that the approach used in this paper can produce systems with clear structural topology and high control performance.

SIMULTANEOUS STRUCTURAL-CONTROL DESIGN

Generally, simultaneous optimization of a structure-control system can be formulated as a nonlinear programming problem in which a certain objective function f is minimized over the structural parameter p_s and control parameter p_c.

$$\min_{p_s, p_c} f(p_s, p_c) \tag{1}$$

It is clear that Equation (1) can be transformed into a main structural optimization nested with control optimizations as sub-processes [2].

$$\min_{p_s} \tilde{f}(p_s) \tag{2}$$

where

$$\tilde{f}(p_s) \triangleq \min_{p_c} f(p_s, p_c) \tag{3}$$

denotes the control optimization. From the point of the main process, the explicit (to a certain extent, only) design variable is p_s, while p_c is implicit or just by-product introduced when p_s is evaluated. In such approach, since the structural and control design variables are treated separately and the control design is well encapsulated, various existing structural and control design techniques can be combined and employed without any difficulty.

STRUCTURAL TOPOLOGY DESIGN

The topology optimization has made remarkable advances and given rise to powerful methods and numerical tools for engineering design since the late 1980s [5-7].

Considering a predefined design domain Ω, the topology optimization problem consists in finding the subdomain Ω_m filled with a limited amount of material (or complementarily, the subdomain $\Omega_v = \Omega - \Omega_m$ occupied by the void) which minimize a given objective function, without any a priori decision on its connectivity. Typically, after the design domain is divided into n finite elements (for simplicity, it is assumed that the finite element mesh is regular and all the elements have the same size and shape), the topology optimization problem can be described as

$$\min_x \tilde{f}(x) \tag{4a}$$

$$\text{s.t.} \quad x_i \in \{0,1\} \quad (i=1,2,3,...,n) \tag{4b}$$

$$V = \frac{\int_\Omega x_i \, d\Omega}{\int_\Omega d\Omega} \leq V_{max} \tag{4c}$$

$$h_j(x) \leq 0 \quad (j=1,2,3,...,n_h) \tag{4d}$$

where x is the vector of design variables and x_i is the component for the ith element. The values 0 and 1 of x_i denote respectively the absence and presence of material in the corresponding element, Equation (4c) sets the volume bound of the material in which V is the volume ratio, while Equation (4d) shows other inequality constraints.

To get reasonable resolution in the domain discretization, the number of elements n is usually very large. In addition, there exists no algorithm applicable to the large-scale 0-1 optimization problem defined by Equation (4). Hence, the binary parameters are relaxed so that they can take all values between 0 and 1.

$$0 \leq x_i \leq 1 \quad (i=1,2,3,...,n) \tag{5}$$

As a result, the relaxed variable x_i has to be associated with some properties of the material or element to make the intermediate values have physical meanings. Since the two-dimensional plate problems are considered in this paper, for simplicity x_i is related to the thickness of the element t_i.

$$t_i = x_i t_{max} \tag{6}$$

Furthermore, since the presence of a large number of "gray" elements having intermediate thicknesses is undesirable, the following penalization is applied to Equation (6).

$$t_i = P(x_i)t_{max} \tag{7}$$

where

$$P(x_i) \triangleq (x_i)^3 \tag{8}$$

By using Equation (7) instead of Equation (6) in the optimization, the intermediate x_i can be forced toward either 0 or 1 for high material efficiency. The theoretical details and several different approaches of the relaxation and penalization can be found in the references [5, 6].

It is known that in addition to the intermediate elements, the topology optimization suffers from the numerical problems such as checkerboards and mesh-dependence. Sigmund and Petersson [6] have a review paper on these phenomena and the corresponding prevention techniques. Two techniques, namely, the filtering and perimeter restriction employed in this paper are discussed briefly here.

First, the filtering function is defined below,

$$F(x_i) \triangleq \frac{\sum\limits_{k \in N} H_{ki} x_k}{\sum\limits_{k \in N} H_{ki}} \tag{9}$$

where

$$H_{ki} = \frac{1}{0.1 + (R_{ki})^4} \tag{10}$$

where R_{ki} is the distance between the centers of element k and i, while the set N includes every element (including element i itself) which has center distance less than R_{min} to element i. The filtering techniques which have been borrowed form image processing can be used to suppress checkerboards and moreover, by defining a constant R_{min} independent to the mesh size, it can be useful for resolving the problem of mesh-dependence. By replacing x_i in Equation (7) with $F(x_i)$, the relation between t_i and x_i can be written as,

$$t_i = P(F(x_i))t_{max} \tag{11}$$

Secondly, for two-dimensional structures, the perimeter is defined as

$$L(x) = \sum_k l_{k,ij} \left(\sqrt{(x_i - x_j)^2 + \zeta^2} - \zeta \right) \tag{12}$$

where ζ is a positive scalar close to 0, while $l_{k,ij}$ is the length of the kth interface (i.e., the common edge) between two neighboring elements i and j. The checkerboards and mesh-dependence can be effectively suppressed by introducing an appropriate constraint on the perimeter.

CONTROL DESIGN

The low-order natural frequencies and mode shapes of the structure obtained by the finite element analysis are used to establish the structural model in state space which has the external disturbance w, control input u, and measured output y. By introducing the feedback controller K, sensor noise signal d, limitation on control input u, and four weighting functions W_w, W_y, W_u and W_d, the augmented closed-loop system can be obtained as shown in Figure 1. For simplicity, H_2 control law is adopted in this paper. Therefore, the purpose of the control design is finding the optimal controller K which minimizes the H_2 norm of the closed-loop system shown in the figure.

$$f_1 = \min_K \left\| T_{(w',d')(y',u')} \right\|_2 \tag{13}$$

NUMERICAL EXAMPLE

One example is used in this paper to demonstrate the effectiveness of the approach for simultaneous topology-control optimization.

As shown in Figure 2, the clamped-free plate structure is represented by a rectangular

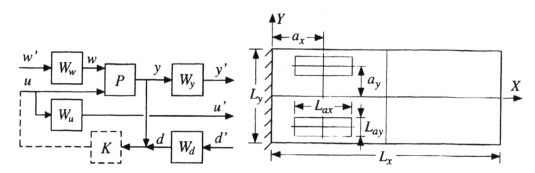

Figure 1. Closed-loop system   Figure 2. Design Domain

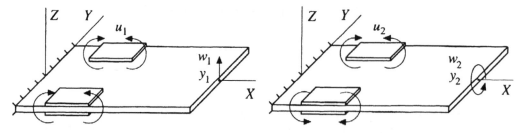

Figure 3. Bending vibration   Figure 4. Torsional vibration

TABLE I. MATERIAL PROPERTIES AND
MESH PARAMETERS

Young's modulus	E	$210 \times 10^9 \text{N/m}^2$
Mass density	ρ	$7.8 \times 10^3 \text{kg/m}^3$
Poisson's ratio	γ	0.3
Domain size	$L_x \times L_y$	$25 \times 10 \text{mm}^2$
Actuator size	$L_{ax} \times L_{ay}$	$5 \times 2 \text{ mm}^2$
Mesh resolution	$n_x \times n_y$	50×20
Max. Thickness	t_{max}	0.4mm

TABLE II. VALUES OF F_i

F_1	F_2	F_3	F_4	F_5	F_6
9. 90	0.50	210	22.6	16.2	0.99

design domain, in which the two small rectangles show the areas for bonding piezoelectric actuators. Four actuators with the same size are bonded on the two surfaces of the structure. Considering the thickness of the plate structure, both the structure itself and the actuator array are assumed to be symmetrical about the *XY* plane and *XZ* plane. Hence, only the topology of the half domain and the *XY* coordinates of one actuator center, i.e., a_x and a_y, are to be determined.

The inputs and outputs of the structure are shown in Figures 3 and 4. The external disturbances w_1 and w_2 are the *Y*- force and *X*- moment acting at the center of the free end, while the corresponding collocated transversal and rotational displacements y_1 and y_2 are used as outputs. By applying different voltages to the actuators, two patterns of control moments u_1 and u_2 can be generated to control the bending and torsional vibrations induced by w_1 and w_2.

Table I shows several constant parameters used in the example.

For simplicity, the piezoelectric characteristics of the actuators are not modeled and the moment combinations u_1 and u_2 are directly used as control inputs. Moreover, since the structural mesh is fixed, if the actuator locations, i.e., a_x and a_y can take continuous values, the four edges of the actuator will have slim chance of precisely lying upon the edges of structural elements. Hence, the actuation moments, stiffness, and mass of the actuators are distributed among the neighboring structural nodes by a procedure which can make the distributed values differentiable with respect to a_x and a_y.

The weighting functions W_w, W_u, and W_d used in the control design are assumed to be constant scalars of 1, 0.001 and 0.01, while W_y is a 3-order low-pass filter

$$W_y = \frac{8 \times 10^{-3} s^3 + 2.513 \times 10^3 s^2 + 3.928 \times 10^8 s + 3.101 \times 10^{13}}{s^3 + 6.283 \times 10^4 s^2 + 1.974 \times 10^9 s + 3.101 \times 10^{13}} \tag{14}$$

In addition, the structural modes which have frequencies higher than 10,000Hz are omitted in the plant modeling for control design.

Finally, the optimization problem can be expressed as

$$\min_{(x, a_x, a_y)} \quad f_1 / F_1 + \max(f_4 / F_4, f_5 / F_5)$$

$$\text{s.t.} \quad 0 \le x_i \le 1 \quad (i=1,2,3,\dots,n)$$

$$3 \le a_x \le 22$$

$$1.5 \le a_y \le 4 \tag{15}$$

$$f_j / F_j \le 1 \quad (j=2,3)$$

$$\max(f_4 / F_4, f_5 / F_5) \le 1$$

$$f_6 / F_6 \ge 1$$

where, f_1, f_2, f_3, and f_6 denote the closed-loop H_2 norm, volume ratio of the whole structure V, perimeter L, and volume ratio of actuator bonding area V_{act}, respectively. While f_4 and f_5 are the two static displacement outputs y_1 and y_2 caused by the *static* external disturbance inputs w_1 and w_2, respectively. The constants F_1, F_2, …, and F_6, are listed in Table II. From the above definition and the table, it can be seen that firstly, the maximum volume ratio is set to be 0.5, i.e., in the final topology the number of solid elements will not exceed 50% of the total element number. Secondly, to assure the actuator bonding quality, the minimum volume ratio of actuator bonding area is set to be 0.99, i.e., the bonding areas are forced to be almost solid. Thirdly, the perimeter, static displacement 1 and 2 are limited to be not greater than F_3, F_4 and F_5. Finally, upper and lower bounds are imposed on the actuator locations a_x and a_y and the topology variable x.

It should be mentioned here that the purpose of introducing the static displacements f_4 and f_5 into Equation (15) is to assure the structure to be continuous and capable of bearing the external loads. Moreover, the constants F_1, F_4 and F_5 are equal to the corresponding f_1, f_4 and f_5 of the *initial* structure *without penalization*. As shown in Figure 5, the initial structure is uniform and has volume ratio of 0.5.

The above-mentioned nested approach is used to solve the simultaneous design problems defined in Equation (15), in which the topology variables and the actuator locations are solved in the main optimization by the method of moving asymptotes (MMA), while the control parameters are designed in the sub optimizations by the Ricatti-based synthesis. The details of MMA and H_2 control design are omitted here and can be found in the references [3, 4].

RESULTS AND DISCUSSIONS

The numerical results are discussed in this part. For saving space, the optimization histories, i.e., the changes of the f_i values with respect to the iteration number are not shown here. The optimization is terminated after 64 MMA iterations.

The initial and optimized structure topologies are shown in Figures 5 and 6, in which the locations of the actuators are indicated by the "X" marks and the values of topology design variables are mapped to the gray-scale bar. As shown in the figures, the initial structure is uniform and has volume ratio of 0.5, while the optimized structure with the same volume ratio looks like a tuning fork with the two branches clamped and the root free. The optimized topology is quite clear and manufacturable since only a few elements remain "gray" while the others become either solid or void. Moreover, by comparing the two figures, it can be found

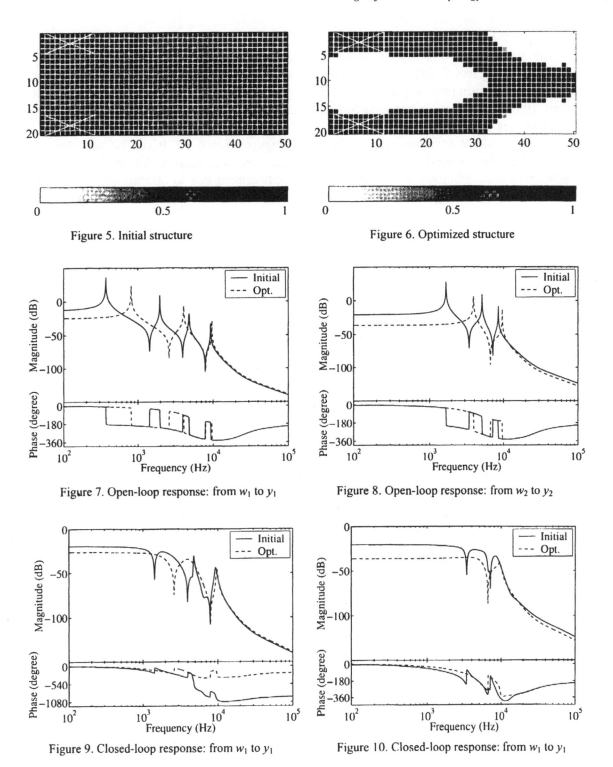

Figure 5. Initial structure

Figure 6. Optimized structure

Figure 7. Open-loop response: from w_1 to y_1

Figure 8. Open-loop response: from w_2 to y_2

Figure 9. Closed-loop response: from w_1 to y_1

Figure 10. Closed-loop response: from w_1 to y_1

that the changes of the actuator locations are very small after the optimization. The result means that for the optimized structure, the actuation will be most effective if the actuators are placed near the clamped side.

The open-loop frequency responses of the initial and the optimized structures are compared in Figures 7 and 8, from which it can be seen that in both bending and torsional directions, the optimized structure are "stiffer" than the initial one. As shown in the figures, below 10,000Hz, the optimized structure only has 3 bending and 2 torsional modes while the initial one has 4 and 3 corresponding modes.

The closed-loop responses are compared in Figures 9 and 10. It can be seen from the figures that the closed-loop responses are improved significantly by the simultaneous topology-control optimization. In addition, the improvement due to optimization in torsional directions is slightly larger than that in bending direction.

CONCLUSIONS

Simultaneous optimization with respect to structural topology, actuator locations and control parameters for a plate structure is investigated in this paper.

The simultaneous design problem is formulated as a main structural optimization dealing with the structural parameters and actuator locations, into which the control optimizations with respect to the control parameters are nested as sub processes. The control optimizations are encapsulated and executed independently to find the control parameters for any given structures produced during the main optimization. Thus, various methods and tools for the control synthesis and structural design including the structural topology optimization can be combined and employed easily.

In the structural topology optimization, the plate is represented by a rectangular design domain which is discretized by a regular finite element mesh and for each element the parameter indicating the presence or absence of material is used as a topology design variable. Due to the unavailability of large-scale binary optimization algorithms, the parameters of the original topology design problem are relaxed to take continuous values. The popular techniques including penalization filtering and perimeter limitation are also applied to obtain clear and manufacturable topology.

The numerical results demonstrate that the approach discussed in this paper can be successfully applied to the simultaneous topology-control optimization and produce systems with clear structural topology and high control performance.

ACKNOWLEDGMENT

The authors are grateful to Krister Svanberg for providing the MMA optimization package.

REFERENCES

1. Onoda, J. and R. T. Haftka. 1987. "An Approach to structure/control simultaneous optimization for large flexible spacecraft," *AIAA Journal*, 25: 1133-1138.
2. Zhu, Y., J. Qiu, J. Tani, Y. Urushiyama, and Y. Hontani. 1999. "Simultaneous optimization of structure and control for vibration suppression," *Journal of Vibration and Acoustics*, 121(2):237-243.

3. Svanberg, K. 1987. "The method of moving asymptotes – a new method for structural optimization," *International Journal for Numerical Methods in Engineering*, 24:359-373.
4. Doyle, J.C., K. Glover, P. Khargonekar, and B. Francis. 1989. "State-space solutions to standard H_2 and H_∞ control problems," *IEEE Transaction on Automatic Control*, 34(8):831-847.
5. Bendsøe, M.P. and N. Kikuchi. 1988. "Generating optimal topologies in optimal design using a homogenization method," *Computational Methods in Applied Mechanics and Engineering*, 71:197-224.
6. Sigmund, O., and J. Petersson. 1998. "Numerical instabilities in topology optimization: A survey on procedures dealing with checkerboards, mesh-dependencies and local minima," *Structural Optimization*, 16:68-75.
7. Du, H., G. K. Lau, M. K. Lim, and J. Qiu. 2000. "Topological optimization of mechanical amplifiers for piezoelectric actuators under dynamic motion," *Journal of Smart Materials and Structures*, 9:788-800.

High-Load / High-Speed Systems

MODELLING AND DESIGN OF HIGH-LOAD ACTUATOR SYSTEMS IN ADAPTIVE STRUCTURES

Elmar J. Breitbach and Harald P. Breitbach

ABSTRACT

Emphasis is placed on the problem of how to optimally design and realize high load smart actuators in adaptive structures. In this general context, primary focus is on concepts of vibration reduction using smart actuator systems made on the basis of piezoelectrica, magnetostrictiva or other high-frequency materials.

The paper deals with the modelling procedure starting with the general and concise identification of all characteristic parameters of the actuator system itself, of the actuator / structure coupling interface as well as of the local structural flexibilities in the close vicinity of the coupling points. With this complete set of characteristics the derivation of the structural dynamics equations of motion is possible. It is shown how these equations can be transformed without any loss of physical information and accuracy into the highly reduced modal space formulated in terms of only a small number of modal DOFs in the problem-related frequency range of interest.

These modal equations establish the best possible and easy-to-use means for the parametric prediction and selection of the final positioning, design and structural embedment of the actuator system. Their formulation takes into account the full range of operation-related static and dynamic loads and of the actuator related requirements such as maximum stroke, force, strength and stiffness as well as minimum weight, size, and energy consumption.

The main advantages of the design concept outlined in the paper have been demonstrated in the past by means of some successfully accomplished industrial projects.

INTRODUCTION

In the context with Smart Structures Technology one of the key questions is how to effectively reduce undesired vibrations and noise radiation in structural systems [1]. Towards this aim, actuator elements made of smart materials such as piezo-ceramics, magnetostrictors or shape memory alloys have to be suitably integrated into the structure. They are activated by

Elmar J. Breitbach, DLR e.V., Director Institute of Structural Mechanics,
Lilienthalplatz 7, 38108 Braunschweig, Germany
Harald P. Breitbach, bec GbR,
Am Hachweg 6, 37083 Göttingen, Germany

means of appropriately conditioned sensor signals representative of the dynamic behavior of the structure. The signal conditioning process is carried out through modern adaptive control systems.

State of the art smart actuators, either commercially available or as lab specimens, are commonly affected by a number of inherent deficiencies frequently impeding their proper operation in the structural compound. So, the most widely used piezo-actuators are characterized by the following disadvantages to be taken into consideration:

- low tensile and shear strength due to high brittleness
  and extremely low fracture strains
- high specific weight ($\sim$ 7...8 g/cm^3)
- low electromechanical strain ($\leq$ 1 μm / mm)
- high-voltage power supply

On the other hand, piezo-based actuators are distinguished by fairly good characteristics such as

- high frequency operation range (up to some kHz)
- high stiffness
- high compression strength
- small size
- high load capacity
- low energy power supply

The main problem is how actuators of this kind have to be designed and integrated in order to operate in the most efficient and best possible way by compensating for their deficits as well as making optimal use of their positive features, both mentioned above.

The following discussions and derivations will be confined to piezo-based actuators because this material presently shows the best potential for applications in light weight structures under high frequency and high load operational conditions.

MODELLING OF THE ACTUATOR SYSTEM

In general, piezo-based actuators are in use either in form of piezo-stacks or in a higher integrated form as distributed sheet-type actuators in the form of thin patches made of piezo-foils or piezo-fibers.

Piezo stack actuator

In the following, special attention will be payed to modelling of highly integrated stack actuators taking into account

- the overall structural dynamic characteristics of the structure to be controlled and
- the active elements integrated into the structure between the interface points i and i+1 (**fig. 1**).

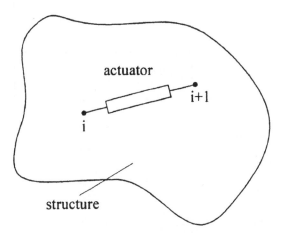

Figure 1: structural system with integrated actuator

By means of a detailed and thorough view of the system (**fig. 2**) the following stiffness elements can be identified:

- stiffness of the piezo-stack
- stiffness of the actuator housing
- stiffness of the tension bolt keeping the piezo-stack under compression load in all operational conditions.
- stiffness of the coupling devices between actuator and structure such as the coupling rods, the lever arms and the local structural stiffness around the lever base.

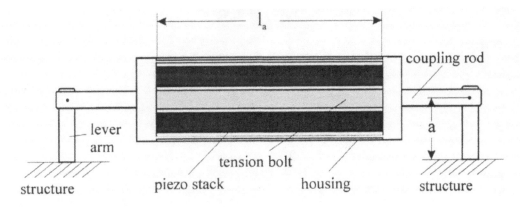

Figure 2: schematic sketch of a stack actuator coupled to a structure

The set-up shown in **fig. 2** can further be reduced to the simplified representation (**fig. 3**) where c_a stands for the stack stiffness, c for the parallel stiffness arrangement of the housing stiffness c_h, the stack stiffness c_a and the tension bolt stiffness c_b whereas c_{c1} and c_{c2} denote the coupling stiffnesses reduced to the lever end points i and i+1.

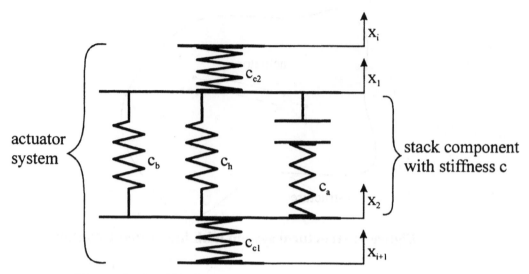

Figure 3: simplified representation of stiffness arrangement

The latter comprise the series arrangement consisting of the coupling rod stiffness, the lever arm stiffness and the local structural stiffness in the close vicinity of the lever arm base.

Force equilibrium at the stack end points

With an arbitrary activable stack displacement Δu the force equilibrium in the interface points 1 and 2 (see **fig. 3**) is given by the equation

$$\mathbf{c}\,\mathbf{x}_s = c_a\,\Delta u \begin{Bmatrix} -1 \\ 1 \end{Bmatrix} + \mathbf{c}_c\,\mathbf{x}_b \tag{1}$$

where

$$\mathbf{c} = \begin{bmatrix} c_{c1}+c & -c \\ -c & c_{c2}+c \end{bmatrix} \tag{2}$$

$$\mathbf{c}_c = \begin{bmatrix} c_{c1} & 0 \\ 0 & c_{c2} \end{bmatrix} \tag{3}$$

and

$$\mathbf{x}_s = \begin{Bmatrix} x_1 \\ x_2 \end{Bmatrix}, \quad \mathbf{x}_b = \begin{Bmatrix} x_i \\ x_{i+1} \end{Bmatrix} \tag{4}$$

Then, the displacement vector $\mathbf{x}_s$ can be expressed as a function of Δu and the displacement vector $\mathbf{x}_b$ by the equation

$$\mathbf{x}_s = c_a\,\Delta u\,\mathbf{c}^{-1} \begin{Bmatrix} -1 \\ 1 \end{Bmatrix} + \mathbf{c}^{-1}\,\mathbf{c}_c\,\mathbf{x}_b \tag{5}$$

Stiffness energy in the actuator system

The stiffness energy stored in the actuator system taking into account all stiffness elements between the coupling points i and i+1 can be written in the form

$$U = \frac{1}{2}\left[c_{c1}(x_i - x_1)^2 + (c_b + c_h)(x_1 - x_2)^2 + c_a(x_1 - x_2 + \Delta u)^2 + c_{c2}(x_2 - x_{i+1})^2\right] \quad (6)$$

as a function of the arbitrary displacement vectors x_s and x_b as well as of Δu.

Partial differentiation of U with respect to x_i and x_{i+1} (Lagrange operation) has to take into account that x_1 and x_2 are defined as implicit functions of x_i, x_{i+1} and Δu according to **equ. (5)**. Hence it follows after some lengthy mathematical operations

$$\left\{\begin{array}{c} \dfrac{\partial U}{\partial x_i} \\[2mm] \dfrac{\partial U}{\partial x_{i+1}} \end{array}\right\} = c_{tot}\begin{bmatrix} 1 & -1 \\ -1 & 1 \end{bmatrix}x_b - \Delta u \frac{c_a c_{tot}}{c}\left\{\begin{array}{c} -1 \\ 1 \end{array}\right\} \quad (7)$$

$$= c_{tot}Tx_b - \Delta u \frac{c_a c_{tot}}{c} t$$

where

$$\frac{1}{c_{tot}} = \frac{1}{c_{c1}} + \frac{1}{c_{c2}} + \frac{1}{c} \quad (8)$$

and

$$c = c_h + c_b + c_a \quad (9)$$

The discretized equations of motion

The equations of motion of the discretized structural system without actuator can be written in the matrix form

$$cx + m\ddot{x} = p \quad (10)$$

Considering a structural system with an integrated actuator leads to the modified equation of motion

$$(c + \Delta c)x + m\ddot{x} = p + \Delta p \quad (11)$$

In this equation Δc stands for the modification of the original stiffness matrix c due to the stiffness effect of the integrated actuator and Δp describes the modification of the forcing vector due to the actuator-induced internal forces.

The matrices **c**, Δ**c** and **m** are of the order m x m with m denoting the number of the physical degrees of freedom of the discretized system. The physical displacement vector **x**, the external force vector **p** and the vector Δ**p** of the internal actuator-induced forces are of the order m x1. For the sake of simplicity, the additional inertia effect of the actuator is assumed to be negligible, i. e. Δ**m** $= 0$. Structural damping is neglected, as well.

Embedding the actuator-related terms as formulated in **equ.** (7) into the order scheme of **equ.** (11) leads to

$$\Delta \mathbf{c} = c_{tot} \begin{bmatrix} 0 & 0 & 0 \\ 0 & \mathbf{T} & 0 \\ 0 & 0 & 0 \end{bmatrix} = c_{tot} \Delta \mathbf{c}^* \tag{12}$$

$$\Delta \mathbf{p} = \Delta u \, c_a \, \frac{c_{tot}}{c} \begin{Bmatrix} 0 \\ \mathbf{t} \\ 0 \end{Bmatrix} = \Delta u \, c_a \Delta \mathbf{p}^* \tag{13}$$

Modal Transformation

Transforming **equ.** (10) into the frequency domain and assuming **p** $= 0$ results in the eigenvalue problem

$$\det \left(\mathbf{c} - \omega^2 \mathbf{m} \right) = 0, \tag{14}$$

the eigensolutions of which are the eigenvalues ω_r^2, $r = 1\,(n)$, with n indicating the number of eigensolutions in a frequency range of interest, and the eigenvectors (normal modes) $\Phi_r, r = 1\,(n)$, comprised in the modal matrix in the form

$$\Phi = \left(\Phi_1 .... \Phi_r .... \Phi_n \right) \tag{15}$$

In general, Φ is a rectangular matrix of the order n x m. Application of the modal transformation

$$\mathbf{x} = \Phi \mathbf{q} \tag{16}$$

to **equ.** (11) along with a left-hand multiplication by Φ^T yields the modal equations

$$\left(\mathbf{K} + \Delta \mathbf{K} \right) \mathbf{q} + \mathbf{Mq} = \Phi^T \left(\mathbf{p} + \Delta \mathbf{p} \right) = \mathbf{Q} + \Delta \mathbf{Q} \tag{17}$$

The implications and advantages of the modal transformation as a powerful mathematical tool are dealt with in detail for instance in [2], [3] and [4].

In **equ.** (16) and (17) **q** denotes the vector of the so-called generalized co-ordinates $q_r, r = 1 (n)$, which can be interpreted as weighing factors accounting for the contribution of the various normal modes in the structural response to an arbitrary for a vector.

The square matrices **K** and **M** are diagonal with the generalized masses $M_r, r = 1(n)$, and the generalized stiffnesses $K_r = \omega_r^2 M_r, r = 1(n)$, as diagonal elements. $\mathbf{\Delta K}$ is a fully occupied matrix taking into account the stiffness effect of the actuator whereas $\mathbf{\Delta Q}$ designates the actuator-related generalized force vector. The elements of matrix $\mathbf{\Delta K}$ are defined as

$$\Delta K_{rs} = c_{tot} \left(x_{i,r} - x_{i+1,r} \right) \left(x_{i,s} - x_{i+1,s} \right) \tag{18}$$

whereas the components of the generalized force vector can be written as follows

$$Q_r + \Delta Q_r = \mathbf{\Phi}_r^T (\mathbf{p} + \mathbf{\Delta p})$$
$$= \mathbf{\Phi}_r^T \mathbf{p} + \Delta u\, c_a \frac{c_{tot}}{c} \left(x_{i,r} - x_{i+1,r} \right), \quad r = 1(n) \tag{19}$$

Equ. (17) showing the equations of motion of the structural system with an integrated actuator have the disadvantage that all equations are coupled to each other through the off-diagonal elements of matrix $\mathbf{\Delta K}$. However, the equations leave the actuator-related stiffnesses c_{tot}, c and c_a as well as the "prescribed" active displacement Δu of the actuator explicitly available. Thus, the equations are best suited for parametric studies in order to find out the optimal position and design of the actuator.

To overcome the disadvantage due to the coupling effect of $\mathbf{\Delta K}$, but to maintain the advantageous availability of the actuator-related design variables a further main-axis transformation has to be applied to equation (17) under the condition that **Q** and $\mathbf{\Delta Q} = 0$, i. e.

$$\det \left(\mathbf{K} + \mathbf{\Delta K} - \omega^2 \mathbf{M} \right) = 0 \tag{20}$$

This mathematical operation entails a new set of modal characteristics $\overline{\mathbf{\Phi}}r$ and $\overline{\omega}_r^2, r = 1,(n)$, the application of which to **equ.** (17) results in

$$\left(\overline{\mathbf{K}} + \Delta \overline{\mathbf{K}} \right) \overline{\mathbf{q}} + \overline{\mathbf{M}} \ddot{\overline{\mathbf{q}}} = \overline{\mathbf{Q}} + \Delta \overline{\mathbf{Q}} \tag{21}$$

where

$$\overline{\mathbf{K}} = \overline{\mathbf{\Phi}}^T \mathbf{K} \overline{\mathbf{\Phi}}, \quad \Delta \overline{\mathbf{K}} = \overline{\mathbf{\Phi}}^T \mathbf{\Delta K} \overline{\mathbf{\Phi}}, \quad \overline{\mathbf{M}} = \overline{\mathbf{\Phi}}^T \mathbf{M} \overline{\mathbf{\Phi}}, \tag{22a}$$
$$\overline{\mathbf{Q}} = \overline{\mathbf{\Phi}}^T \mathbf{\Phi}^T \mathbf{p}, \quad \Delta \overline{\mathbf{Q}} = \overline{\mathbf{\Phi}}^T \mathbf{\Phi}^T \mathbf{\Delta p} \tag{22b}$$

The eigenvectors $\overline{\Phi}_r$ are comprised in the square matrix $\overline{\Phi}$ which is of the order n x n.

The matrices $\overline{K}$, $\overline{M}$ and $\Delta\overline{K}$ are all diagonal with the special feature of $\Delta\overline{K}$ that the actuator design variable c_{tot} is still explicitly available because $\Delta\overline{K}$ can be written

$$\Delta\overline{K} = c_{tot}\overline{\Phi}^T\Phi^T\Delta c^*\Phi^T\overline{\Phi} \tag{23}$$

Herewith, the equations of motion are available as a set of n equations completely decoupled from each other, thus greatly scaling down further analytical investigations to the treatment of simple SDOF equations of motion. The same holds true for the forcing terms $\overline{Q}$ and

$$\Delta\overline{Q} = c_a\frac{c_{tot}}{c}\overline{\Phi}^T\Phi^T\Delta p^* \tag{24}$$

OPTIMAL ACTUATOR DESIGN

According to **equ.** (21) the best possible controllability of a structural system for a given external excitation $\overline{Q}$ requires the actuator-related internal forces $\Delta\overline{Q}$ to be conditioned such that the structural response due to the actuator forces must be equivalent to the structural response due to the operation-related forcing vector $\overline{Q}$. This is a short form description of the general principle that full cancellation of operation-related vibrations requires complete interference with artificial actuator-induced vibrations. At first glance, this condition of complete interference could theoretically be achieved if

$$\overline{Q} = -\Delta\overline{Q} . \tag{25}$$

However, it is hardly possible to directly derive from the generalised force vector $\Delta\overline{Q}$ the corresponding actuator force vector Δp only sparsely occupied in a few structural positions with internally acting force pairs (actio = reactio), whereas the operation-related forces p are in general external forces often continuously distributed throughout the structure. Typical examples are bridges under wind loads or aircraft under aerodynamic loads.

In view of this general situation it seems to be much more efficient and easier to achieve the above mentioned interference condition by minimising the complete structural response rather than to fulfil **equ.** (25). This concept requires the extension of the undamped system as defined in **equ.** (21) to a damped system which can be described by the equation

$$\left(\overline{K}(I + j\Gamma) + \Delta\overline{K} - \omega^2\overline{M}\right)\overline{q} = \overline{Q} + \Delta\overline{Q} . \tag{26}$$

With the diagonal matrix $\Gamma = diag(\gamma_r)$, damping can be approximately defined in form of the imaginary part $j\overline{K}\Gamma$ of the complex stiffness matrix $\overline{K}(I + j\Gamma)$, ($I$: Identity matrix, γ_r: modal loss factor). Hence the overall structural response can be achieved by the retransformation

$$\mathbf{x} = \mathbf{\Phi}\,\overline{\mathbf{\Phi}}\,\overline{\mathbf{q}}\,. \tag{27}$$

By taking into account **equ.** (26) it follows from **equ.** (27)

$$\mathbf{x} = \mathbf{\Phi}\overline{\mathbf{\Phi}}\left(\overline{\mathbf{Q}}+\Delta\overline{\mathbf{Q}}\right)\overline{\mathbf{H}}\,, \tag{28}$$

where

$$\overline{\mathbf{H}} = \left(\overline{\mathbf{K}}(\mathbf{I}+j\mathbf{\Gamma})+\Delta\overline{\mathbf{K}}-\omega^2\overline{\mathbf{M}}\right)^{-1}. \tag{29}$$

stands for the generalised admittance matrix. Fulfilling the above mentioned cancellation condition means that in the search for the best possible modal interference $\Delta\overline{\mathbf{Q}}$ has to be varied with the objective that $\mathbf{x} = \min$.

Towards this goal, state-of-the-art control technology offers a great variety of different control concepts appropriately adaptable to each practical case. Especially the small size of the modal structure / actuator model considerably minimises the effort necessary to realise an optimally controlled structure.

Once the mathematical model of a vibration-controlled structure has been established on the basis of the above described modal formulation, a sequence of further steps has to be carried out to establish the final actuator design:

- Identification of the responding normal modes to be controlled.
- Calculation of the optimal number and positions of the actuators based on the mode shapes as well as on the special design allowables and restrictions.
- Determination of maximum force and displacement amplitudes of the actuators attributed to the different normal modes.
- Deduction of the actuator size (length and cross section) from the electromechanical strain as well as from the required stiffness and strength.
- Determination of the above-described stiffness values c_a, c_h, c_b, c_{c1} and c_{c2}.

To get a first impression of how a real actuator behaves in comparison with the best possible theoretical actuator, the corresponding limit value actuator-induced generalized force vector can be determined as follows

$$\Delta\overline{\mathbf{Q}}_{lim} = \lim_{\substack{c_h, c_b \to 0 \\ c_{c1}, c_{c2} \to \infty}} \frac{c_a\,c_{tot}}{c}\,\overline{\mathbf{\Phi}}^T\mathbf{\Phi}^T\Delta\mathbf{p}^{\bullet}$$
$$= c_A\,\overline{\mathbf{\Phi}}^T\mathbf{\Phi}^T\Delta\mathbf{p}^{\bullet} \tag{30}$$

Dividing $\Delta\overline{\mathbf{Q}}$ by $\Delta\overline{\mathbf{Q}}_{lim}$ leads to the efficiency factor

$$\eta = \frac{c_{tot}}{c} \tag{31}$$

which simplifies for the general case $c_{c1} = c_{c2} = c_c$ to

$$\eta = \frac{1}{1 + 2\frac{c}{c_c}} \tag{32}$$

illustrated in **fig. 4**.

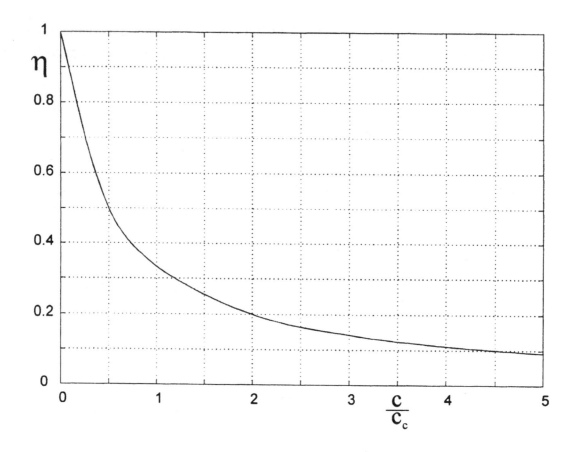

Figure 4: η acc. to **equ. (32)** as a function of c/c_c

This stresses that the best possible efficiency requires $c_c = 0$ whereas c has to be a finite value such that between the coupling points i and $i+1$ sharp stiffness changes can be avoided.

CONCLUSIONS

It has been demonstrated that the modal transformation is a very powerful tool for establishing quite small-size mathematical models of structures with embedded actuators. This considerably scales down the prediction effort towards determining optimum actuator positions, establishing efficient control systems and identifying the best possible actuator design.

It seems to be quite easy to use the same modelling concept for structures provided with either very thin 2D foil-like actuator systems or complex 3D-actuators.

It has to be mentioned that the modal design concept implies no simplifications and approximations hampering the efficiency, accuracy and speed of the design process.

REFERENCES

1. Breitbach, E., Lectures on Smart Structures Technologie at the University of Braunschweig.

2. Breitbach, E., Treatment of the control mechanisms of light airplanes in the flutter clearance process, *NASA CP 2085, Part II*, 437, 1979.

3. Breitbach, E., Recent Developments in Multiple Input Modal Analysis, *Journal of Vibration, Stress and Reliability in Design*, 1988.

4. Breitbach, E., Experimentelle Simulation dynamischer Lasten an Raumfahrtstrukturen mittels modaler Erregerkraftkombinationen, *Habilitationsschrift, RWTH Aachen*, 1989.

DEVELOPMENT OF SMART MISSILE FINS WITH ACTIVE SPOILER

Seung Jo Kim, Chul Yong Yun, Seong Hwan Moon, and Sung Nam Jung

ABSTRACT

For conventional missiles, electric or hydraulic actuators are mounted inside the missile fuselage to activate the aerodynamic control surfaces. These internally mounted actuators occupy considerable volume which otherwise can be used for payload or additional fuel. Reducing the size of the internal actuators and hence lowering the total actuator weight may improve the overall performance of missile significantly. The goal of this research is to develop a light-weight, low cost smart missile fin capable of surviving the supersonic operating environment while providing necessary performance comparable to existing missile fins. In this study, in an alternative to facilitate realistic design concepts and to generate enough actuation forces required for missile controls, smart fins with trailing-edge-mounted retractable wedge are considered. The retractable wedge stretches in or out appropriately to decrease the applied pitching moment of the missile fin. For the theoretical calculation, a commercial CFD software package is used to obtain the forces and the moments generated by the fin with retractable wedge. The piezoelectric actuation mechanism that is applicable to the wedge type actuator is also investigated.

INTRODUCTION

So far, due to the extreme operating environment encountered during the supersonic flight, successful supersonic smart missile fin has been more difficult to develop than

S.J. Kim, Professor, School of Mechanical and Aerospace Engineering, Seoul National University, Seoul, 151-742, Korea

C.Y. Yun, Graduate Research Assistant, School of Mechanical and Aerospace Engineering, Seoul National University, Seoul, 151-742, Korea

S.H. Moon, Graduate Research Assistant, School of Mechanical and Aerospace Engineering, Seoul National University, Seoul, 151-742, Korea

S.N. Jung, Professor, Department of Aerospace Engineering, Chonbuk National University, Chonju, 561-756, Korea

subsonic counter part of the smart missile fins. Currently, research is being conducted to develop a light-weight smart missile fin capable of surviving the supersonic operating environment. Since a missile fin actuator takes up considerable available volume, the performance of the missile is limited. Most of the actuators are of pneumatic, hydraulic or electro-mechanical and occupy 2 to 6% of fuselage volume. Since these actuators are mounted within the fuselage, they cannot be placed adjacent to critical components like rocket motors or seeker assemblies and they require bearings and linkages. Small missiles must use an extension tube aft of the rocket motor to accommodate these internal actuators. They are also not amenable to some fin folding arrangements such as those used on several types of unguided munitions. Accordingly, there is a need for a type of inexpensive fin that occupies less fuselage volume and is capable of generating the forces and moments required for the controlled flight. Therefore reducing the size of the internal actuators may improve the overall performance of the missile and allow smaller missiles to carry a larger payload.

Barrett[1,2] demonstrated concepts for smart missile fin design without using fuselage mounted actuators. The design consists of a torque plate that was bonded at the wing root to a fixed base and to the shell at the wing tip. This design demonstrated static deflections of 8.5 deg at low airspeeds but is prevented from use at high airspeeds due to a tendency to become divergent. August and Joshi[3] conducted preliminary design research on the use of smart structures to control missile fins at supersonic speeds. Seven distinct control schemes have been identified which use aerodynamic forces to rotate or assist in the rotation of an all moving control surface. The results have shown that altering missile fin hinge moments using aerodynamic control is possible using the forces and displacements that smart structures are capable of generating, but the effects on pitching moment are small to reduce the size of traditional actuator.

In this study, in an alternative to facilitate realistic design concepts and to generate enough actuation forces and displacement required for missile controls, smart fins with trailing edge attached wedge are considered. The trailing wedge of the fin moves inward or outward to decrease the pitching moment, which is applied to actuator. Also, a compact smart actuator driven by piezoelectric ceramics was preliminarily designed for applications in missile fin control. The actuator using oscillation motion of the one way clutch bearing which is forced by vibrating piezoelectric ceramics was presented to develop high force and long stroke smart actuators. In the CFD calculation, the entire 2-D domain is gridded using CFD-GEOM, the grid generator package. CFD-FASTRAN code, developed at CFDRC, is utilized to calculate the lift, drag and the resulting pitching moment generated by the missile fin with wedge and no wedge. Flow condition at Mach 2.0 and angles of attack from 0 to 20 degrees is chosen and Navier-Stokes flow solver with k-ε turbulent model is used in the present computations. The Roe's upwinding differencing scheme with the minmod flux limiter is employed and steady state solution is obtained using fully implicit scheme.

AERODYNAMIC ANALYSIS

Problem Description

The problem to be investigated is a diamond airfoil with the thickness-to-chord length ratios of 10%. The simulated altitude is 10,200 meters (initial ambient condition with $P=2.5701\times10^4 N/m^2$, $T=221.97^\circ K$) and the flow has a free-stream Mach number 2. Calculations were carried out at angles of attack from 0 to 20 degrees. The schematic of the missile fin with retractable wedge is shown in Fig. 1.

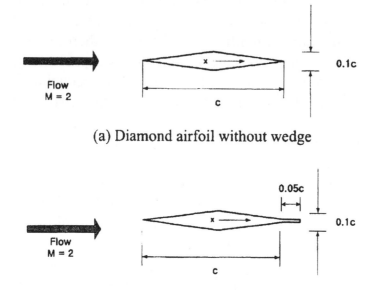

(a) Diamond airfoil without wedge

(b) Diamond airfoil with retractable wedge

Fig. 1 Schematic view of the missile fin.

Grid Generation

The grid for the problem is generated using the CFD-GEOM[4]. The total number of meshes used in the computation is 9720 for the fin without wedge and 12000 for the fin with wedge. Fig. 2 shows the meshes for both the cases.

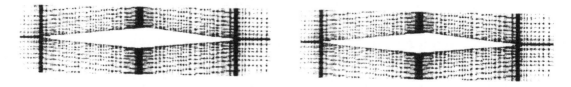

(a) Diamond airfoil with no wedge   (b) Diamond airfoil with wedge
Fig. 2 Computational grid for supersonic flow past a 2-D diamond airfoil.

Spatial differencing scheme

The numerical results are obtained by using the Roe's upwind differencing scheme[5], which is a second order method, with the minmod flux limiter. All the Roe simulation used an entropy fix of 2.0 to achieve a higher order spatial accuracy for both linear and nonlinear waves. Steady state solution is obtained using fully implicit integration scheme.

Numerical Results

Fig. 3 shows the pressure distribution obtained for both the diamond airfoils at Mach No. 2 and at an angle of attack of 10 degree.

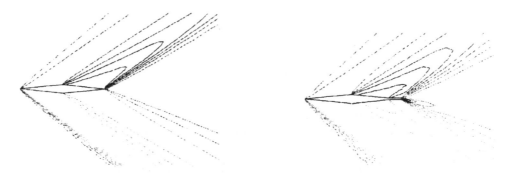

(a) Diamond airfoil with no wedge   (b) Diamond airfoil with wedge

Fig.3 Pressure Distribution for Supersonic Flow Past a 2-D Diamond
Airfoil operating at M=2 and angle of attack of 10 degree.

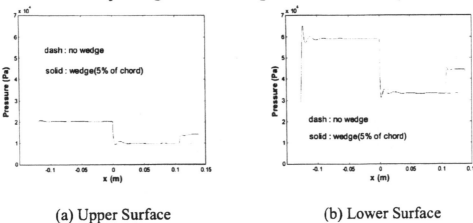

(a) Upper Surface   (b) Lower Surface

Fig.4 Pressure Distribution at the Surface of an Airfoil
Operating at M=2 and Angle of Attack of $10°$.

Figure 4 shows the pressure distribution at the top and bottom surface for the diamond airfoil at Mach No. 2 and an angle of attack of $10°$. Due to the actuation of the wedge type actuator, the pressure distribution at the upper and lower surfaces of the airfoil is increased simultaneously but the amount of increase in the pressure is more pronounced at the lower surface. This results in a substantial reduction of hinge moment for the airfoil.

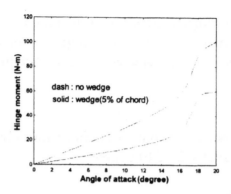

Fig.5 Change in Pitching Moment vs. Angle of Attack for a missile
fin having a 0.254m chord and 0.254m span operating at M=2.

Fig. 5 shows the effect of the actuation of the retractable wedge on the hinge moment of
the diamond airfoil as a function of the angle of attack of the fin. The angle is varied
from 0 to 20 degrees. As expected from Fig. 4, the hinge moment of the diamond airfoil
using the wedge became reduced significantly than that without using the wedge. The
above results indicate the feasibility of the smart fin proposed in this study.

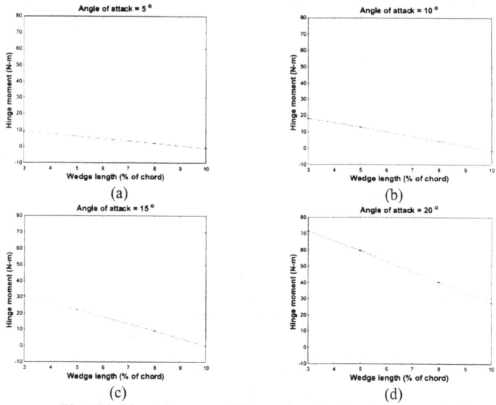

Fig.6 Pitching Moment at different length of wedge for a missile
fin having a 0.254m chord and 0.254m span, M=2

For practical situation, the relationship between the wedge length and the hinge moment should be known priori to control the missile fin in a desired manner. Figure 6 shows the effect of the wedge length on the hinge moment at several angles of attack; 5, 10, 15, 20 degrees. The hinge moment becomes reduced as the wedge length increases, as is seen in Fig. 6. These results demonstrate the feasibility of the proposed actuating method in this study.

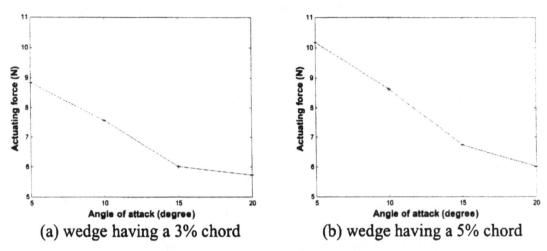

(a) wedge having a 3% chord   (b) wedge having a 5% chord

Fig.7 Actuating force for a missile fin having a 0.254m chord and 0.254m span operating at M=2.

Figure 7 shows the calculation result for the actuating force of the missile fin having a chord length (c=0.254m) with changing angles of attack form 0 to 20 degrees. The actuating force becomes reduced in response with the increase of the angle of attack, as presented in Fig.7.

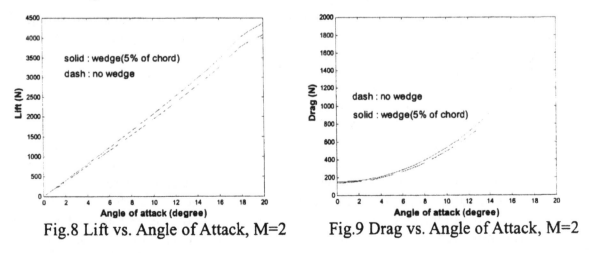

Fig.8 Lift vs. Angle of Attack, M=2   Fig.9 Drag vs. Angle of Attack, M=2

Figures 8 and 9 show the lift and drag forces obtained for the diamond airfoil, respectively.

CUMULATIVE ACTUATOR

A compact smart actuator driven by piezoelectric ceramics was preliminarily designed for applications in missile fin control. The actuator using the oscillatory motion of the one-way clutch bearing which is forced by the piezoelectric ceramics was used to develop high force and long stroke smart actuators. The actuator consists of the one-way clutch bearing, driving shaft with thread and nut parts, and piezoelectric ceramics. The one-way clutch bearing which is free to rotate in one direction and locked in the other direction is used to rotate the shaft and nut in the desired direction. The piezo-element is used to oscillate the shaft and the nut that contain clutch bearings. When the piezo-element is driven with a voltage signal at a certain frequency, the piezo-element bends back and forth. This bending action induces an angular input to the shaft, which is then rectified with the clutch bearing to rotational output of the shaft. This actuator can work as a bi-directional actuator.

Fig. 10 shows the schematic of the one-way clutch bearing. The counter-clockwise rotation of the outer frame forces the circular ring into the tapered wedge angle, which results in locking the clamp and forcing the shaft to rotate with the outer frame. Clockwise rotation of the outer frame unlocks the rollers and allows the outer frame to rotate back relative to the shaft. In this way, the oscillatory motion of the outer frame is rectified to constant counter-clockwise rotation of the inner shaft. If one driving shaft is used with the one-way clutch bearings, two one-way clutch bearings are needed. One roller clutch is served as the oscillatory input motion rectifier and the other roller clutch, which is in line with the support bearing and is fixed to the ground, prevents the shaft from rotating backward during the returning phase of the input clutch housing. Fig. 11 is the conceptual drawing of the actuator. In this design, two piezo-bimorphs are used to actuate the wedge. One is used for operating the shaft to pull the wedge, and the other is used for operating the nut to push the wedge. The fixed piezo-bimorph is used to operate the shaft so that the retractable wedge moves toward the preloading direction. The piezo-bimorph is clamped from the base and the free end is connected to the housing of the clutch bearing. When the piezo-bimorph bends back and forth by the application of the voltage, the shaft is rectified with the clutch bearing to the rotational output of the shaft. The moving piezo-bimorph rotates the nut part which is directly connected to the retractable wedge. It is composed of nut, clutch bearings and housing. When the shaft is rotated in one direction through the clutch bearing mechanism by using the piezoceramics and the nut part is constrained to rotate, the wedge moves into the fin along the shaft. On the other hand, when the shaft is constrained to rotate and the nut part is rotated in one direction, the wedge moves out the fin.

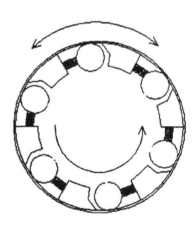

Fig. 10 One-way clutch bearing

When a pre-load P is applied to the retractable wedge, the moving distance of the wedge by working cumulative actuator can be calculated. In this actuator, the distance that the actuator pulls the wedge is different to the distance that the actuator pushes the wedge. Fig. 12 shows the actuation mechanism with bimorph element. Bimorph element which bends at a frequency f forces connecting arm to apply the torque in shaft in order to move the wedge. Considering friction coefficient of thread, μ, thread lead, R, and effective diameter of thread, d_e, the torque necessary to rotate the shaft can be calculated by the relation

$$T_1 = P \cdot \tan(\rho - \lambda)\frac{d_e}{2} \tag{1}$$

where $\rho = \tan^{-1}(\mu)$, $\lambda = \tan^{-1}(R/\pi d_e)$

If the torque applied by bimorph element is greater than the torque, T_1, the actuator can work. The load applied to the end of piezoceramics is $F_r = T_1/l_1$, and the displacement by the piezoceramics under this loading is obtained as

$$u_w = u_0(1 - F_r/F_b) \tag{2}$$

where F_b is the block force of the piezoceramics and u_0 is the tip displacement when no

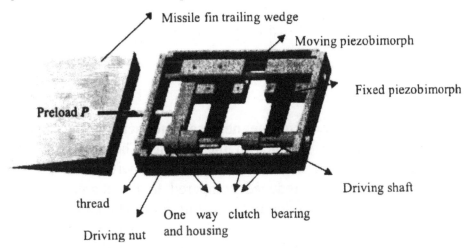

Fig. 11 Schematic drawing of the actuator

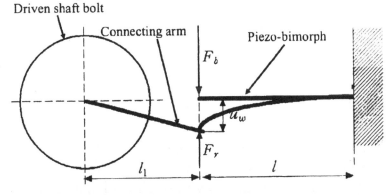

Fig. 12 Actuation mechanism with bimorph element

load is applied. The piezoceramics are applied by the A.C. voltage with frequency f so that the driving shaft is rotated through an angle Θ. The amount of the rotation of the shaft is

$$\Theta = 2\frac{u_w}{l_1}f \tag{3}$$

Then, moving distance of the wedge is

$$\Delta_1 = \Theta\frac{R}{2\pi} = \frac{R}{l_1\pi}f \cdot u_0\left(1 - P \cdot \tan(\rho - \lambda)\frac{d_e}{2l_1} \cdot \frac{1}{F_b}\right) \tag{4}$$

Similarly, the torque necessary to rotate the nut part is obtained by the relation

$$T_2 = P \cdot \tan(\rho + \lambda)\frac{d_e}{2} \tag{5}$$

Since the rotating nut part makes the missile fin wedge move outward, the torque required is higher than that required in the case of rotating shaft. In this case, the displacement is amplified as

$$\Delta_2 = \Theta\frac{R}{2\pi} = \frac{R}{l_1\pi}f \cdot u_0\left(1 - P \cdot \tan(\rho + \lambda)\frac{d_e}{2l_1} \cdot \frac{1}{F_b}\right) \tag{6}$$

CONCLUSION

The method of reduction in torque necessary to actuate the missile fin was investigated to improve the performance of the missile. The CFD simulation results show that the pitching moment of the diamond airfoil with trailing-edge-mounted retractable wedge is significantly reduced compared to the diamond airfoil without wedge while lift and drag forces applied to the airfoil are little changed, and the amount of reduction is not anomalous but in proportion to the wedge length. These results demonstrate the feasibility of the proposed actuating method in this study.

The actuator which consists of the one-way clutch bearing, driving shaft with thread and nut parts, and piezoelectric ceramics has been preliminarily designed to operate the retractable wedge proposed for the missile fin control. Though the design of the actuator is completely design and demonstrated, it has the potential to be applicable to long stroke actuator.

ACKNOWLEDGEMENT

This study has been supported in part by a grant from the BK-21 Program for Mechanical and Aerospace Engineering Research at Seoul National University and in part by the Ministry of Science and Technology through National Research Laboratory Programs.(Contact No. 00-N-NL-01-C-026)

References

1. Barrett, R., 1994."Active Plate and Missile Wing Development Using Directionally Attached Piezoelectric Elements," *AIAA Journal*, Vol. 32, No. 3 : 601-609.

2. Barrett, R., Gross, R. S. and Brozoski, F., 1996."Missile flight control using active flexspar actuators," *Smart Materials and Structures*, Vol. 5 : 121-128.

3. August, J. A. and Joshi, S. P., "Preliminary Design of Smart Structure Fins for High-Speed Missiles," *Proc. of SPIE*, Vol. 2721 : 58-65.

4. *FASTRAN User Manual Version 4.0* March 2001.

5. Roe, P.L. 1981."Approximate Riemann Solvers, Parameter Vectors and Difference Schemes", *Journal of computational Physics*, 43 : 357-372

6. Agnone, Anthony M. 1983."Supersonic Flow over a Cruiciform Configuration at Angle of Attack," *Journal of Spacecraft*, 23(5) : 143-181.

7. Watson, K.P. "Prediction of Fin Loads on Missiles Incompressible Flow," Naval Coastal System . Center, 88-2543-CP.

8. Cayzac, R. and Carette, E. 1994. "Kinetic Energy Projectile(APFSDS) Aerodynamics Numerical Conceptual Approach," AIAA 94-1939.

9. Frank, J., Koopmann, G. H., Chen, W., Mockensturm, E. and Lesieute, A., 2000."Design and performance of a resonant roller wedge actuator," *Proc. of SPIE*, Vol. 3985 : 198-206.

AN ADAPTRONIC SOLUTION TO INCREASE EFFICIENCY OF HIGH-SPEED PARALLEL ROBOTS

D. Sachau, E. Breitbach, M. Rose, and R.Keimer

ABSTRACT

The demands on industrial robots regarding speed and precision are increasing. To come up with more productive systems new innovative concepts are necessary. To reduce inertia forces extreme lightweight structural components are designed for those parts which are moved with high speed and accelerations. The disadvantage of conventional lightweight structures is the lower stiffness and therefore these mechanism tend to vibrations and to larger deviations from the planned path. The problems can be solved by using active materials in the lightweight members. In the paper a new design of an adaptive rod is shown. In order to develop control strategies, simulation tools as explained in this paper are needed.

INTRODUCTION

During the past years machine tools were developed which use sensors and electronics to increase accuracy. In the field of advanced robots e.g. artificial hands [1] large numbers of sensors and highly developed control strategies are introduced. In such kind of applications sensors and actuators are integrated into the robot, therefore they have to be very small. The mechanical part of these systems can be mathematically modelled by discrete nonlinear differential algebraic equations. In the research field of "Adaptronics" [2], the development is focusing on a new class of so-called smart structures. These are lightweight structures which are capable to self-adapt to different operating conditions. This approach requires optimum integration of sensors and actuators within the flexible parts of the system on the basis of functional materials, such as piezoceramic fibers, patches and stacks with adaptive controllers [3]. Such multifunctional elements satisfy both load-bearing and actuatoric/sensoric requirements at a time. Adaptronics allows new approaches by engineers as, apart from classical stiffness and strength considerations, they may also include in their design 'virtual' properties such as changing stiffness, damping or mass distribution. In this way, structural elements can be developed which are not subject to any deformation as a result of external

Delf Sachau, DLR e.V., Inst. of Struct. Mech., Lilienthalplatz 7, 38108 Braunschweig, Germany
Elmar Breitbach, ditto
Michael Rose, ditto
Ralf Keimer, ditto

forces and, consequently, exhibit an apparently 'infinite' stiffness. With this approach it becomes possible to equip optimized mechanical structural systems with structure-conforming integrated actuators and sensors as well as adaptive control systems offering real-time capability.

CONCEPT OF ADAPTIVE PARALLEL ROBOT

With the increasing requirements in machine speed and accuracy, reduction of structural vibrations becomes more and more important. The vibrations make the process results worse and shorten the lifecycle [4]. Especially the need to reach both contradictory goals of high speed and high accuracy makes Adaptronics necessary to extend the performance level [5]. Good experimental results have already been achieved by introducing an active interface at a high precision form measuring machine [6]. A similar concept has been used in a turning machine tool [7].

Looking at a high speed flexible parallel robot in a first investigation a redundant drive was successful used to increase the performance [8]. Starting from this point a German DFG Sonderforschungsbereich, consisting of some institutes of the Technical University of Braunschweig and the Institute of Structural Mechanics at DLR, was set up [9], with the goal of using adaptronic methods to further increase the performance of high speed parallel robots.

Two work packages will be described in this paper. On the one hand the design of a demonstrator including adaptive measures and on the other hand simulation and control. For the modelling, the multibody simulation programm SIMPACK [10] is used. This commercial multibody simulation code is highly developed to consider the deformations of flexible bodies in the simulation [11]. This offers potentialities to simulate flexible bodies with distributed actuators and sensors as needed for the adaptronic design process [12]. Also the kinematical and dynamical equations of motion which are necessary for enhanced control concepts shall be generated in an efficient way.

REALISATION OF DEMONSTRATOR

Rough Design

For first tests of adaptive measures a five-joint structure is planned. The demonstrator will initially be a robot with two degrees of freedom x,y. For further tests regarding robot-control an effector will be applied, giving extra degees of freedom (e.g. rotary motion around z, or movement in z direction). In order to avoid the need to stiffen the structure in z direction a vertical configuration is chosen.

The goal of the demonstrator is to reach accelerations of 10g to 20g at the effector. With the given direct-drive-motor this just can be achieved by limiting the mass of the structure. Regarding the accelerations the pose in Figure 1 is worsed-case and the maximum acceleration is given by equation (1).

Table 1: Data of demonstrator

Name	Description	Value
b	Distance between axes of motors	300 mm
l_C	Distance between axes of crank-joint and axes of motors	300 mm
l_R	Distance between joints of a rod	500 mm
F	Force in direction of rod	ca. 230 N
M	Maximal Torque of Direct-Drive	70 Nm
I_M	Moment of Inertia of Direct-Drive	0.036 kg m^2

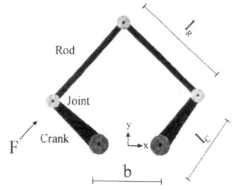

Figure 1: Main dimensions of demonstrator

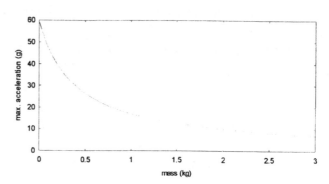

Figure 2: Acceleration over mass

$$a_{max} = \frac{\frac{M}{l_C}}{\frac{I_M}{l_C^2} + m_{joints} + m_{rod} + m_{effector}} = \frac{233\,\text{N}}{0.4\,\text{kg} + m} \tag{1}$$

Figure 2 shows the plot of this function and it yields the result that the mass of the demonstrator must be limited to 2 kg or below. As device for embedding adaptive measures an active rod is chosen. Using the main dimensions shown in Figure 1 and Table 1 the demonstrator is designed as given in Figure 3.

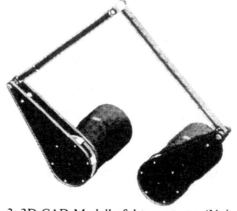

Figure 3: 3D-CAD-Modell of demonstrator (Unigraphics)

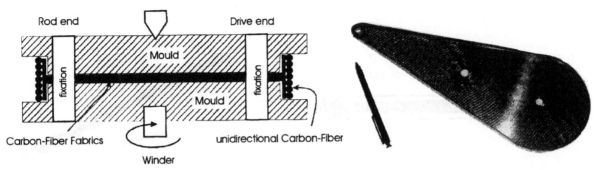

Figure 4: Principle of Crank-Manufacturing   Figure 5: Picture of CFRP panel

Lightweight Crank

The crank consists of two CFRP panels (see Figure 5). The manufacturing is done by using a mould in wich carbon-fiber fabrics is placed (see Figure 4). The fiber-filaments are oriented in 45°/45° regarding to the joint to joint axis of the crank in order to bear the shear in the panel. To get a high bending-stiffness in direction of the motion, a belt of unidirectional carbon-fiber is winded.

Active Rod

The active rod for the demonstrator is designed with the possibilty for changing parameters in mind, so that it can be used in other parallel-robots. The active rod is at the same time actuator and sensor. This is done by using a piezocermic stack actuator with one layer connected as sensor. It is well known that the piezostack as a ceramic component is not able to bear tensile loads, thus arising the need for mechanical prestressing the actuator.

There are 5 functional parts to the active rod, which are exchangeable in a modular way (see Figure 6):
1. the connectors are housing the joint-bearings,
2. the piezostack is actuator and sensor (stiffness c_a, free stroke l_0)
3. the rod is a tube of unidirectional carbon-fiber (stiffness c_r) transmitting the stroke from the actuator to the connectors
4. the belt is built of unidirectional carbon-fiber (Stiffnes c_p) and provides the stiffness for prestressing the actuator
5. the mechanism for prestressing.

The stroke x_2 of the active rod is dependent on the stiffnesses and the free stroke of the actuator and is given by equation (2).

$$x_2 = \frac{c_a c_r l_0}{c_p c_r + c_a c_r + c_a c_p} \tag{2}$$

By changing the stiffnesses (through changing cross-sections of belt, rod or actuator) and changing the length of the actuator (thus changing the free stroke) the weight and the stroke of the complete active rod is determined. In order to obtain the best ratio between stroke and weight an optimisation is done.

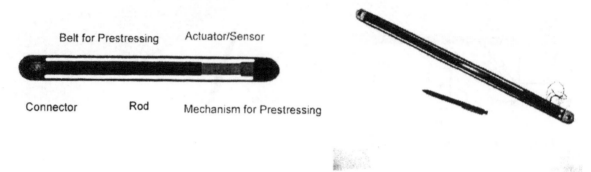

Figure 6: Principle of active Rod   Figure 7: Picture of active Rod

SIMULATION

Software aspects

In order to get a usable model of the five-joint gear, we have to consider, which software we will use to obtain suitable results. There are some possibilities, which we will discuss now.

DIRECT MODELLING WITH LAGRANGE EQUATIONS OR PRINCIPLES OF VIRTUAL WORK

This method is quite good to obtain direct equations of motion for rigid body systems with simple joints. Though our robotic platform has only revolute joints, the elasticity of the cranks and rods complicates the resulting equations and changes in the model become very time consuming. For some systems however, such a direct approach can be used to avoid algebraic constraint equations. This is especially possible for the five-joint gear, and the dynamical equations of motion are give by a proper state space model.

USING A MULTIBODY SIMULATION PROGRAM

The use of a multibody simulation system, which can also deal with elasticities, is very promising to get quick models. But the export facilities of such software products has to be examined carefully, if the model itself has to be embedded in other environments. The use of absolute or relative coordinates is also of some importance for the size of the created underlying differential system. If the equations are exported as FORTRAN-code for example, the dynamic system can be solved numerically, but the understanding of the system is greatly reduced. Incidentally, the automatic generation of models with closed kinematic loops leads to differential algebraic equations of motion. The additional constrained equations must be solved with iterative procedures, which is inacceptable for real time applications.

HYBRID APPROACH

To combine the advantages of both methods, we can use the exported model of the MBS-

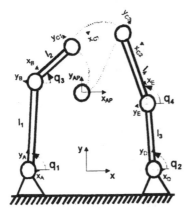   

Figure 8: Open tree structure for the hybrid approach   Figure 9: SIMPACK-model with rectangular trajectory

system, but substitute the generated constraint equations by explicit equations, derived from a direct modelling approach. This avoids the iterative solution of implicit constrained equations, but is only applicable in certain applications. Figure 9 shows the essential elements, which has to be modelled in the case of the five-joint gear platform. Here the actor and sensor elements have been neglected.

Software used for the five-joint gear platform

Our goal is to develop control laws in the MATLAB/SIMULINK environment. Because of the relative coordinate approach and the capability to integrate elastic bodies, the MBS-system SIMPACK was choosen to buildt an elastic model of the five-joint gear structure (see Figure 10). In order to model the elastic bodies, a finite element model for the cranks has been buildt. The FEMBS-preprocessor is able to transform such FE-structures in a special elastic file format, which contains a modal representation of the elastic body. The rods are created by a special preprocessor called BEAM.

The modal description of the elastic bodies implies the need to select the important eigenmodes of each elastic body. With respect to the interesting frequency range for the whole structure, it turns out, that it is sufficient to select the first one or two bending modes of the cranks and the first longitudal mode of the rods.

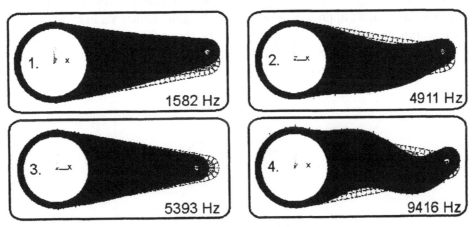

Figure 10: Mode shapes of the cranks

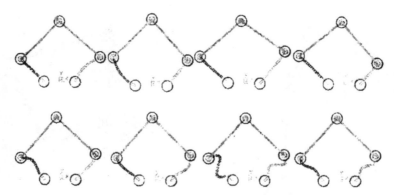

Figure 11: Examples of mode shapes of the five-joint gear with fixed drives

In Figure 10 the first four bending modes of the crank body are shown. Though the eigenfrequencies are relatively high, the complete system has some eigenfrequencies below 1000Hz. Examples of these mode shapes are displayed in Figure 11.

There are several steps to be considered for the complete control strategy of a robotic structure. Given the position, velocity and acceleration values of the target platform from the trajectory planning module, the drive control directs the drives to achieve these requirements. Typically this control strategy is based on a rigid body model. The active elements within the elastic rods are used by the structural control, to reduce vibrations of the structure due to elastic deformations. The frequency range with respect to elastic deformations is higher than in the drive control case. Figure 12 shows the principal combination of these two control aspects.

The robotic structure is realized in MATLAB/SIMULINK by using the interface SIMAT from the SIMPACK environment. Because of the closed kinematic loops a cosimulation block was needed. Therefore both software packages MATLAB and SIMPACK must run simultaneously during the simulation. The next planned steps are as follows:

1. Linearization of the nonlinear equations of motions leads to an A,B,C,D state space model of the structure in several distinct positions. These linear models will be used to derive first control strategies for the active vibration control in different positions. The possibility to interpolate the control laws for intermediate positions will be examined.

2. The nonlinear behaviour of the structure due to the rigid body motion will be eliminated from the dynamic equations. This gives symbolic equations for the A,B,C,D matrices. This step needs a direct approach to obtain the dynamic equations in symbolic form.

3. Other research involves the effective control of the stiffness of the structure for contact problems, as well as indirect measurement of forces by the piezoceramic elements to improve force control applications.

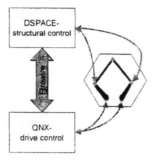

Figure 12: The combination of drive control and adaptive control

CONCLUSION AND OUTLOOK

The design of the five joint parallel robot including the CAD model is completed. The manufacturing of the structural members, based on the manufacturing concepts is in progress and will be finished until the end of 2001. The development of special simulation interfaces and tools which are necessary for the application of the control strategy are almost completed.

After the evaluation of the control strategy by simulation, the experimental testing of the complete structure can start in early 2002.

The new and challenging approach in this research can be seen in the inclusion of adaptronics in all design phases of parallel mechanisms. Therefore the time spend so far on the design, manufacturing and modelling is well invested, because basic tools are generated, enabling the design of more general adaptive parallel structures than considered in this paper.

REFERENCES

1. Hirzinger, G., Brunner, B., Butterfaß, J., Fischer, M., Grebenstein M., et al.. 2001. „Space robotics - towards advanced mechatronic components and powerful telerobotic systems". i-SAIRAS 2001, 6th International Symposium on Artificial Intelligence and Robotics & Automation in Space, St-Hubert, Quebec, Canada, June 18-22,2001, Canadian Space Agency, CD i-SAIRAS 2001.

2. Breitbach E. 1997. „Adaptive Structural Concepts: State of the art and future" Proc. CFAS Intenational Forum on Aeroelasticity and Structural Dynamics, Rome.

3. H. Hanselka, D. Sachau. 2001. „An Overview to the German Research Project ADAPTRONIK", SPIE`s 8th Annual International Symposium on Smart Structures and Materials, Newport Beach, CA.

4. Ohira, G. 1993. „Active Control of Machine Tools". Japanese Journal of Tribology 38 / 8, pp. 1037-1046.

5. Hesselbach, J., Helm, B. 2000. „Adaptronics in Machine Tools". Production Engineering, Vol. VII/1, pp. 83-86.

6. Campanile, F., Lammering, R. Melcher, J. 1996. „High Precision Form Measurements Through the use of adaptive structures Technology". Third European Conference on Smart Structures Technology, Lyon.

7. S. Olsson. 2001. „Active Vibration Control of Cutting Operations", Adaptronic Congress 2001.

8. Kock, S., 2001. „Parallelroboter mit Antriebsredundanz" VDI Fortschritt-Berichte, Düsseldorf.

9. TU Braunschweig, DLR Braunschweig. 2000. „Robotersysteme für Handhabung und Montage – Hochdynamische Parallelstrukturen mit adaptronischen Komponenten". Antrag Sonderforschungsbereich 562.

10. Kortüm, W., Sharp, R. S. 1993. „Multibody Computer Codes in Vehicle System Dynamics". Volume 22 of Supplement to Vehicle System Dynamics. Swets & Zeitlinger.

11. W. Kortüm, D. Sachau, R. Schwertassek. 1996. „Analysis and Design of Flexible and Controlled Multibody Systems with SIMPACK", Space Technology, Vol. 16, No. 5/6, pp. 355-364.

12. M. Rose, D. Sachau. 2001. „Multibody Simulation of Mechanism with Distributed Actuators on Lightweight Components", SPIE`s 8th Annual International Symposium on Smart Structures and Materials, Newport Beach, CA.

Author Index